Wheeler's Wake

Volume I

A Biographical Novel

Andrew Clyde Little

Order this book online at www.trafford.com
or email orders@trafford.com

Most Trafford titles are also available at major online book retailers.

Printed in the United States of America.

ISBN: 978-1-4269-0679-4 (sc)

Our mission is to efficiently provide the world's finest, most comprehensive book publishing service, enabling every author to experience success. To find out how to publish your book, your way, and have it available worldwide, visit us online at www.trafford.com

Trafford rev. 08/31/2010

www.trafford.com

North America & international
toll-free: 1 888 232 4444 (USA & Canada)
phone: 250 383 6864 • fax: 812 355 4082

BY THE SAME AUTHOR

On the Road Again . . . Again, 2001
Before Whispers Become Silence, 2003
Time Exposures, 2005
A Way With Words: One Writer's Journey, 2007

For Shiseido, my faithful muse.

Acknowledgements

Few authors work in a vacuum. Once again I had the help of many people bringing this story to life.

First, there were the individuals who read and provided valuable input into the many versions of the manuscript:

Nick Haramis and Lionel Lumb, my first draft readers, veterans of this process from previous books, and Joe Brita for his dedication and diligence with the manuscript. Tom Henighan for his encouragement.

Then, there were those who provided background information - Joe Rothberg, Andrew Webster, Vince Carlin, Mark Bulgutch and Bert Titcomb.

Dozens of early radio enthusiasts guided me through a maze of technical detail.

There were library archivists in Kingston, Windsor, Toronto, Ottawa and Montreal.

Queen's University and The Canadian Broadcasting Corporation cooperated.

And newspaper editors across Canada responded promptly to requests for reprints of articles.

And finally, my faithful editor, Doug Campbell. This is our third joint venture and Doug deserves two awards this time – one for patience, the other for perseverance.

About the Title

Wheeler

The nickname for someone who frequently uses a bicycle as a means of transportation.

Wake

1) To bring somebody back from a dormant or inactive condition.
2) A celebration of a life.
3) A track, course, or condition left behind something that has passed.

Cover Art

Front – Photograph of Clyde Little's pool hall (1925)
Back – Photograph Wheeler with the author (1936)
Design – Andrew Clyde Little

Introduction

Dad lay wasted and dying in his private hospital room. The insidious smell of cancer was inescapable. It was as if his diseased lungs had processed each molecule of air within these walls and tainted it with the unmistakable scent of death itself. He was propped up in his bed, his skin sallow from jaundice, and he tried to smile when I entered. The smile, like his body, was shrunken, more a grimace than a smile.

"Hello, Pop," I said as I sat on the chair beside his bed.

"Hi, Bub," he replied, raising a hand from the sheet in an attempt to wave.

"How's the pain?" I asked.

"It's there," he acknowledged, "but manageable."

I handed him a folded copy of *The Montreal Star*. "Something to read," I suggested.

"Later," he said, his voice a hoarse whisper. He placed the newspaper on his lap, still folded. He motioned to a glass by the bedside. "Some water." I took the glass and raised it to his lips. He sipped weakly from the bent straw. When he'd had enough, I returned the glass to the small table and sat back down. I noticed the radio on the table, its dial glowing faintly in the dying afternoon light. Its volume had been turned down.

"Some music?" I asked. It had once been my radio; Dad gave it to me after winning it in an office raffle. It had graced my nightstand during my high school years. On evenings when the reception permitted I'd listen to Yankee baseball games, pulling in the signal from a station in upstate New York. Many nights I fell asleep with the radio on, becoming aware of the fact in the morning only because I couldn't recall the final score. It was always mute by then, and I knew that either my mother or my father had turned it off when they went to bed.

"No," said my father, "no music. Nothing but Christmas carols at this time of the year and they get tiresome." He smiled an apology.

I could hear the faint sound of an instrumental version of "White Christmas" playing ever so softly over the hospital's public address system. I wondered if Dad could hear it or if the morphine had silenced that as well as his pain.

With an effort, Dad unfolded the newspaper and glanced at the front page. This gave us an excuse to avoid conversation.

What do you say to a dying man? What do you say when the hospital staff has cautioned you against mentioning the fact that he's dying.

"We don't want to rob him of his willpower," they had explained. "We want him to fight the cancer."

Did Dad know he was dying? Were we, all of us, actors in a play, performing our assigned roles in a drama written by the doctors? It would be years before I realized how foolish this bit of theatre was, but I do recall thinking in those last days how silly it would be if Dad did know he was dying but was keeping up the fiction that recovery was possible in order to spare our feelings.

There are many questions I wish I had asked my father back then—every time I think of them I feel again that familiar flare of anger at the prohibition the hospital had thrust upon me. But then I ask myself, would I, at eighteen, have asked those questions? Would they have seemed important to me back then? I don't know. Perhaps if my father had realized his end was near he might have set the agenda, and told me things that, although they might not have seemed important to me at the time, would have become a treasure in my later years.

Dad folded the paper and looked down at me. "How's school?"

I was in my second year at Bishop's, a residential university in the Eastern Townships of Quebec.

"It's good," I said, suddenly enthusiastic. I'd enrolled the previous year as a science student, at Dad's request. He was an electrical engineer with a degree from one of the top schools in Canada, and he thought a B.Sc. carried more weight than a B.A. It had been a mistake, and it had almost cost me my year. After I registered a zero on the Christmas math exam, my father had reluctantly accepted the school's recommendation that I switch to the Arts program. "I haven't got my marks for this term yet," I said, "but I'm sure you'll be pleased when they do come in." I wondered to myself whether he would still be alive when those marks arrived in the mail.

"Good," he said, and for the first time that afternoon I heard in his voice a hint of the father I had once known. This was the old Willis, upbeat, optimistic, and sure of himself. Then, just as suddenly, his voice softened to a whisper: "Your mother would have been proud of you, you know."

"I know," I said, and the room once again fell silent. It had been little more than a year since my mother had taken her own life. She shot herself with a 22-calibre rifle in the basement of our home. My mother was dead, and Dad and I were left, still trying without success to fill the void that her loss had left in our lives.

When I left the hospital that day, wet snow was falling, and the only sound I remember is the hiss of automobile tires on the slick pavement. My father died two weeks later—and, yes, he lived long enough to savour the fact that I'd done well in my mid-term exams.

It has been more than fifty years since his death, and now it is the void that *he* left that I am seeking to fill. My father's story deserves to be told, because in many ways I owe my life to him. Had it not been for his calm, steadfast presence, I most assuredly would have been swept out to sea, beyond saving, by the deadly undertow of my mother's tortured life.

I was fortunate that my grandmother lived on after the death of my father, her only child, and was able to provide some of the answers to the questions I'd like to have asked him.

I
Anna Maria and Clyde

Chapter 1

Aboard the SS Canadian, November 1887

Eight-year-old Anna Maria Anderson stood in line at the serving table with other steerage passengers aboard the SS *Canadian*, waiting for food. The steward was ladling out split pea soup, shifting his balance from time to time to accommodate the ship's movement. There might be enough soup for seconds today, because seasickness had robbed many of the immigrants of their appetites. Anna Maria was lucky—she had suffered a brief wave of nausea when the ship moved into the open sea, but that was all. In fact, the smell of the soup was making her hungry. She gave the steward her bowl.

"Careful," he said as he handed it back to her. "It's hot."

"Thank you," she said. The steward didn't realize that Anna Maria spoke almost no English. She had just used two of the five words she did know. The others were "please" and "excuse me."

"My pleasure, miss." He motioned to a plate of hardtack biscuits. "Help yourself." In his navy blue uniform with the Allan Line crest on its cap, he reminded Anna Maria of the policeman in her hometown.

She was suddenly sad. She missed her friends, and her family's snug little cabin on the outskirts of Stockholm. But she didn't miss the hunger. It had been part of her life for as long as she could remember. She made her way along the passageway, careful not to spill her soup, as the SS *Canadian* pushed through the uncharacteristically tranquil waters of the North Atlantic.

When Anna Maria emerged into the daylight from the ship's interior, she paused for a moment to let her eyes adjust. Before her in the forward section of the ship was a recessed area, open to the sky, and cluttered with things that were not needed when the ship was at sea. Coils of thick rope, cargo nets, and drums of kerosene were lashed to the deck. It was the only place where the men and women steerage passengers could mingle freely. Anna Maria had spent most of the daylight hours here, away from the sleeping quarters below—away from the smell of unwashed bodies, the sound of babies crying, and the sight of too many people occupying too little space.

It was chilly today, so she was glad she'd worn her heavy winter coat. Her hair was tied back in a pigtail and hidden beneath a white kerchief. She leaned against the capstan, took a spoon from her pocket, and began eating. The bowl warmed her hands and comforted her. It had come from home, and so had the cutlery.

After a few minutes her father and her younger brother, Ernest, joined her. Andrew Anderson was six feet tall and slim. His full beard masked his gaunt cheeks, but his hazel eyes took in everything. Ernest was two years younger than Anna Maria, and had the soft features and gentle disposition of their late mother. The three remained silent as they ate their meal. Around them other steerage passengers were talking in a babble of Greek, Italian, Russian, and English.

Some of them seemed to be complaining about the food; it was the third day in a row they'd been served split pea soup. The Andersons weren't complaining, though. Their crops in Sweden had failed for three consecutive years, and that was why they had decided to leave. Andrew was sure that his late wife, Greta, would be alive today if food hadn't been so scarce. She often gave her share to the children and did without herself. She had lost strength, and when the epidemic hit she was no match for it. Andrew bought tickets for the trip to Canada the week after Greta's funeral. The eighteen krona he paid for them wiped out all the savings he'd accumulated during ten years of work as a tanner.

He looked at the sky, fearing that the run of fine weather was too good to last. It was almost clear, with only a few light wispy clouds to the west. Beneath his feet he felt the steady, reassuring pulse of the ship's engines.

"What are those?" asked Anna Maria, pointing to the mast. "Those things up there." She'd finished eating, and now that her hunger was at rest her curiosity had awakened.

"Those are sails, Anna Maria." Andrew Anderson smiled and waited. He knew this simple answer wouldn't satisfy his daughter.

"They don't look like sails to me."

Andrew bit into his hardtack. "They're furled."

"Furled?"

"Tied up . . . like a bedroll."

"That's silly," said Anna Maria disgustedly."Why don't they unroll them? Wouldn't we get to Canada faster that way?"

"We might," said Andrew, "but most of the time these ships only use their engines."

"Then why bother with sails?"

Andrew scraped the last of the soup from his bowl. "I suppose . . ." He paused, considering his words. "I suppose it's in case the engine breaks down."

"Like the lifeboats?" She cocked her head. "In case we sink?"

Andrew wished he'd invented a different answer to his daughter's question. "I don't think that's very likely." He rose. "Let's go wash our dishes."

He moved across the deck to a barrel of seawater in a corner, where a bulkhead butted against the side of the ship. Anna Maria and Ernest followed. They waited as a young couple rinsed their dishes. The man was wearing a worn three-piece suit and a striped shirt, and his wife, a long dress and high button shoes. She was one of the few women on deck not wearing a kerchief. Her hair was pinned up, and it reminded Anna Maria of her mother. Greta haunted Anna Maria's dreams. This morning she had awakened from a dream-memory of her mother calling from a distant shore, calling to them to come back to Sweden. But Anna Maria had shaken off the dream. It was no use. Andrew wanted a new life in a new country. A cousin in Montreal had promised him work.

After rinsing their bowls they sat down on the deck, their backs resting against the side of an open hatch. Andrew had no wish to return to the cramped sleeping quarters. He shared a narrow berth, just eighteen inches wide, with Ernest. Like most of the other immigrants, they had bought a straw mattress at the dockside. There were one hundred eighty-eight steerage passengers aboard the SS *Canadian*, and as nearly as Andrew could tell there were only a half-dozen other Swedes. He filled his pipe and lit it, shielding the flame of the match with his hands. He puffed contentedly.

A group of people on his right were gathering round the couple they'd seen washing dishes earlier. The man had taken a button accordion from a black leather case and was playing a few tentative notes. Then he smiled at the woman and began to play a tune. After a few bars the woman began to sing in a clear contralto voice. Conversations halted; card games stopped. Soon the dozen or so people around the couple joined in for the chorus:

On a fine Allan Liner
We're sailing in style,
But we're sailing away
From the Emerald Isle.

The Andersons couldn't understand the words, but Anna Maria had no trouble keeping time, drumming her fingers on the deck to the rhythm of the song. They might be travelling in steerage, but this was first-class entertainment.

Conditions might have been a lot worse. The SS *Canadian* was new, one of a fleet of ships founded by Sir Hugh Allan, a Scotsman who had settled in Montreal. The relative comfort these passengers enjoyed was a tribute to him. At one time, steerage, or "third-class" passengers as they were sometimes known, had had to provide their own food, and in the days before steam the trip might take six weeks, not twelve days. Under Sir Hugh, food and service for steerage passengers were improved. He made sure a doctor sailed with each ship. Word quickly spread throughout Europe, and soon the Allan Line was attracting thousands of immigrants each year.

After the company was commissioned to carry the Royal Mail, Sir Hugh Allan built a mansion on the east slope of Mount Royal and named it Ravenscrag, after a similar piece of land near his birthplace in Aberdeen. It had a spectacular view of the St. Lawrence River and the busy port. Sir Hugh always watched for the arrival of the SS *Canadian* with more than routine interest. Two previous Allan Line ships bearing that name had been lost at sea.

With nightfall the temperature dropped, and after a dinner of bread, salt pork, and tea, Andrew and Ernest crawled into their lower bunk. Ernest snuggled in his father's arms, his head brushing Andrew's beard, and was soon asleep. Andrew, though, was not so fortunate. He lay listening to the muttered curses, the snoring, and the occasional startled cry, born of a nightmare. But it was not these sounds that kept Andrew awake. It was the smell. Portholes were always shut tight against the night cold. The hatches were left open, but the odour of tobacco, sweat, and rancid food remained. He longed for the scent of pine from the woods near his home in Sweden. It was with that in his mind that he finally dozed off.

When he awoke in the morning, the gentle swells had given way to whitecaps, and by noon a slashing rain had arrived, forcing passengers from the open steerage area into the ship's cramped interior. The hatches were closed, and it was not long before there was an overwhelming smell of vomit.

No longer a vast open stretch of sea, the North Atlantic had become a moving mountain range of waves that sent avalanches of water cascading down upon the SS *Canadian*. It seemed to Anna Maria that she retched each time a wave hit. Her stomach had given up the last of its food long ago; her spasms now produced nothing more than occasional traces of bile. Looking out of a porthole only made things worse. She was sure they were going be swallowed up by the sea.

In the men's sleeping quarters Andrew sat on the edge of the bunk, his head down, cradling a bowl in his lap. He and Ernest took turns using it. There were moans and muttered curses all around them, and occasional screams of fear when the ship dipped at a particularly alarming angle. In the few moments of stability between waves they could hear voices murmuring prayers in counterpoint to the howling wind. A salt spray whipped from the crests of the waves seeped in through the hatches and drenched the people below.

"Are we going to sink, Papa?" Ernest asked, his lower lip trembling. "Will we all drown?"

"No, no . . . don't be silly," Andrew said. "We're bobbing on the waves like a cork. We'll be fine." Andrew wasn't as confident as he sounded.

He was reaching for his son to give him a reassuring pat when the ship pitched so violently that they were both thrown from the bunk. They slid across the floor in a tangle and slammed into the opposite berth. The hold filled with shouts and the sounds of things breaking. Then, suddenly, it was very quiet. They could no longer hear the wind howling. The ship was listing at a forty-five degree angle. For what seemed like a very long time Andrew and his son remained pinned against the berth, unable to move. Then, slowly, the ship righted itself.

"Jesus," Andrew whispered, as he helped Ernest back into the bunk. He wasn't a religious man, but he got down on his knees and prayed silently for deliverance. Not for himself, but for the children.

On the bridge, Captain Greg Titus and his first officer had also been thrown to the floor. The captain knew his ship had been hit broadside by a wave. He also knew they were incredibly lucky they hadn't capsized. If the cargo had shifted, even slightly, the SS *Canadian* would certainly have rolled over. He struggled back to his feet and regained control of the wheel. There were charts, navigational instruments, and broken china scattered on the floor. An empty bottle rolled back and forth with the motion of the ship.

"You all right?" the captain asked, looking down at the first officer, who was rising unsteadily.

"Fine, sir," the man said.

"Stand by for a course correction."

"Aye aye, Captain," said the first officer as he took over the wheel.

"Take her to port forty degrees." The ship's owners might not like the delay, but he wasn't going to battle the storm any longer. The sea had spoken, and Greg Titus was listening.

The first officer wrestled the wheel to the new setting. The ship shuddered slightly and began its slow turn. Both men were silent for a time—the course change wasn't without danger of its own. Bringing the SS *Canadian* about in the storm meant exposing her starboard side to the pounding waves for at least part of the turn. Another direct hit and they might not be so fortunate; the cargo had held once, but there was no telling whether it would hold again. The captain clenched his fists as the first officer kept the wheel steady. When they were broadside to the wind the ship dropped into a trough; then it rose slowly with the next wave, hung for a moment on the crest, and slid back down into another trough. The next breaker took them up again, but this time their descent was slower—they were no longer moving at right angles to the wind. The captain unclenched his fists. Each successive wave found them less and less at risk, until finally they had come about fully and were headed south. Now they would run with the wind instead of against it. The SS *Canadian* might not make Montreal on schedule, but at least it would arrive.

The passengers in steerage were unaware of the course change, so to them the waves were still filled with menace as the deck continued to pitch from side to side beneath their feet. But gradually the buffeting subsided; the pendulum swings were no longer so extreme. Running with the storm, the SS *Canadian* was taking on less water now, and the waves were working for them rather than against them. A Lutheran minister knelt and led members of his emigrating congregation in a prayer from *The Sankey Hymn Book*, which was damp in his hands:

Eternal Father, strong to save,
Whose arm hath bound the restless wave,
Who bids the mighty ocean deep
Its own appointed limits keep,

Oh, hear us when we cry to
Thee For those in peril on the sea.

When the storm finally blew itself out, the SS *Canadian* was considerably off course. The winds had taken them southeast; the nearest port was Philadelphia. The Allan Line had offices there, so the captain set a new course and informed his passengers. He told them they'd be in port for a day or so to take on coal and provisions, then they'd be off again, up the coast. He expected to make Montreal within a week.

A pilot boarded the ship at the mouth of the Delaware River. His name was Lars Svensen. He was part of a large Swedish community in Philadelphia, a descendent of the Swedes who, two centuries earlier, had first settled in what was then called the New World. As deckhands scrambled to tie up the ship, Andrew noticed him in conversation with a passenger. They were speaking in Swedish. When the men parted Andrew approached the pilot.

"Pardon me, but I couldn't help overhearing you. It was good to hear my language again."

"Second generation Swedish," he answered. "My name's Lars." Then he continued, with a note of pride, "I'm an American. But my folks never learned much English." He shrugged. "I'm afraid my Swedish is a little rusty. Where you headed?"

"Montreal. I have relatives there I can stay with until we can afford a place of our own." He tried to strike a match, but the head was damp and it wouldn't catch.

Lars took a small box of wooden matches from the pocket of his pea jacket and handed them to Andrew. "Do you speak any English?"

"I'm afraid not."

"It can be hard when you don't speak the language."

Andrew lit his pipe.

"I noticed your hands," said Lars. "You are a tanner, yes?"

Andrew nodded. Over the years the tannic acid had left its mark.

"I have some friends who could use a man like you. They have a tannery in Wilcox. They'd pay your train fare from Philadelphia and give you a house to live in."

His eyes drifted to the stevedores who had secured a gangplank for the passengers to disembark. "Keep the matches. They have the name of an agent who might be able to help you."

Andrew noticed Anna Maria looking up at him. He had always

been shy; he'd spoken very little to the other steerage passengers on the *Canadian*. Most of his time on board had been spent caring for Anna Maria and Ernest. But he realized now that if he was to continue caring for them he would have to overcome his shyness.

Chapter 2

Wilcox, Pennsylvania, 1897

Anna Maria Anderson deftly slipped the spatula under the eggs that were sizzling softly on the grill. She prided herself on the quality of her breakfasts. She never browned the bottoms of her sunny side ups. Her toast, done one side at a time on the same gas grill, had a delightful almost-burnt quality. She used her own recipe for pancakes, and knew how to crisp back bacon without overdoing it. This Friday morning had been busy for Anna Maria, but the breakfast rush was almost over. She loved the feeling that came to her at moments like this, the knowledge that her carefully cooked food was warming men on their way to work in the tannery.

The bells on the restaurant door in Grant House tingled. Anna Maria looked up from the eggs and saw a man standing in the doorway. He was wearing a collarless dress shirt, open at the throat, a dark blue vest, matching trousers, and scuffed black boots. He was carrying a small canvas bag and his jacket was folded over his arm. He looked around and took in the restaurant interior with a quick glance, then headed for the counter. He was a small man, shorter by a few inches than Anna Maria, but he moved confidently, and seated himself easily on one of the stools.

"I'd like some breakfast," he said, smiling.

"What will it be?" She was already filling his cup with coffee.

"Eggs, I think. Easy over. And toast."

Anna Maria moved to the grill and reached into a wire basket for the eggs. As she cracked them open she was aware, even with her back to the man, that he was watching her. She put two slices of bread on the grill, then quickly took a folded newspaper from a shelf and turned to face him.

"Something to read?" she said, extending the paper.

He dropped his eyes and took it, embarrassed that he'd been caught watching her. "Thank you, yes."

Anna Maria couldn't place the man. He wasn't one of the regulars. She was sure she hadn't seen him before. As she turned the eggs over, she glanced back. He was reading the paper now, the sports page. She buttered the toast and cut the slice in two. When the eggs were ready she slipped them onto the plate, added the toast, and turned to the man. He looked up from his paper and saw that she was waiting with his plate.

"I'm sorry," he said, folding the paper. "I missed the ball scores and I was just catching up."

"You like baseball," she said. It wasn't a question. She put the plate down where the paper had been.

"You bet," he laughed. "Better than just about anything."

"Than anything?" she teased.

He shrugged, embarrassed. "Well, almost anything," he said, reaching for the salt and pepper. "What about you?"

"I might like it if I understood it."

"Ever play baseball?"

"No, I was born in Sweden." Anna Maria was aware of her accent for the first time that morning. "We didn't play baseball there."

"I could teach you," said the man, his eyes fixed on his eggs.

"Maybe you could," she answered, and moved down the counter to serve someone else. As she poured a cup of coffee, she thought about how little she really cared about baseball, but how much she liked the idea of having this young man teach her.

By the time Thomas Clyde Little had finished his breakfast he was the last customer in the restaurant. He'd planned it that way.

"Can I get you anything else?" Anna Maria asked. "More coffee?"

"No, nothing more, thanks." Clyde took a dollar from his wallet and laid it on the counter. "Have you ever been to a baseball game?"

Anna Maria shook her head, took the dollar, and reached into her apron for change.

"Would you like to?"

"I don't know . . . I guess so," she said, pushing the change across the counter.

Clyde put two tickets down beside the money.

Anna Maria felt her cheeks flush. "What are those?"

"Tickets for the game. The Leamington Lions are playing a team from the tannery tomorrow afternoon."

"I'd have to ask my father," said Anna Maria, feeling her excitement rising.

"Bring him along," said Clyde. "There are two tickets."

"But . . ." Anna Maria hesitated, confused. "But then we'd need three tickets."

"I won't need one," said Clyde, laughing. "I'll be on the field. I play second base for the Lions."

"Second base?" Anna Maria shook her head, puzzled. "What's second base?"

"Why don't you come out tomorrow and see for yourself?" He picked up his bag and moved to the door. "I'll bet your dad would enjoy it." He walked out onto the street. When the door closed behind him the tingling of the bells sounded louder in the emptiness of the restaurant.

Anna Maria looked down at the tickets. Beside them lay a quarter; the man had forgotten his change. Or had he? No one had ever tipped this much before. She smiled to herself and slipped the tickets and the money into her purse.

Chapter 3

Wallaceburg, Ontario, 1898
Detroit, Michigan, 1899

That fall, after the World Series, the Detroit Tigers held their annual tryout for semi-pros. This gave athletes who were talented enough to earn money barnstorming in the summer but required other work in the off-season a chance to become full-time professional ballplayers. The odds were slim, but the Tigers were always interested in players from the Detroit-Windsor area. Local heroes boosted attendance, and Clyde's hometown of Wallaceburg was considered part of this region.

Clyde crouched on the hardened infield dirt of Navan Field, his slate-grey eyes fixed on the pudgy man with the baseball bat standing at home plate.

"Throw to first, then cover second," the man called to Clyde.

Clyde wasn't worried. He had good range, quick hands, and a strong arm. When the coach swung he darted to his right, backhanded the sharply hit grounder, pivoted, and threw. The first baseman relayed the ball to the catcher, who stepped on home plate and rifled it back to second. Clyde took the return throw just inches off the bag.

At twenty-eight Clyde was the oldest player on the field, and at five foot six and one hundred forty pounds he was also one of the smallest. But he'd led his league in stolen bases that summer and had played errorless ball. Although he was from Wallaceburg, he played for a team from Leamington, forty miles north of Detroit. Scouts had spotted Clyde in games he'd played against teams in the U.S.

"Come home this time, then cover first," the coach shouted. Another ground ball, this one to Clyde's left. He short-hopped it, planted his foot, and fired home. His throw was precise, low, and on the plate. He took off for first and caught the catcher's return throw as he crossed the bag. His timing was perfect.

"Next," called the coach.

Clyde retrieved his light blue cardigan and slipped it on. Navy blue letters on the back spelled out the word "Lions" and, in smaller print, "Champions–1898." They'd won it all this season, and Clyde hoped his luck would hold. He gazed at the empty seats of the stadium. He'd been here many times, but always as a fan. Now he was actually down on the field, on the same diamond where Ty Cobb, the Georgia Peach, had run wild. As he looked around, Clyde felt like a pilgrim visiting one of the world's great cathedrals.

At home plate the coach spat tobacco juice into the dirt, then hit a ground ball to the nineteen-year-old who'd taken Clyde's place on the infield. The kid caught the ball cleanly on the second hop and threw to first in one smooth easy motion. There were thirty-one young men trying out on this cool October morning, each hoping his talent would get him to the warmth of Florida for spring training the following year.

Clyde's love of baseball had been with him as far back as he could remember. His father, Clayton, was a cooper. He made barrels on the family farm. Clyde's first bat was a barrel stave and his first ball a stone lobbed to him by an older brother. He played his first game at a family picnic and, unsure of the rules, threw the ball at his father, who was trying to stretch a single into a double. It hit Clayton on the head. Fortunately it wasn't much of a throw, and his father was more surprised than hurt.

Clayton rubbed the spot where his son had beaned him. "What were you trying to do?"

"Get you out!" Clyde said indignantly, convinced he hadn't done anything wrong.

Clayton laughed. "A little harder and you might have knocked me out," he said, then patiently explained to his five-year-old son how to put out a base runner.

That was twenty-three years ago, and Clayton was no longer patient with Clyde. "It's a tomfool idea," he had said. "Time you got this silliness out of your head and settled down. There's a better future for you in barrels than baseballs."

The rest of the morning at the stadium was taken up with more fielding drills. Each time Clyde was called he acquitted himself well. Pop-ups, line drives, turning the double play, taking the cutoff throws from the outfield, relays, run-downs—he handled them all without an error.

When it was time for lunch Clyde sat in the dugout eating a ham sandwich. He was joined by Ray Phelps, a tall, slender pitcher from a rival team in a town a few miles down the road from Leamington.

"How did you do?" Phelps asked, sipping a bottle of pop.

"So far, so good," said Clyde. "How 'bout you?"

"Lousy." Phelps kicked at the dirt with his cleats. "I was wild . . . couldn't find the plate."

"Nerves," said Clyde. "You'll settle down this afternoon." Clyde had hit against Phelps. The guy had a good fastball but had trouble controlling it.

"Hope so." Phelps unwrapped the waxed paper from a sandwich. "Think any of us will make it this year?"

"You never know. Always a chance, I guess."

"What would you say the odds are?"

Clyde wasn't sure. "A hundred to one?" He poured coffee from a thermos. "I know one thing. If I don't make it I'm going to pack it in. I promised my coach I'd play for Leamington next year, but after that I'm hanging up my spikes." He sipped his coffee. "The old man has been after me to help out with the business. Not that I want to be a cooper for the rest of my life, mind you, but maybe I could save a little money and start up something of my own."

"What sorta thing?" Phelps balled up the waxed paper and tossed it into a garbage pail in the corner of the dugout.

"I'm not sure. Open a tavern . . . or sell something, maybe baseball equipment. I know I'd like to be my own boss."

"How about a billiard hall?" Phelps asked.

Clyde cocked his head. "Billiards?" He hadn't thought of that, but the idea was worth considering. "I wouldn't mind."

"Fanshaw is getting on; maybe you could buy him out."

Clyde laughed. "With what?" He thought of the six dollars he'd managed to save that was hidden in a preserve jar on the top shelf of his bedroom closet.

"Wouldn't hurt to ask," said Phelps.

"I'll think about it." Clyde was a Friday-night regular at Blair Fanshaw's poolroom. He liked a friendly game of snooker, and all the baseball talk that persisted there throughout the winter months. "What about you?" Clyde asked. "You gonna keep playing ball?"

"Oh, I guess," said Phelps with a laugh. "Till the old arm gives out."

Clyde finished his coffee, put his thermos into a small metal lunch pail, and snapped the lid closed. "All set?" he asked as he climbed the steps of the dugout.

"Set as I'll ever be." He followed Clyde onto the field.

In the afternoon the players were tested for hitting. Clyde was nervous now. Since his size ruled out power, he used speed to get on base—walks, bunts, the occasional scratch base hit. He kept his batting average at a respectable .300, but he had to work at it. When it was his turn to hit he selected a light bat from the rack, a thirty-ounce Louisville Slugger. The man on the mound was Abner Benson, the Tiger pitching coach, a former major-leaguer.

"Take some swings to loosen up," he called to Clyde. "The first couple will be easy, down the middle."

Clyde lined the first pitch sharply to left field. It felt good. His contact with the ball had been solid. He undercut the second pitch and popped it up.

"OK," said Abner, "these next ones will be for real." He wound up and fired home.

Clyde had barely begun his swing when the ball rocketed past him into the catcher's mitt, where it made a sound like the crack of a whip.

"Shit," Clyde muttered, and stepped out of the batter's box to knock some imaginary dirt from his cleats. So that was "high heat," the major league fastball he'd heard so much about. The fact that it had been thrown by a coach long past his prime, a man pushing fifty, didn't escape Clyde. How would he do against a younger man, a real major league pitcher? Next came the curve ball. Clyde choked up on the bat. He knew it was a curve from the way it left the coach's hand, but knowing it and doing anything with it were two different things. He tried to cut down on his swing this time, but the ball kept tailing away; he found himself flailing at it, and missed again.

The coach threw three more pitches, two fastballs and a change-up. The best Clyde could do was to dribble a foul ball down the third-base line.

"Next," the coach called.

Clyde walked back to the bat rack, his head down and his shoulders slumped, not bothering to hide his disappointment.

When he reached the dugout he dropped his infielder's glove into his canvas equipment bag. He knew he was putting away more than a glove; he was putting away a dream. Baseball was full of slick-fielding, weak-hitting second basemen—and they were all in the minor leagues. He looked down at his hands. They were not the hands of professional ballplayer.

But Clyde's disappointment that fall was tempered by something entirely new. He was, he realized, in love for the first time, and not with baseball. He simply couldn't get the image of Anna Maria Anderson out of his mind. But that image always came accompanied by another image, that of her taciturn father, big and raw-boned, asking the hard question after the game in Wilcox.

"How do you earn your living?" Andrew Anderson's Swedish accent was much more pronounced than his daughter's. "Surely not by playing a child's game."

It was clear to Clyde that Andrew Anderson saw baseball as an amusement, not a profession.

Sensing this, Clyde gave a safe answer. "I'm a cooper." He knew it sounded a lot better than "semi-pro ballplayer."

Andrew relaxed. "Ah," he said, "you make barrels?" He was only too familiar with barrels, working as he did in the Wilcox tannery.

"Yes, my father has a farm back home, and we make barrels after the crops are harvested."

Clyde had watched as Andrew Anderson processed this information. He hoped it would be enough, because he wanted to see more of Anna Maria, and that meant winning the approval of her father.

He had known when he met Andrew Anderson for the first time and shook hands with him that he was in the presence of another Mason. No words were spoken on the topic, but the secret handshake signalled mutual membership in the order. It had been a promising beginning. Clyde's father and grandfather had been Masons, and he had followed the same path. But these were facts he would withhold from Andrew Anderson until the time was right.

Clyde began courting Anna Maria by mail. It wasn't easy for a man whose schooling had ended when he was sixteen. Writing skills hadn't been important when it came to helping out on the family farm, assembling barrels, or fielding baseballs. At times, as he wrote, he felt as if he was back in the classroom—an experience he hadn't relished.

Anna Maria, with even less education than Clyde, found that writing came easily. She liked the fact that when she put words down on paper they bore no hint of her Swedish accent. It wasn't that she was ashamed of that accent, she just wanted to put Sweden behind her, to embrace life in the new world—and, she thought with a smile, to embrace the handsome Canadian who had appeared in Wilcox so unexpectedly that morning. Until then she had thought that her future would be spent with one of the many young men who worked alongside her father in the tannery. Now she was looking north, to Canada, where they had originally intended to settle, before the storm on the North Atlantic had intervened. Anna Maria wondered whether it was fate—whether she'd been meant to live in Canada all along, and had only been marking time in Wilcox.

Clyde, for his part, realized that the way to Anna Maria's heart was through her father. Andrew Anderson needed reassurance that his daughter would be improving her lot in life if she married Clyde.

Chapter 4

Leamington, Ontario, 1899 Wilcox, Pennsylvania, 1900

Clyde began frequenting Blair Fanshaw's pool hall more regularly. He was already a fair snooker player, but he soon improved his game and began to win more and more often. Winning meant increasing the nest egg he was building for what he hoped would be a change in his fortune.

Blair noticed this pattern and befriended Clyde. He'd been a baseball fan all his life, and had followed Clyde's career with interest. He had even hoped that the young man would make it to the major leagues some day. It would be a first for a local kid, something they could brag about when the occasional American happened into the pool hall. When Clyde gave up on his dream and packed it in, Blair felt a surge of sympathy for him. With no sons of his own he felt a special bond with Clyde, and soon learned about Anna Maria Anderson.

One day Clyde sat reading his latest letter from Anna Maria on a bench in the poolroom, and Blair sat down beside him.

"Been thinking," he said, shifting the plug of chewing tobacco from one side of his mouth to the other.

Clyde looked up from the letter.

Blair spat into the brass spittoon at the end of the bench. "Seems to me a young man like yourself might improve his chances with that young lady if he had something a little more regular in the way of employment."

Clyde nodded and remembered Ray Phelps's words. He had said, "Blair Fanshaw's getting on," and that was what had encouraged Clyde to spend more time in the poolroom, refining his game. But he wanted to accumulate enough money that he could make a reasonable offer for the business.

"Been wondering . . ." Blair paused again, spat, and continued. "I could use someone here, someone to spell me so I could spend more time at home."

Clyde looked up, waiting."And . . ." he said, hesitantly.

"And," Blair continued, "if I had someone here to spell me—an assistant, say—it might give me the chance to take a break, put my feet up so to speak."

"Who'd you have in mind?" Clyde asked, hoping he wasn't pushing things.

"Fella like yourself," Blair answered, "fella looking to impress a young lady and her father."

Clyde smiled. "That would do it," he said, unable to keep the enthusiasm from his voice. "That would surely do it."

Blair Fanshaw put his hand on Clyde's shoulder. "Forget the 'assistant' idea. How 'bout we call you the manager? That'll go over better with the folks in Wilcox."

At that moment Clyde Little knew that the second great dream in his life, the dream of winning Anna Maria Anderson's hand, might now come true. The dream of making the major leagues had been cut short, but this new dream was far more important than the game of baseball.

The wedding party was an odd mix. The front pews on the groom's side of the Lutheran Church in Wilcox, Pennsylvania, held the Canadian ballplayers, a dozen young men in their twenties and early thirties. On the bride's side the Swedish families, friends of the Andersons, were assembled. They'd brought their young children, adding energy to the usually sedate surroundings. It soon became clear that there wouldn't be enough room on the bride's side of the church for the Swedish contingent, so after a hurried consultation with the minister it was decided that the overflow should be seated behind the ballplayers.

Clyde wanted to celebrate both of his two great loves on the occasion of his wedding. He had taken time off from the pool hall to join the Leamington Lions for one last barnstorming trip through upstate New York and Pennsylvania. The trip had wrapped up in Wilcox, but the team had thrown the traditional stag party for Clyde the night before the game—and before the wedding—at the local hotel, with the result that the hung-over Lions were soundly trounced by the local team.

Clyde's parents had been invited to the wedding, but the demands of the farm back in Wallaceburg had made the long trek to Wilcox impractical. Instead, it was decided that Andrew Anderson would travel to Canada in the fall to meet his son-in-law's family.

At the reception after the ceremony, the shy ballplayers were warmly welcomed by the Swedish families. Food and drink were plentiful. The players were seen as celebrities, men who had mastered a game that the Swedes were still learning. For most of the immigrants, soccer had been the sport of choice. But this was America, and baseball ruled.

Many of the women envied Anna Maria. To them, life in a tannery town was only marginally better than what they'd left behind in Sweden. It seemed to them that Anna Maria was fortunate to be moving on, and to be marrying a handsome husband who, most important of all, didn't have to make a living working with his hands.

II
Early Years

Chapter 5

Wallaceburg, 1905 Leamington, 1911

In her exhaustion Anna Maria breathed a sigh of relief. Although the price of the sigh was a spasm of pain, it was a different kind of pain, one that was mixed with a profound sense of pride. She'd done it. After a difficult delivery, she'd brought a healthy baby into the world. She'd been in labour more than a day, and the new arrival's large head had taxed the midwife's talent, but after one final push the baby had suddenly been there. Anna Maria remained tense as she waited for the first cry; it was only when it came that she relaxed. It was her first child, and as the midwife carried the infant into the next room she resolved that it would be her only one. She had endured poverty and hardship, and was going to make sure her son didn't have to do the same. One of her father's favourite phrases came to mind: she would put all her eggs in one basket.

After a moment Clyde shyly entered the room with his newborn son in his arms. The baby was swaddled in blankets, and Clyde held it awkwardly, not quite sure yet how to manage this new little bundle of life. Hands that could field a line drive flawlessly, almost by instinct, suddenly threatened to fumble when the baby stirred. Clyde moved quickly across the room and laid the child at Anna Maria's breast. Then, embarrassed by the sight, he turned and sought refuge in a chair. He pulled it to the side of the bed and sat down.

"How you feeling?" he asked tentatively.

Anna Maria smiled at her husband. "I've felt better," she admitted. She noticed that Clyde's dress shirt was wrinkled and collarless and his vest was unbuttoned. His clothes gave off the unmistakable scent of the coarse pipe tobacco he favoured, and there was a light stubble on his face. It was clear to her that he hadn't managed to get much sleep in the day and a half since the contractions had begun.

He reached out and patted his wife's arm. "Doc says the pain should ease up in a day or so."

Anna Maria looked at Clyde's hand and then the baby, which was now nursing. "We did it, didn't we!" she said. "We've got ourselves a son, a new beginning."

Clyde smiled. "You did it, Anna Maria, not me. I'm so proud of you."

Anna Maria arched one eyebrow." Are you saying you had nothing to do with it?"

Embarrassed, Clyde shrugged. "My part was easy. You did all the hard work."

"Well," said Anna Maria with a chuckle, "you'll have your work cut out for you in the future, raising our boy."

As he often did, Clyde began thinking of baseball. Here was a son who could carry on the tradition; he might even become a major-leaguer.

"I've been thinking about a name," said Anna Maria, interrupting Clyde's thoughts. "Now that we know it is a boy . . ."

"And . . . ?" said Clyde. He waited, knowing that Anna Maria's thoughts were never idle.

"It came to me back in Wilcox, at our wedding reception."

Clyde was confused. "I don't understand."

"Dad had hired a local musician to play the piano. I noticed that it was called a 'Willis' piano, and for some reason I said to myself then and there that if I ever had a son I'd call him Willis. Not William—there are too many Billys in the world. But 'Willis': now that's a name with a difference."

Clyde tried it out, feeling a little uncertain. "Willis ..."

Anna Maria sensed his uncertainty and quickly added, "Willis Clayton, after your dad."

Clyde beamed at his wife. She could always surprise him; that was one of the things he loved about her. "'Willis Clayton.' That has a fine ring to it. And Dad will be honoured."

Clyde Little did not give up baseball entirely. When he was thirty-three they moved from the family farm in Wallaceburg to a rented home on Talbot Street in Leamington in order to be closer to his work, and at that time there were two teams in Leamington competing for the same fans.

Clyde managed to convince them to combine their talents, and offered to manage the new amalgamated team.

The Leamington Lions soon became the class of the league, and they reflected Clyde's own strengths as a player. Earl Robson wrote in *The Leamington Post and News*, "The Leamington players never rated as sluggers. They held their heavy-gunned opponents off the scoreboard by fielding and pitching." As for Clyde's starting pitchers, "Besides control, George Fry used a curve and knuckle ball. Earl Taylor's spitball was even effective against professionals and Pick Pickford could rifle a speed ball that would leave the marks of its stitches on a pine board." According to Robson, Clyde's dedication was complete: "He was proud of his team and backed it with money, time and energy and in return there was not one player who did not honour and respect him."

However, he wasn't as successful coaching his son.

It was the first really warm day of spring, and Clyde had taken six-year-old Willis to the baseball diamond across the railway tracks, in spite of the boy's protests.

"It's time the boy learned a little ball," Clyde had said over breakfast. Anna Maria had only nodded as she poured another cup of tea into her chipped blue mug. She set the teapot down and wiped absently with her napkin at the moist ring it had left on the oilcloth table cover.

"He really shoulda started last year," Clyde continued.

Anna Maria started to protest. "He couldn't . . ."

"I know, I know," said Clyde. "But he ain't sick this year."

Anna Maria parted the lace curtains to watch Clyde, carrying his baseball glove and a sawed-off bat, leading Willis across the railway tracks to the ball field. Willis was carrying the ball. When they approached home plate Clyde took the ball from Willis and handed him the bat.

"Now let me see you swing," Clyde said to Willis. The boy hoisted the bat onto his shoulder and made a feeble and reluctant cut through the spring air.

"You're batting cross-handed. Switch your hands, the right on top of the left." Clyde leaned over his son and moved his hands onto the correct position on the bat.

"There, that's better. Give us another swing." Willis swung again, harder this time, putting his resentment into it.

"Much better; much, much better," his father said, then walked toward the pitcher's mound. He stopped a few feet short of the mound and turned. Willis was standing at home plate with his bat on his shoulder.

Clyde faced the boy. "I'm going to pitch a few to you nice and easy, and I want you to hit them as hard as you can." He lobbed the first pitch in. It was low and outside, and Willis waved his bat at it, missing by a good foot.

Anna Maria let the lace curtain fall back into place. Clyde and Willis, tiny in the distance, were dwarfed by the red brick high school that formed the backdrop to the baseball diamond. To Anna Maria, her husband and son seemed somehow lonely, as if more players were needed to make the scene look right. She sighed and turned back to the kitchen, where the breakfast dishes waited. She had seen enough. She knew there was trouble ahead. She had tried to tell Clyde that their son wasn't cut out for baseball, but her husband wouldn't have it. He was determined to see Willis share his love for the game.

"You don't have to swing at every pitch, son. Only swing at the ones that are right over the plate, between your armpits and your knees. That's the strike zone. If it's outside that zone, don't swing at it."

Clyde lobbed the ball again. This time it cut through the heart of the plate, belt high. Willis hadn't moved the bat from his shoulders.

"Now that was one you should've swung at. That was a strike." A touch of frustration had crept into Clyde's voice. Willis retrieved the ball from the backstop and, still holding the bat in his left hand, rolled the ball back to Clyde with his right hand, bowling style.

"Another thing, Willis: never, never roll the ball. Throw the ball. Never let me see you roll the ball like that again. That's bowling, not baseball." Clyde picked up the ball and threw it gently to Willis. The pitch cut the middle of the plate again. This time Willis swung as hard as he could—so hard, in fact, that after he missed the ball the bat slipped out of his hands and down the third base line.

Willis watched the bat skitter to a stop in the infield dust. He looked sheepishly back to his father. "It slipped, it just slipped out of my hand, Dad."

Clyde swallowed his impatience. "That's OK, son. It's kinda hard to hold onto that bat. When I sawed it down I shoulda stuck some tape on it to take the place of the knob."

Willis trotted down the third-base line and retrieved the bat. When he got back to home plate he picked up the ball and threw it back to Clyde, underhand.

"Overhand, Willis. Always throw the ball overhand." With the ball in his hand he pantomimed the motion of an overhand throw. "This isn't softball, you know." He looked down at the ball in his hand, as if to confirm the statement. "This is baseball; we don't throw a baseball underhand. You throw a softball underhand." He said the word softball with scorn. To him, softball was an aberration. All right for girls, maybe, but never to be confused with the real thing, with baseball.

Clyde pitched the ball. Willis swung and missed again, although this time he managed to hold onto the bat. The ball hit the backstop. Willis picked it up and threw it back awkwardly, but overhand this time. For the next half-hour the two played out this ritual, Clyde pitching and Willis missing, each with a mounting sense of frustration and fear. The frustration for both was the same: no matter how carefully Clyde pitched, no matter how hard Willis concentrated, the boy simply couldn't hit the ball. Their fear was different: for Clyde it was the fear that his son would never be the major-league ball player he'd wanted so desperately to be himself. For Willis the fear was simply that he'd never be able to please his father.

Chapter 6

Leamington, September 1913

Willis came home from school carrying a note from his new grade three teacher, and dutifully handed it to his mother.

"What's this?" Anna Maria asked. She saw the name Willis Little neatly inscribed on the envelope.

"Miss Standish asked me to give it to you," Willis explained. The envelope had been sealed and Willis had no idea what was inside. He wasn't concerned. He was a quiet boy, a decent student.

Anna Maria arched an eyebrow. This was a first. She tore open the flap and sat down at the kitchen table to read. In the same concise hand, Miss Standish had written a short, one-page note:

> It has come to my attention that Willis has difficulty reading the homework assignments written on the blackboard. His seat is at the back of the classroom and yesterday, after the children were dismissed, I noticed that he approached the blackboard to write down his assignments. It is possible he needs spectacles. He is no trouble in class, his work is always done on time, and he seemed genuinely interested in his studies, particularly arithmetic. I trust you won't take offence at my bringing this to your attention. Please feel free to contact me if you have any questions.

Suddenly things fell into place: so that was why it was so difficult for Willis to play ball with Clyde! If the boy needed glasses, if he couldn't see well enough to read what was on the blackboard, no wonder he'd found catching or batting a baseball a challenge. Anna Maria knew there was an optometrist in town, and she resolved to take Willis to have his eyes checked the next day after school. But she decided she would withhold the news of the note from Clyde until they'd seen the optometrist.

"What does it say?" Willis asked. He was sitting across the table from his mother eating one of the doughnuts from Anna Maria's weekly batch. He sipped some milk from his glass.

Anna Maria wondered what to tell her young son. Was the teacher right? And if she was, how would Willis feel about the prospect of wearing glasses? She decided to keep it simple.

"Miss Standish noticed that you seem to have trouble reading assignments on the blackboard."

"Yes, that's right," Willis acknowledged, unconcerned, "but I can always copy them down after class."

"Maybe you should sit closer?" Anna Maria suggested. "Naw," said Willis scornfully, "that where the troublemakers sit—so Miss Standish can keep an eye on them." He smiled.

"Well," said Anna Maria, feeling her way, "tomorrow we're going to walk downtown after school and let someone have a look at you."

"Someone?" Willis was confused.

"An optometrist."

"What's an . . . ?" Willis couldn't pronounce this new word.

"He's a kind of doctor . . . an eye doctor."

"But I feel fine," Willis insisted, not yet grasping the significance of the conversation.

Anna Maria rose from the table and moved behind her young son, wrapping her arms around him. "You're fine, you're fine," she murmured, "but we want to make sure you stay fine, that's all. It's just a . . ."—she searched for the right word—"... a checkup."

"Okay." Mollified, Willis shrugged and reached for another doughnut.

The checkup confirmed Miss Standish's suspicions. Willis would need glasses. He took the news with surprising ease.

"If I can see the blackboard better, I guess that's a good thing."

Anna Maria wondered if this attitude would change when he actually had to wear the glasses. It would be several days before they were ready.

That night she stayed up to wait for Clyde, who usually finished his shift in the pool hall at midnight.

"Glasses?" he said, shaking his head. "How do you know?"

Anna Maria explained the teacher's note and the results of the examination. "We should have realized earlier," she said, with a touch of annoyance in her voice. "Willis had so much trouble playing ball with you."

Clyde took his pipe from his cardigan pocket and filled it with tobacco from a brass humidor. He was digesting the information, trying to recall how many major-league baseball players wore glasses. He couldn't think of any. "When will his glasses be ready?" He tamped down the coarse, locally grown tobacco and then lit a kitchen match with his thumbnail.

"In a few days," Anna Maria said. "Then we'll see the optometrist again to make sure they fit properly and are the right strength."

"Well," said Clyde, "I guess things could be worse. At least it's something we can put right." He had always prided himself on his own keen eyesight, but knew that Anna Maria wore glasses for reading and sewing. He was about to suggest that when his son did have his glasses he might take him back to the baseball diamond and see if they made a difference, but then he thought better of it and remained silent.

Anna Maria knew what Clyde was thinking, and she resolved to discourage it. A boy with glasses could be hit in the face with a ball, and that could inflict damage they couldn't correct. But like her husband, she kept her thoughts to herself for now.

"I'm off to bed," she said, heading for the stairs. "Be sure to lock up when you finish that pipe."

Clyde sat quietly, the smoke forming a small cloud that hung in the still air. His thoughts remained on his son and this latest news. He wondered, as he puffed on his pipe, how Willis would look in glasses.

Chapter 7

Leamington, 1913–14

Willis had little trouble adjusting to his new glasses. He endured a few days of teasing from his school chums, but the clarity the glasses provided more than made up for it. He hadn't been aware of how much he'd been missing. His grades immediately improved, and Anna Maria sensed a subtle change in him. He seemed to gain a measure of self-confidence.

The baseball season was over, so there would be no immediate confrontation with Clyde. In the meantime her son seemed to have taken up another sport, soccer, and Anna Maria encouraged it. Willis didn't need his glasses to play soccer, so there was no risk. Soccer had been popular in Sweden, and Anna Maria wrote proudly to her father in Wilcox that his grandson had begun playing a game he would understand. Andrew Anderson had never taken to baseball. The rules seemed unnecessarily complicated. In fact the only games he'd actually seen were the ones that Clyde had played in Wilcox.

When Willis was eight, Andrew Anderson finally managed to visit Leamington. He had received letters from his daughter describing the town and her new life, but he wanted to see things for himself. Clyde gave him a brief history of the region. The discovery of oil and natural gas deposits in the area at the turn of the century had led to a modest economic boom, and the town's population had grown to five thousand. However, the reserves were quickly depleted, and Leamington might have returned to its sleepy pre-boom status as a farming community if it hadn't been for the Heinz canning company. In 1908 Heinz chose the town as the location for its Canadian plant. The rich soil in the surrounding farms was ideal for growing tomatoes, one of the staples of the Heinz line of products. The weather was favourable too, and Point Pelee, a nearby spit of land extending into Lake Erie, had the distinction of being the southernmost point in Canada. Two slogans emerged to describe the town. The local chamber of commerce dubbed it "The Tomato Capital of Canada," and later the town sought a broader appeal by advertising itself as "Canada's Sun Parlour."

The Leamington Andrew Anderson encountered, then, was a small but vibrant town. Clyde insisted that they visit Windsor as well. He wanted to show his father-in-law, now an American citizen, that his daughter would be living close to the United States. Andrew spent a pleasant hour in the billiard hall where Clyde worked, but pool, like baseball, was a game he had never played. There was no poolroom in Wilcox; until this visit Andrew Anderson had thought of the game as something played by wealthy gentlemen. But he came away with a sense of pride in his son-in-law's profession. It was clear from the conversations he overheard that Clyde was respected in his community. When they were on their way back to the house on Talbot Street Clyde also mentioned the fact that he'd been putting money aside, and that he hoped some day to own the pool hall outright.

Andrew Anderson described himself as a lapsed Lutheran, but he was a little disappointed to learn that in Leamington his daughter had shed her religion entirely. He worried that she might be missing out on the social life the church provided. However, she assured him that she and Clyde had found a pretty fair substitute for organized religion: Clyde was a Master Mason, and Anna Maria was a prominent member of the Order of the Eastern Star, the female side the organization.

Andrew, also a Mason, took comfort in this. As a widower, he had become heavily involved in the Wilcox chapter. When he saw Anna Maria carrying out the many responsibilities she'd assumed in the Eastern Star, he came to understand that there was another side to the organization. It became clear to him that the Masons were providing Anna Maria and Clyde with a social life that was both rich and rewarding. It was more than adequate as a substitute for membership in a church congregation.

As he prepared to leave Leamington Andrew Anderson decided that his daughter had chosen well. Andrew only wished, as he often did, that his wife had survived to see the life the family was building in this new world.

On his ninth birthday Willis woke while it was still dark, slipped out of bed, tiptoed along the upstairs hall, and went silently downstairs. He had told his parents he only wanted one present that year, but it was a big one. He shivered at the foot of the stairs and made his way across the cold linoleum kitchen floor. He unlocked the door to the back porch and unhooked the screen. There it was, leaning against the outside wall, a shiny new bicycle. It was dark blue, and Willis ran his hand over the embossed silver letters: "CCM." He didn't know what these letters meant, only that the brand was highly coveted among his friends. Later Clyde would explain that they stood for "Canada Cycle & Motor Co. Ltd."

The grey of dawn was beginning to appear in the sky, so Willis returned into the kitchen, carefully closed the door, made his way upstairs and into his bedroom, and snuggled beneath the covers. His father liked to sleep in, and Willis didn't want to wake him. He would wait until his parents were up before he'd venture back down. He hoped he could act surprised, as they'd expect him to be, when he went downstairs again. He soon dozed off. Sleep had never been a problem for Willis.

He woke again when he heard sounds from below. He could hear the creaking of the pump at the kitchen sink as his mother primed it, and his father feeding the wood stove. He knew that in a matter of minutes the chill would be banished from the kitchen. He waited for the smell of bacon frying to waft upstairs, and then he rose from his bed, dressed, and made his way down.

Anna Maria was at the stove, turning the bacon. "Hello there, lazybones. We wondered if the birthday boy was going to sleep his special day away."

Willis rubbed his eyes. "Morning," he said, taking a seat opposite his father at the kitchen table.

Clyde looked up from his newspaper. "How does it feel?" he asked, smiling. He was wearing a collarless dress shirt open at the throat, and suspenders. His hair was slicked back and parted in the middle, and there was a fine grey stubble on his face.

"Feel?" Willis wasn't sure what he meant.

"Well, you're nine now . . . a whole year older."

Willis laughed. "I feel okay. A little excited, I guess."

Anna Maria put plates of bacon and eggs on the table and returned to the stove for toast. "Excited?" she teased, as she sat down with them.

Trying to hide his embarrassment, Willis dipped a piece of toast into the egg yolk and popped it into his mouth.

"Well," Anna Maria continued, "it isn't every day a young man turns nine." Willis noticed that his mother had taken extra care with her appearance this morning. No dressing gown. She wore a floral-patterned dress, and her hair was tied in a tight formal bun. The family ate in silence, and when the bacon and eggs were finished Anna Maria cleared the table. She returned with a coffee pot, poured Clyde a cup, and watched as he added cream and sugar. Her own choice in the morning was tea, but that could wait.

She glanced at the kitchen door. "Have you looked outside? It's a lovely day."

Taking his cue, Willis rose from the table with a glass of milk in his hand. He pushed open the screen door, as he had done earlier, and gave what he hoped would be a genuine whoop of surprise. He waited a few seconds—the length of time he felt was appropriate—then retraced his steps back into the kitchen.

"Mom!" he said. "Dad!" His mother was standing, teacup in hand, and his father was still at the table. "It's perfect!" He gave Anna Maria a hug and turned to Clyde. "Just perfect! Exactly what I wanted."

"Just be careful," his father warned, always embarrassed by shows of emotion. "We don't want any accidents."

"I'm sure he'll be careful," said Anna Maria. "Willis is a good boy."

The bicycle added a whole new dimension to Willis's life. He already knew how to ride, having learned on his best friend's bike, but he wasn't prepared for the sense of freedom the gift provided. Now he could go anywhere he wanted in a quarter the time it used to take, and within days he was riding through areas of the town he'd never seen before. There was a gently sloping hill that led to the beach, and Willis loved to coast effortlessly down it, the wind in his hair. Even his glasses took on a special value when he was riding. They kept the shadflies from his eyes.

He had always been curious about how things worked, and the bike offered him a new challenge. His first test occurred when he had a flat tire and had to patch the inner tube. That was a simple task. Within a month, though, he'd learned how to take the bike apart and reassemble it. When he understood the mechanics of the bicycle he gained an appreciation of its efficiency. This marked the beginning of a passion he would retain for the rest of his life.

Anna Maria watched this new development in her son with pride. She never let on that she had heard her son slipping down the stairs that April morning before they were up. And she liked the way Willis had handled the situation. She knew that Clyde would never guess that Willis had already seen his birthday present, and Anna Maria had no plans to tell him.

Chapter 8

Leamington, 1918

The First World War, the "war to end all wars," took a heavy toll on the youth of Leamington. When hostilities ended there was a profound sense of relief in the town, but there was also a wrenching sadness, because of the many lives that had been lost. Clyde and Willis had been fortunate in that Clyde was too old for service and his son was too young.

However, Willis's youth did not protect him from one of the indirect consequences of the war. Anna Maria knew something was wrong when he failed to appear at his usual time one morning. He had seemed lethargic the previous day—he'd even passed up a chance to join his pals on a bike ride to the beach. It sometimes seemed to Anna Maria that her son spent more time on that bike than he did walking. But she didn't fret. It was good exercise and got her shy young son out with his friends.

When he didn't appear for breakfast she climbed the stairs to his bedroom with a mounting sense of dread. With the war winding down and talk of an imminent armistice, the mood in the country had been upbeat. But then, suddenly, the word on everyone's lips was not "peace," but "influenza." The so-called "Spanish influenza" had cropped up first in Europe, and had spread to Canada and the States only when the troops had begun to return. This was not the flu Anna Maria was familiar with, involving an occasional week in bed with a fever. This flu was a killer. It took the lives of a high percentage of those it infected.

Anna Maria crossed the floor of her son's darkened bedroom and spoke gently, trying to keep her voice light. "Hey, there, lazybones, you don't want to be late for school."

Willis turned amid the tangle of his sheets, and when he answered his voice was a croak. "I don't feel so good, Ma."

Anna Maria reached out and placed her hand on her son's forehead. It was hot, and Anna Maria felt a stab of fear. She tried to keep the fear at bay as she asked her son if he wanted anything—a glass of water perhaps. He muttered that he just wanted to sleep. Anna Maria moved to the bathroom and took a thermometer from the medicine cabinet. She shook it down and returned to the bedroom.

"Open up," she said, and as Willis groaned a feeble protest she placed the thermometer beneath his tongue. Then she filled a glass with water from the pitcher on the bureau, placed it on the nightstand, and set about straightening the bedsheets. When a minute had passed, she took the thermometer from her son and moved out of the darkened bedroom into the hall, where the light was better. His temperature was 103 degrees. Anna Maria thought back to the various illnesses her son had contracted and realized that he'd never had a fever this high. She also knew that fever was usually lowest in the morning, and rose as the day wore on.

She debated as to whether to rouse her husband. Clyde worked late at the poolroom, and usually slept in, rising at about ten on most mornings. She wondered what to do. They had no telephone, so someone would have to go down to the doctor's office and see if a house call could be arranged.

Anna Maria went back downstairs to the bedroom she shared with Clyde. She reached out and gently shook his shoulder. He muttered something indistinguishable and pulled the covers closer. Anna Maria persisted, and Clyde eventually turned to face her. "What is it?" he said irritably, his voice thick with sleep.

"It's Willis," she said; "he's not well. I think we need a doctor."

Her words had the desired effect. Clyde threw back the covers and sat on the edge of the bed. "Is it that bad?" He knew that if she thought Willis needed a doctor, it must be serious.

"He has a fever," said Anna Maria. "103 degrees."

"Damn!" Clyde looked at Anna Maria and they shared a common thought. "Spanish flu?" he asked.

"Get dressed," said Anna Maria as she moved to the door. "I'll make you a cup of coffee, and then I'd like you to head downtown and see if Dr. Brown will come by and take a look at Willis."

"Right." Clyde rose from the bed and opened the wardrobe.

From the kitchen Anna Maria called, "And I'd like him to come by today, not tomorrow."

Within a few days their suspicions were confirmed. Willis had the Spanish influenza. For the next two weeks it was touch and go. His fever reached as high as 105 degrees, and Anna Maria and Clyde took turns at his bedside. Much of the time Willis was delirious, shouting names and thrashing about as if he was possessed. Dr. Brown dropped by when he had time, but the epidemic was spreading and there was really very little he could do. Anna Maria and Clyde encouraged their son to drink water, applied cold compresses to his forehead, and sponged down his fevered body with rubbing alcohol.

Clyde and Anna Maria were not alone in their concern. Many of Willis's schoolmates had also contracted the flu. Classes were suspended in an effort to contain the epidemic. There seemed no logic to it. The young and the old, the affluent and the poor, the strong and the weak—all were stricken. Deaths were reported daily. Blair Fanshaw even closed down his pool hall, freeing Clyde to stay home with Anna Maria and Willis.

No one was quite sure what to make of this deadly new illness. It hit entire families, but it also spared entire towns. Doctors were at a loss when it came to treatment—all they could suggest was to rest, plenty of fluids, and hope for the best. Clyde and Anna Maria settled into a routine, sitting in shifts by Willis's side. They managed meals when they could, and they tried to get some sleep, but there were many times when sleep wouldn't come, when fear made it impossible.

One night about three weeks after Willis had taken sick, things began to change. It was after midnight, and Clyde was alone with his son, dozing in the chair beside the bed while Anna Maria was catnapping downstairs. Something, he wasn't sure what, roused him. Willis lay on his back snoring softly, and in the faint moonlight that streamed through the window Clyde detected a fine sheen of perspiration on his son's forehead. He reached for the face cloth, rinsed it in the bowl of water at his feet, and applied it carefully to Willis's forehead. It was then that he realized that the forehead was no longer hot. His son was sweating, and the sheets were damp. Clyde felt a surge of hope. He knew that when the fever broke the worst was usually over.

His mind raced. Should he take his son's temperature? Should he wake Anna Maria? He decided to wait. He removed the face cloth and replaced the damp sheets that covered Willis. His son shivered slightly, then rolled over onto his side.

Clyde was not a religious man, but there had been times in these recent days when he had wished he was. Anna Maria had been raised a Lutheran, and he knew she prayed daily for her son's recovery. He wondered if it helped. He reached out again and felt his son's forehead. Definitely cooler. He sat back in his chair and waited. Anna Maria needed her sleep. He didn't want to raise her hopes unnecessarily. Eventually he dozed off himself.

A faint voice woke him. "Dad?" Had he been dreaming or had his son spoken?

"Dad?" The voice came again, and this time Clyde knew it had been no dream.

He reached out and placed his hand on his son's shoulder. "Yes, son?"

"I'm thirsty."

Clyde lifted the glass from the bedside table, propped his son in his arms, and watched Willis drink. A faint glow from the dawn suffused the room. When Willis had finished, Clyde gently lowered him back and put down the glass.

The voice came again. "Dad?"

"Yes, son?"

"I'm hungry."

At that moment Clyde knew his son would make it, and he was filled with joy.

This was too good to keep to himself. He patted Willis again on the shoulder, the touch no longer tentative. "I'll be right back," he said as he hurried downstairs to waken Anna Maria.

Chapter 9

Leamington, 1919

Willis slowly regained his health, but Anna Maria wasn't happy with his recovery. His fever had subsided and he was regaining the weight he'd lost, but he seemed lethargic, and slept for a good part of each day. Although the schools had reopened, Anna Maria was reluctant to send her son back to classes. His teachers sent work home so that he wouldn't fall behind.

One evening in the pool hall Clyde was discussing his son with a friend, Sparks Nichols. The man's real name was Norman, but he'd served in the army as a teletype specialist, and like so many others filling that role he'd been nicknamed Sparks. When Norman returned home he became fascinated with radio, the newly emerging form of home entertainment, and so the nickname stuck. He was the first man in Leamington to fashion his own crude crystal set.

"You know," he said to Clyde, "the day is coming when you'll be able to listen to your Detroit Tigers on the radio. You won't have to wait for the morning papers to get the results."

Until then Clyde had thought of radio simply as a newfangled gadget; it was of little interest to him. But when he heard it linked with baseball it took on a new meaning.

"Explain that to me, Sparks," he said. "Wouldn't I have to run wires to the house?" Clyde was thinking of the teletype machines that used Morse code to transmit signals.

"Nope," he said, "no wires. The radio station sends out a signal through the airwaves, and if you've got the right equipment you can pick it up."

"And what about that equipment? Isn't it expensive?"

"It would be, at first, if you bought it from a store. But I made my own set at a fraction of the cost."

"And what do you hear on it?"

"Depends on the weather. Sure, a thunderstorm can ruin the reception, but if it's clear I can pull in stations from all over."

Clyde persisted. "But what do you actually hear, when the reception is good?"

"Music, mostly. It's like having your own Victrola, but without the records. And some stations are broadcasting news, crop reports, weather."

The excitement in Sparks's voice was contagious. It occurred to Clyde that this might be something Willis would like, now that he was spending so much time at home. Maybe a crystal set would help his son pass the time as he regained his strength.

"You know my son had the Spanish influenza?" he asked Sparks.

"Sure . . . and you're one of the lucky ones . . . your son survived."

"Yeah, he did, but it took a lot out of him. He's still not back at school." Clyde thought Anna Maria was pampering the boy, but he kept this to himself. "I was wondering if a crystal set . . . a radio . . . might be something he'd enjoy."

Sparks sensed a prospective customer for one of his homemade crystal sets. He'd been trying to tempt Clyde with his talk of baseball, but he could see that he might do better through Willis. "Sure, I could fix you up with a radio if you like. See if the boy likes it."

Clyde pondered the idea. "Let me talk it over with my wife," he said, "and see what she thinks."

Sparks was pretty sure he'd made a sale, but just to reinforce the pitch he added, "Even if Willis doesn't take to it, it won't be long before they'll be broadcasting the ball games. They're already doing them in New York and Chicago."

When Clyde walked home that night, the town was quiet. He liked this time, after his late shift, when most folks were home asleep. It was when he did his best thinking. And tonight his thoughts were about radio and his son. He had no idea that he was about to introduce something that would not only amuse Willis, it would captivate him, and lead to a career no one could foresee in 1919. No one, perhaps, except Sparks Nichols.

Sparks was burdened with a box, a suitcase, and a roll of copper wire as he carefully made his way up the path to the house on Talbot Street. It had rained continuously the previous day and the temperature had dropped overnight. The result was a thin film of ice, and Sparks didn't want to risk a fall that might damage the fragile material he was carrying. His progress up the slippery path was slow, and Anna Maria, who had been watching for him, opened the front door and came out on the porch.

"The steps are clear of ice," she called. "Clyde scraped them on his way out this morning. He didn't have time for the path, though."

It was an overcast Saturday, and Sparks had arrived to set up a crystal receiving set for Willis, as he'd promised Clyde.

"I'm almost there," he said as he took his last hesitant steps.

Anna Maria came down the steps to greet him. "Here, give me the box," she said, extending her arms.

Sparks handed it to her as he reached the foot of the stairs. "Be careful with it."

"I will," Anna Maria assured him. Behind her, framed in the door, her son was watching. Anna Maria turned and mounted the steps, the box cradled in her arms. "Willis, put the kettle on," she said. "Mr. Nichols could use a nice cup of tea. He's probably chilled to the bone."

Willis retreated into the house, followed by Anna Maria and Sparks. The man put down his suitcase and the roll of wire in the hall. "Whew, that path was tricky," he said, laughing, as he removed his cap.

Anna Maria was still holding the box. "Where shall I put this?"

He gestured to the case and the wire at his feet. "Right here for now. I think it might be best to set it up upstairs, but we can probably put things together in the kitchen first, if you have the room."

Anna Maria placed the box gently on the floor, then took Sparks's cap and hung it on a peg in the hall.

Willis reappeared from the kitchen. "Can I help?" he asked. He was still in his pyjamas and was wearing worn-down slippers.

"In a minute." Sparks removed his windbreaker and hung it beside his cap. He smiled at Willis. "I'd like to have a look at the kitchen first, to see if we have enough room to put this stuff together."

Willis moved into the kitchen, and Anna Maria and Sparks followed him. "Oh, this should do," said Sparks. "This should do nicely."

Anna Maria had cleared the table, and there was lots of light from the window above the sink. The kitchen was at the back of the house, and it was noticeably warmer there. Wood crackled in the large black cookstove, giving off heat and the faint smell of creosote. Sparks recalled that Clyde used old railway ties as fuel. In the past he'd helped him cut them down to size in the backyard with a crosscut handsaw.

They heard a shrill whistle, and Anna Maria took the kettle from the stove and poured the boiling water into a teapot beside the sink. "Sit yourself down," she said, motioning Sparks to a chair by the table. "We'll let this steep a minute or two."

Willis was already sitting, /and Sparks joined him. "All set, young man?" he asked, winking at Willis.

"I hope so," said the boy, trying to sound confident.

Anna Maria brought a tray to the table and sat down. "We can't thank you enough," she said." A wireless will be just what the doctor ordered. Willis is still weak, but he needs something other than schoolwork to give his time to."

"Well, the wireless should do it," said Sparks. "It's a marvellous invention. And we've just scratched the surface. No telling where it will all lead."

"How much do you know about it?" he asked Willis.

"I've been reading up about it since Dad told me you were coming. I learned Morse code as a cub scout so I know a little bit."

"That's a good beginning," said Sparks with enthusiasm. "Why don't you bring my stuff in here and we can get started? You can leave the roll of wire where it is for now."

"He's a smart boy," said Anna Maria when her son had left the kitchen. "He loves to tinker with things." She poured the tea into the cups, and Sparks spooned sugar and milk into his, then took a sip.

Willis returned with the box and the suitcase.

"Leave the case on the floor," said Sparks. "It's got all my tools. If we can clear some space on the table you can put that box down right here." Anna Maria removed the tray to make room for the box.

Willis watched with fascination as Sparks carefully removed the electronic components. Some he recognized, but others were new to him. He'd used headphones for his Morse code transmissions. There was a coil on a small wooden frame, and something that looked a little like a teletype machine, something he was familiar with.

"This is the tuning coil," said Sparks. The he pointed to a small knob above the coil. "You move this to determine which signal you pick up."

"What's this?" asked Willis, pointing to a smaller piece of apparatus.

"That's the detector," Sparks explained. "We'll hook that to the coil and the earphones, then ground the wire and look for a suitable antenna." He reached down, opened the case, and took out a pair of pliers and a soldering iron. "Can you put this on the stove, Anna Maria? Heat the metal part and leave the wooden handle away from the stovetop, the way you would with a frying pan."

Anna Maria did as she was told. All this was a complete mystery to her, but she had seen how Willis's mood had brightened when Clyde had suggested a wireless set, and she was willing to do anything that would help her son recover some of his enthusiasm. "Now we need to fashion an antenna," said Sparks, and rose from the table. "I'll rig up something temporary outside with that copper wire. Later, if you like, we can install something more permanent on the roof. For now, I'll string a wire from that old apple tree in the backyard. He rose from the table and went back into the hall to get his jacket and cap and the roll of copper wire. He moved to the kitchen door and handed Willis the end of the wire. "Hang onto this, Willis," he said. "I'll just spool it out." He opened the door and backed out, allowing the wire to unspool, then very gently closed the door so that it was open just a crack. "We don't want to pinch the wire," he explained to Anna Maria.

Willis and Anna Maria watched through the window as Sparks moved across the yard to the tree, then climbed the lower branches and stretched the wire above the ground like a clothesline. He threw the coil higher so that it looped over an upper branch and unspooled as it fell to the ground. He snipped the wire with his pliers, freeing it from the coil, and wound it around the trunk of the tree.

Back in the kitchen he attached the antenna wire to the little post on the side of the tuning coil. Then he cut another section of wire, wound it around a second post on the coil, and ran that wire to the sink, where he looped it through the pump handle. "This should ground it," he explained, as much to himself as to Anna Maria and her son. "All this is just temporary," he said apologetically to Anna Maria, looking at the wires on the floor. "It's so that Willis can get the idea."

Back at the table he picked up the earphones and took a length of solder from his case. He gestured to the stove. "That iron should be hot enough." Anna Maria took the iron gingerly by the wooden handle and brought it to the table. Within minutes Sparks had soldered the wires that connected the earphones to leads from the detector and the ground. "We're just about ready," he said as he donned the earphones. "Cross your fingers." He began to slide the knob slowly along the slot in the tuning coil. Then he smiled, removed the earphones, and handed them to Willis. "Hear for yourself."

Willis donned the headset. At that moment Anna Maria saw a smile on her son's face that she hadn't seen in months. Through the phones came the tinny but clearly recognizable sound of another human voice. Willis didn't know it at the time, but it belonged to Bill Scripps, a young Detroit ham operator who had convinced his father, the wealthy publisher of *The Detroit News*, to finance a radio station, WWJ. The station signal was broadcast from the newspaper's own building, which was more than fifty miles from Leamington.

Chapter 10

Leamington, 1919

Within days, Sparks had attached a proper radio antenna to the chimney and had run wires down the side of the house to the ground and into the boy's bedroom window. Willis had found a place for the crystal set on his desk, where the reception was best. He soon made contact with some local hobbyists, most of whom were capable of only limited transmission. Willis soon learned that these pioneers were known as "ham operators." The word had originally been applied to rookie telegraph operators, who were called "ham-handed" because they lacked skill in the use of the Morse code, which required an operator to depress a key for a shorter or a longer period in order to indicate a "dot" or a "dash"; specific combinations of these dots and dashes represented specific letters of the alphabet. Later the term "ham" would be adopted by those in commercial radio to refer to all those amateurs who were felt to be jamming the airwaves. Soon these amateurs accepted the name with some pride, ignoring its negative connotation. Willis soon learned to identify the various hams by the sound of their voices.

Although Willis had become acquainted with Morse code as a cub scout, Sparks insisted that he learn the code in earnest. "You practise, now," he insisted. "Every radio man worth his salt knows it. And you don't want to be thought of as ham-handed."

Contact with this new world transformed Willis. His colour suddenly improved and his energy returned; before long Anna Maria decided he was well enough to go back to school. When he returned to the classroom that winter he was a different student. He'd not only grown a good deal taller during his convalescence, but he seemed to have gained self-confidence as well. No longer the shy, quiet young teen who rarely spoke in class, he became much more involved both in his studies and in extra-curricular activities. While he had no steady girlfriend, he was a good-looking teen, and had no trouble finding an occasional date for a school dance.

Behind these changes was a passion to learn everything he could about radio, an innovation that was about to change the lives of millions of families in Canada and the United States. When he couldn't find the information he was seeking at the local library in Leamington, he'd take the hour-long trolley ride into Windsor. There was a bigger library there, and the bookstores had a good deal more information about radio. Willis set about educating himself.

During these visits he struck up a friendship with a Windsor librarian named Doris Black, who was relatively new to her work. She was a tiny woman whose hair was cropped short and whose wire-rimmed reading glasses hung around her neck by a delicate silver chain. One afternoon when he brought an armful of books to the check-out desk, she commented, "You seem to be particularly interested in radio."

"Yes, I am," Willis answered politely.

"Me too," said the librarian enthusiastically, "but I don't understand how it works. It seems to me like a miracle. My father has a crystal set at home and in the evening he sometimes lets me listen with the earphones."

Willis laughed. "Well, I know how it works, but it still seems a bit of a miracle to me, too."

"It's a wonderful hobby."

"Oh," said Willis, "I hope to make it more than a hobby. I'd like to work in radio when I finish high school."

"Surely you'll need more training for something like that," said Miss Black. "Are you off to university when you graduate?"

"I hope so," said Willis, although this was not an ambition that he'd shared with his parents.

"Where would you go?" the librarian asked.

"I'm not sure," Willis conceded.

Another patron interrupted their conversation, and Willis hurried out of the library to catch a Leamington-bound streetcar. He didn't want to be late for dinner.

A week later he was back at the Windsor library, and Miss Black had a surprise for him. She beckoned him to her desk and handed him a sheet of paper. "I did some research," she whispered. "I made a list of some of the universities that offer courses in radio."

Willis looked down at the paper, a little surprised at this development. "Gosh, thanks," he said. "But you didn't have to . . ."

"Don't be silly," countered Miss Black. "It was no trouble. It's part of my job."

"Well, thanks anyway," Willis said. He glanced at the list. Miss Black's handwriting was neat and precise, like the woman herself. At the top of the list he noticed Queen's University, in Kingston.

"Are these in any particular order?" he asked.

Miss Black blushed. "I did take some liberties there," she admitted, and slipped on her reading glasses. "According to my research, Queen's seemed the best fit. McGill is known for its medical degree, and Varsity for law, but Queen's has a reputation for engineering."

Willis thanked her again and turned away from the desk, but she called him back. "I thought you might like to see these," she said, and pushed several books across the counter. "They're from some of the universities. You can't check them out, but you're free to read them here."

Willis took the books and smiled. "This is terrific," he said. "I wouldn't have known where to look."

"Never underestimate a library," said Miss Brooks.

"Or a librarian," Willis countered.

Willis wondered whether his parents could afford to send him to university. Tuition at Queen's was more than $500, and there was also the cost of room and board. But when he approached Anna Maria, she responded without hesitation. "If it's the best, then that's where you'll go."

"But the tuition is awfully expensive," Willis said. "And it's in Kingston. That would mean I'd have to leave Leamington."

Anna Maria pondered this. As much as she wanted her son at home, though, she was determined to see that he got the education that was denied to her father and her husband.

"You take care of your studies," Anna Maria said with conviction, "and your father and I will make sure you have the money you need. If you say Queen's is the best, then Queen's it will be." When she had decided to have no more children after Willis was born, she had done so on the understanding that as a result of that decision she would be able to provide her son with whatever he needed to make a success in life.

As it turned out, Clyde could afford to send his son to Queen's. He took formal ownership of the poolroom in 1923, and changed its name to Clyde Little's Billiard Hall and Tobacco Store.

Willis made the high-school soccer team—although it wasn't baseball, it was at least something Clyde could tell his pool hall customers about. During the summer, he hired his son to clean up the place in the morning before it opened—he swept the floors, emptied the spittoons, and dusted the felt tables. One of the benefits of this job was the free use of the pool tables when there were no customers waiting to play. Willis suddenly found himself popular with a new group of friends. He learned to play a reasonable game of snooker, although he never took to skittles or billiards.

Leamington was a small town, but because it was close to both Windsor and Detroit it was caught up in the decade that became known as the "Roaring Twenties." Prohibition had been introduced in the United States, and although similar legislation had been passed in Ontario, it was not illegal to manufacturer liquor in the province. The result was a thriving smuggling business between Windsor and Detroit. Beer, wine, and hard liquor were much more readily available in Canada, and as a teen Willis was, like most of his friends, no stranger to beer and rye whiskey. While never a serious drinker, he did fall under the influence of tobacco, something that was both popular and locally grown.

He smoked his first cigarette with his high-school friend, Bud Hancock, when they were fourteen. Clyde's pool hall stocked a complete inventory of cigarettes, cigars, pipes, and tobacco—the smoking, chewing, and snuff varieties. So Willis had easy access to Buckinghams, Turrets, and Players, the popular Canadian cigarettes of the day, as well as imported American brands such as Camels, Lucky Strikes, and Pall Malls. Bud became violently ill after his first cigarette and never again tried tobacco, but Willis had no such problem, aside from feeling slightly light-headed, and within a year he was smoking regularly. It was something he hid from his parents, or so he thought. But Anna Maria's gossip grapevine soon relayed the information, and she confronted her son.

"Don't see the harm," he blustered. "Dad smokes a pipe."

"Well, he's a man and you are still a boy. If you want to smoke when you're a man, that's fine, but it will stunt your growth."

Willis, who was already taller than both his father and his mother, reluctantly agreed to stop. But the habit was too strong for him to resist, and he soon resumed smoking. Anna Maria must have known, but she didn't raise the subject again. Willis, for his part, practised greater discretion and never smoked at home.

Willis may have disappointed Clyde when it came to athletics, but on May 6, 1923, a month after his eighteenth birthday, he redeemed himself by becoming a Mason. The formal ceremony was held in a large room above Knowlton's Jewellery Store, home of the Leamington Lodge of the Freemasons.

Clyde's eyes filled with tears as he watched the initiation. There had been a brief time when he'd hoped Willis might become a major-league baseball player, succeeding where he himself had fallen short. He had no illusions about Willis's joining him in the pool hall business. No, his son was destined for bigger things. The problem arose when Willis became captivated with radio. Whether it was Clyde's lack of a formal education or mere stubbornness on his part, he couldn't bring himself to share his son's passion, and the two had grown apart. Now, though, it was as if the gap between father and son had been bridged.

The Masons had long been central to Clyde's life. His father and grandfather had been Masons. In the Leamington Lodge, he'd risen through the ranks to the very pinnacle, eventually becoming Worshipful Master, and Anna Maria was equally involved in the activities of the Eastern Star, so Willis's membership in the Masons was in many ways inevitable. It was something Clyde had looked forward to from the day of his son's birth. While Anna Maria had told Willis the traditional bedtime stories, Clyde had held forth on the colourful early history of the Freemasons.

Later, it became a natural fit for a young man caught up in a world where science seemed poised to eclipse religion. Even the order's symbol, the square and compass, were significant to the future engineering student. And Willis liked the secrecy that shrouded the organization. He was now a member of a powerful worldwide brotherhood. It included prime ministers, presidents, kings, and many of the country's leading industrialists, scientists, and inventors. Heady stuff for a young man.

As owner of his own poolroom, Clyde had begun to receive gifts from the various tobacco companies whose products he sold. One firm's idea of a suitable gift was unique. They sent him books. Billiards and books might seem an unlikely combination, but Clyde dutifully took the volumes home. Included were the collected works of such authors as Robert Louis Stevenson, Daniel Defoe, Alexander Dumas, and Victor Hugo. Clyde and Anna Maria were not serious readers, but from time to time Willis availed himself of this steadily growing library.

One Sunday morning Willis had his nose buried in a book at the kitchen table when his father appeared for breakfast.

"What are you reading?" Clyde asked.

"*The Hound of the Baskervilles*," Willis replied, "a Sherlock Holmes mystery."

"By Conan Doyle, isn't it?"

Willis looked up. "Yes," he said, trying not to sound surprised.

Clyde smiled. "Conan Doyle. A fine writer. And did you know he was a Mason?"

"Really?" Willis was intrigued.

"Yes," said Clyde as he poured himself a cup of coffee. "Quite a few famous authors were Masons." He sat down opposite Willis. "Artists, musicians. Did you know that Mozart was a Mason?"

"No, I didn't," Willis admitted.

"Well, as a Mason you're now a member of a very elite group," said Clyde as he sipped his coffee. "A very elite group," he repeated.

Willis knew at that moment that the note of pride in Clyde's voice expressed more than just pride in the Masons. It expressed a father's pride in his son.

"I wonder," Clyde mused, "whether Sherlock Holmes was a Mason."

While Willis liked Conan Doyle well enough, reading fiction was something he did in his spare time. With a focus on Queen's, he gave particular attention to his studies in mathematics and science. His marks were solid and he was accepted.

Clyde was no engineer, but the growing popularity of radio had convinced him that Willis would need a university degree if he were to succeed in this rapidly changing world. He might not be able to brag to his friends about his son's baseball prowess, but how many of them had boys who were going off to college?

III
Queen's University

Chapter II

Kingston, Ontario, October 1925

When Willis got off the train in Kingston in the fall of 1925, he was excited and more than a little apprehensive. It was the first time he'd been away from home for any length of time. He was comforted, though, when he saw the waters of Lake Ontario in the distance. They reminded him of his hometown on the shores of Lake Erie.

He found a cab, loaded his bulky wardrobe trunk and leather Gladstone suitcase into the back seat, and climbed in beside the driver.

"Where to?" the man asked.

"Frontenac Street," said Willis. "Number 118."

The driver put the car in gear and they drove away from the station. "Glad to be back?" he asked Willis.

"Back?"

"Back at university," the driver explained.

Willis was flattered that the man had mistaken him for a returning student. "It's my first year," he said.

"Really? You don't look like a freshman. Want me to give you the guided tour? Get the lay of the land?"

"No, but thanks anyway," said Willis. "I'd rather get settled in first." He wasn't sure how much the guided tour would add to the fare.

"Suit yourself." The driver shrugged, clearly disappointed, and lapsed into silence.

Later, when the first-year students assembled in the university auditorium, Willis understood why the driver had assumed he was a freshman. The Spanish flu had cost him a year in high school, and now, at twenty, he was older than the majority of his classmates. He was also taller, having inherited his mother's lanky frame. He stood an even six feet, and weighed one hundred seventy-five pounds.

The boarding house on Frontenac Street was an old red-brick home with twin gables. Willis had picked it from an accommodations list provided by Queen's. He'd been surprised to learn that the university had no residences for men, although it did provide one for women in the newly opened Ban Righ Hall.

Willis used the knocker on the front door of the house on Frontenac Street and was greeted by an older man. He was wearing a clean white shirt open at the collar with sleeves rolled up to his elbows, and over it a vest.

"I'm Mr. MacKinnon," he said, his tone brisk. "And you are?"

Willis introduced himself.

Mr. MacKinnon stood for a moment, sizing up Willis. "My wife and I run a tight ship here," he said, "so we ask that you mind your p's and q's."

Willis indicated his trunk. "I have some things."

"Well, since you're the first to arrive," said the man, "you can take your pick. We have two rooms, both on the second floor." He indicated the stairs. "Let me know when you've unpacked and we'll go over some of the house rules." With this, he turned and retreated down a hall into what Willis took to be the kitchen.

Willis chose the room at the back of the house. There were two cots, and bedding was folded neatly on the bare mattresses. Beside each one there was a nightstand with a reading lamp. There was a single window that looked out on a small yard of patchy grass bordered by cedar hedges.

Willis retrieved his trunk, hefted it up the stairs, and placed it upright in a corner of the room. He smiled and ran his hand over the stencilled initials, "W.C.L.," that his father had insisted he paint on its side. A veteran traveller from his days as a barnstorming baseball player, Clyde had had a good deal of luggage go astray because it lacked proper identification.

Willis kept the trunk in the corner, opened at a right angle, and it became his closet. It was just over four feet high. One side held a rod and hangers for ties, jackets and pants, and the four drawers on the other side would be used for shirts, socks, sweaters, and underwear.

Willis found the bathroom down the hall, and after splashing water on his face and running a comb through his hair he decided it was time to learn the "house rules" from Mr. MacKinnon.

They weren't complicated. No lady guests, no booze, and no "carousing," to use his landlord's word. Breakfast at was at seven-thirty, supper at six.

Room and board—six dollars and fifty cents a week—was due on Monday morning.

"Any questions?" Mr. MacKinnon asked.

"No, sir." Willis was fairly sure the man was a veteran.

"Fine. We'll see you at six."

Briefing concluded, Willis decided it was time to explore, to get what the cabbie had called "the lay of the land." Outside, he lit a cigarette and set out to walk the campus. He met a few young men along the way and exchanged polite hellos, but he was too shy to initiate any further conversation. He assumed that they were mostly freshmen, since freshmen were required to register early, two days before those in the sophomore, junior, and senior years. The idea was to give the newcomers a chance to familiarize themselves with university before the others arrived.

One of his first stops was the university library, where he decided it would be a good idea to learn a little more about Kingston. A helpful librarian recommended a book, and Willis settled down to read. Some of it was familiar from his high school history. He knew that hostilities with the United States in the previous century had led to the construction of the Rideau Canal, a waterway that linked Kingston and the Great Lakes with Ottawa. Those hostilities also explained the presence of the city's historic fort, the site of the Royal Military College, where young men trained to become officers in the Canadian army.

He learned that the city had once been favoured to become Canada's capital, but had lost out to Ottawa in 1857. He was surprised to find that there remained a lingering sense of resentment about this decision, almost seventy-five years after the fact. While Kingston had lost some of its strategic importance in the wake of improved relations with the United States, it still boasted a decent port, which moved grain, lumber, and limestone along the Great Lakes. However, more and more of that business had shifted to the rapidly expanding city to the west, Toronto.

In 1925 Kingston had a population of twenty thousand. Canada's largest penitentiary was located here, and it boasted a fine hospital and an asylum. Willis wondered whether, in the future, the city's prosperity might depend more on its institutions than its trade. Queen's was certainly one of those institutions. The university was founded in 1845 to prepare young men for the Presbyterian ministry. It still fulfilled this role in 1925, but its mandate had expanded dramatically over the years. There were slightly fewer than three thousand students, and they were enrolled in the faculties of science, arts, engineering, medicine, and law. Women had begun attending, but they were outnumbered ten to one by the male students. This inequality was balanced somewhat by the presence of the young women at the hospital who were training to enter the nursing profession.

When he returned for his evening meal Willis met the three other students who would be rooming at the MacKinnon house. They were all freshmen. Mike Phillips and Don Curry, from Toronto, had gone to the same private school and wanted to room together. This meant that Hugh Wells, from Ottawa, would bunk in with Willis. Over supper Willis learned that Mike and Don were aspiring engineers, while Hugh, a self-admitted bookworm, was enrolled in the arts program.

On Willis's first walkabout of the campus he had noted a lot of bike racks outside the buildings, and he decided then and there to acquire a bike of his own. Two days later, using the map provided to freshmen, he strolled downtown. He stopped for a coffee and bought cigarettes at a local restaurant, where a waitress informed him that a bicycle repair shop was just a few blocks away.

Jed's Cycle was a two-storey clapboard building in the section of Kingston that ringed the port. It was between a hardware store and a shoemaker's. Willis pushed open the door and a tiny bell tingled. The familiar smells of oil, glue, and leather hung in the air. A variety of bicycles hung from the ceiling and there was a glass counter to the right of the entrance. A middle-aged man emerged from the back of the shop wearing a leather apron. He had a handle-bar moustache, and a cigarette dangled from his lips.

"Hello, young man," he said, wiping his hands on rag. "What can I do for you?"

"Well . . ." Willis started, but the man raised a hand to silence him.

"Let me guess. You're a student. Probably a freshman. Right?"

"Right," Willis echoed, and wondered if it was that obvious.

"And you're looking for a bike?" He took the cigarette from his lips and flicked the ash on the floor.

"Yep," said Willis with a smile, "but nothing fancy."

"Where ya from?" The man's bushy eyebrows, flecked with grey, mimicked the contours of his moustache.

"I'm from Leamington," said Willis and, in response to the proprietor's puzzled look, he added, "It's near Windsor."

"Long way from home," the man mused as he dropped his cigarette to the floor and ground it out with his boot. "Well," he said, "let's see what we can find for you. Name's Jed, by the way."

"I'm Willis, Willis Little."

"You're a big fella," said Jed. "You'll need something solid." He turned to the hanging bikes.

"But nothing fancy," Willis repeated. "Maybe something second-hand?"

"Right. And you're in luck. The graduating students usually leave their bikes with me to sell, and since you're my first customer this fall the selection is good. Some of them are in pretty fair shape."

"I'm pretty handy at repairs myself."

"Now don't go advertising that fact," said Jed with a laugh. "Wouldn't want you stealing my customers."

"No, no, sir. I meant . . ." Willis searched for a way out. "I meant I can easily repair my own bike."

"Sure," said Jed. "I was just teasing. When Queen's fills up I'll have more business than I can handle."

Willis narrowed down his choice to two used bicycles. One was similar to the CCM model he'd left in Leamington, while the other was a Raleigh, a fancier bike, with gears, imported from England. The Raleigh was more expensive and Willis decided it was beyond his means.

"I'll take the CCM," he told Jed, and the man took the bike down from the rack.

He checked the tires. "Pressure's fine," he said, and rang the small bell on the handlebars. "This could use some oil." He reached for a can on the countertop and squirted a shot into the bell's mechanism, then tried it again. The sound was louder. "That should do it," he said, and moved behind the counter.

Willis handed him the two dollars they'd agreed was a fair price. Jed deposited the money in a cash register and then moved to the door and held it open so that Willis could wheel his bike out onto the sidewalk.

"It was a pleasure doing business with you, Willis," said Jed. "If you need anything—spare parts for the bike––you know where I am."

"I do." Willis heard the door to the shop close as he mounted the bicycle and turned up Queen Street on his way back to the university campus. Some of the apprehension he'd felt about his new surroundings subsided. It felt good to be back on a bicycle.

Chapter 12

Queen's University, October 1925

When Willis got down on all fours the smell made him gag, but he swallowed hard and looked at the peanut, still in its shell, that lay nestled in fresh horse manure on the turf at the edge of the football field. The entire field was spread with manure, and he, along with forty-odd other freshmen, had been assigned the task of pushing a peanut across the field using only his nose. Next to him Hugh Wells was already retching, but Willis was determined to keep his own breakfast down. He'd learned the trick of holding his nose without pinching it while jumping from the Leamington Pier into the waters of Lake Erie. He employed this technique as he began the tedious task. It was the last day of initiation rituals, and he tried to console himself with the fact that once he reached the other side the humiliations of a week of hazing would be at an end.

The upperclassmen were jeering from the sidelines, and when Willis reached the halfway point he paused and glanced around. He was pleased to see that while he was not at the head of the line, he was among the leaders. He wiped at his nose with the back of his hand and was surprised to see blood. Apparently his efforts had rubbed the skin raw. No matter. He bent down and resumed pushing. After a few more minutes he crossed the finish line. He breathed a sigh of relief and rose to his feet, then saw that his roommate still had a few feet to go.

"Now, pick up the peanut," said a senior who was squatting on the sideline.

Willis did as he was told.

"Now, eat it," the voice commanded.

One last humiliation, Willis thought, and cracked open the shell. He popped the two nuts into his mouth, again resisting the urge to vomit, and swallowed them. He looked back at the field and saw that most of his group had made it across, although a few stragglers brought up the rear. A large cheer went up from the upperclassmen as the last freshman crossed the finish line.

There were congratulations all around. Even the seniors, juniors, and sophomores who had turned out to watch the spectacle good-naturedly slapped the backs of the freshmen.

On the way back to residence, Hugh told Willis that the high point for him had come after the event. "This senior came up to me," laughed Hugh, "and, to prove there were no hard feelings, shook my hand." He held his up manure-stained palm to show Willis. "The look on that guy's face when he realized the mistake he'd made . . . well, it almost made it worthwhile."

That night, wearing a Band-Aid across the bridge of his nose, Willis joined Hugh and the other freshman at the King's Head tavern, located off-campus in downtown Kingston. Although he had never been much of a drinker in his high school days, Willis was soon caught up in a ritual of chug-a-lugging draft beer. The idea was to drink the entire contents of a glass without pausing, after which the next student would be required to do the same. By midnight Willis had lost count of how many glasses he'd consumed. When the floor of the tavern appeared to tilt and dizziness overcame him, he stumbled to the door and pushed out into the cool night air. At this point he lost the battle with his stomach that he'd so steadfastly controlled earlier in the day. He threw up on the sidewalk, and retched until there was nothing left but bitter bile. As he wiped the sweat from his forehead he saw that Hugh had joined him, and together they made their way unsteadily back to the rooming house.

When he woke up the next morning he couldn't remember how he had managed to make it back. He was still dressed, but at least he had discarded his shoes beside his bed. When he propped himself on one elbow his head seemed to explode. He'd had mild hangovers in the past, especially when he'd sampled rye whiskey, but they were nothing like this. He slowly rolled out of bed and waited for the wave of nausea to subside. He noted that his roommate was still sound asleep. He grabbed a towel from his wardrobe trunk, made his way quietly to the bathroom down the hall, and turned on the tap at one of the sinks. He put his head down and let the cold water sluice through his hair and down his face. Then he put his mouth to the tap and gulped the water. God, he was thirsty! Finally, he towelled himself off and went back to his room, where he stripped off his clothes and got under the covers. He wouldn't die, he knew, but it sure felt as if he would. Within minutes he lapsed into a deep sleep. It was an ability he'd picked up during his fight with influenza and would maintain for the rest of his life. No matter how difficult the situation, Willis could always put his worries aside and sleep.

One of the first things Willis did once his hangover subsided was to seek out the campus radio station. It was located in Fleming Hall, an imposing four-storey structure of grey limestone named after Sir Sandford Fleming, a former Queen's chancellor. Willis paused at the entrance, trying to recall what he knew about Fleming. He remembered that he was the man who'd come up with the idea of standard time, and also that he had been involved in the construction of the railway across Canada. What he didn't know at the time was that the former chancellor had also been a Mason. If he'd been aware of this, he might have seen it as a promising omen.

Inside Fleming Hall he encountered a young man carrying an armful of books and asked him where he might find the radio station.

"Oh, it's in the basement," came the reply. "Through those doors and downstairs."

As Willis descended the steps he wondered why the station wasn't on the top floor—the higher elevation would be a better location for an antenna. He had never been in a radio station before, and when he pushed through another set of doors in the basement he was surprised at the informality of the setting. The room was divided into two distinct areas, one crammed with electronic equipment and the other looking a little like a kitchen, complete with table and chairs. There were no place settings, though; instead, at the centre of the table there was a microphone, and wires ran from its base to the equipment side of the room.

There was no one at the table, but a man with his back to Willis was bending over a bench that held an array of coils, tubes, and wires. He was so intent on his task that he hadn't noticed Willis. A faint wisp of smoke rose from the bench, and Willis recognized the smell. The man was soldering a connection. Willis cleared his throat.

"Damn!" said the man, startled by the interruption. He turned, accusingly. "Don't you know any better than to sneak up on someone like that?" He was of medium height and slight build. His hair, neatly parted, had a touch of grey. With his even features and thin moustache he reminded Willis of a silent film star.

Willis mumbled an apology. "I'm sorry sir."

"Let me guess," the man said, with a note of resignation. "You're a freshman?" He wore no jacket but his vest was buttoned and his tie was knotted tightly against a starched white collar.

Willis nodded.

"And you're interested in radio?" the man continued.

"Yes, sir."

"Well, I'm Professor Bain, and as long as you're here you might as well make yourself useful." He motioned Willis to the bench. "Sometimes four hands are better than two." He extended two wires to Willis and reached for the soldering iron. "Hold these together, steady, and we'll see if we can get them properly connected."

Willis took the wires, rested his hands on the bench, and watched as Professor Bain heated beads of solder and delicately placed them where the wires met.

"Don't move," he said. "Give it a moment to set."

Willis remained motionless, determined to make up for his earlier interruption.

"There." Professor Bain took the newly joined wires from Willis and gently tested the connection. "Well done, young man." He smiled. "Now perhaps you can tell me your name?"

"Willis, sir. Willis Little. I'm from Leamington."

"Well, Willis, welcome to my domain. This is Radio Station CFRC and it is my responsibility to make sure it runs with the precision of a Swiss clock." The smile had faded from the professor's face. "I'll excuse you this time because you are new to Queen's and don't know any better, but entry into this inner sanctum is by invitation only."

This new curt tone in his voice suggested to Willis that the man had a military background.

"Now, if you'd be so good as to leave, I'll get on with my work," said Professor Bain, and turned back to his workbench.

Willis left the room, careful to close the door quietly behind him. Not a promising start at the university radio station, he thought, as he made his way up the stairs. But then again he had been of some assistance. He hoped that would count for something.

Willis began to spend more time in Fleming Hall, watching the people who manned the university radio station. It was clear that members of the electrical engineering faculty were still filling the more demanding positions, but by his mere presence Willis began to win their trust. During this time, Professor Bain was putting in long hours upgrading the station's Mark II transmitter to the more powerful Mark III model. Willis was always available, more than ready to provide an extra hand, run an errand, and stand as a listener when testing was being carried out.

Willis had by no means been a "mama's boy" during his high school years, but there was little doubt that Anna Maria had played a large part in her son's development. He was her boy, and she had made sure he grew up with a strong sense of his own worth. Clyde may have been disappointed that his son wasn't more athletic, but because he worked nights he was rarely around when Willis needed advice on dating or homework. It was Anna Maria who shaped her son. She did not have a formal education, but she was bright and had an inquiring mind. She instilled in Willis the drive to do his best at all times and to play by the rules. She stressed that it didn't matter what others did or said, it was what one thought of oneself that counted.

At Queen's Willis blossomed socially. Word spread quickly that he could do minor bicycle repairs, and a good many of his early friendships were forged as he patched tires or replaced faulty pedals. He fulfilled his athletic requirement by playing intramural soccer—a plus with the girls. He would never attain the adoration on campus attained by the star football players, Pep Leadly and Harry Batstone, but he did shed his inherent shyness, and began to emerge from his mother's shadow as a young man in his own right. Along with his fundamental sense of his own worth, he knew what he wanted to become in life, an advantage that a surprising number of undergraduates did not share.

His hopes to become involved in the university radio station didn't materialize. Freshmen were occasionally allowed in to watch some programs as they were broadcast, but because they were in their first year they weren't deemed experienced enough to merit more than observer status. And it seemed to Willis that the various professors who were involved did most of the interesting work. The more senior students were allowed to participate, but from what Willis could pick up in casual conversations with them they were usually assigned routine tasks.

The radio station was gaining a reputation, however, and this fact was not lost on the editors of *The Queen's Journal*. On the front page of the February 5 edition the following article appeared:

On the Air

> Station CFRC—Queen's University—is becoming very popular with dial twisters. Under the management of Prof. Bain of the Electrical Dept., Queen's has taken a place among the leading broadcasting stations in Canada. The popularity of CFRC is shown by the many letters received weekly from members of large radio audiences. Queen's broadcasts at 267.7 metres. Everything worthwhile is "put on the air" and the varied programme of Queen's winter life assures versatile radio entertainment. All games are broadcast from the "ringside." And every concert and lecture is sent out from the stage. In addition to this Queen's has a well-equipped studio in Fleming Hall from which concerts are broadcast. The management of the radio department are always pleased to receive written word from their radio friends, and will make every effort to fulfill any desire they may express.

Willis may have been relegated to the sidelines of CFRC in his freshman year, but he was an avid listener to just about all the programs the station produced. The Queen's football team was a dominant force on the intercollegiate scene, and the home games were broadcast from Richardson Stadium. Things didn't always run smoothly, though. Transmitter trouble cut short the broadcast of the October 10 game against Toronto. In his room Willis took off his headset and wished he had been there in the studio to see what had gone wrong.

During the following weeks the station was more successful, and the games against the Royal Military College and McGill came off without a hitch. The hockey team wasn't on a par with the football squad, but games were broadcast from Jock Harty Arena nonetheless. These were never very successful, because within the close confines of the rink the announcer could scarcely be heard over the cheering of the crowd. Willis also listened to concerts that were broadcast from Grant Hall. He had very little appreciation for the classical music that was played, but dutifully stayed tuned in order to judge the quality of the sound.

One freshman did get the chance to work at the station. George Ketiladze was a Russian immigrant and an accomplished pianist. He was often called upon to accompany vocalists in the recitals the station broadcast. His musical talent gave him an entrée to CFRC that Willis couldn't match. George was big, gregarious, and loud. His distinct Russian accent seemed out of place on the staid Queen's campus.

One day he approached Willis outside Fleming Hall after a lecture. "You fix bikes?" he asked.

"Depends," Willis answered.

"Depends?" George looked perplexed.

"Depends on the problem," Willis explained.

"I have problem." The Russian pointed to a rusty old bike that was leaning against a nearby brick wall.

"Let's have a look," said Willis, and they walked over to George's bike.

Willis saw immediately that there was more than one problem. There were, in fact, several, not the least of which was two flat tires.

"I pay," said George, reaching into his pocket.

"Hang on," said Willis. "First let's see if these tires will hold air." There were several other bicycles occupying space along the same wall. Willis quickly scanned them and found what he was looking for—a small pump clamped to the frame of a nearby bike. "We'll just borrow this," he said, and was soon filling the tires, listening intently for the telltale hiss of escaping air. There was none. Then he moistened a fingertip and rubbed his spit over the valves. "No bubbles," he said, almost to himself, "and no leaks." He looked up at the Russian. "That should look after the tires," he said, "but this chain is pretty rusty." He rose and pushed the bike a few feet, producing a grating squeak of protest. "And the wheel bearings could do with some oil."

"I can ride now?" George asked.

"I wouldn't advise it," said Willis. "But I have some machine oil back in my room. I live pretty close."

"You sure?"

"It's really no trouble," Willis assured him. "Let me just return the pump first, then we'll walk the bike over to my place." On their way to his room Willis asked George why he'd come to Canada.

"Is long story," said the Russian. "Back home revolution changed everything in my country. No future for me there. I had to get out."

He explained how he'd signed on as a deckhand aboard a Soviet freighter. When it docked in Dartmouth he jumped ship, walked to the nearest police station, and turned himself in. During his detention by

Canadian immigration officials he not only worked on his rudimentary English, but entertained his fellow detainees by playing the piano. Word of his musical talent soon spread.

"One day," George continued, "man from Halifax showed up to hear me. I play a few classical pieces for him. I learn he is music professor. I guess he likes, because he offers to help."

"And what did you say?" asked Willis, clearly intrigued by the story.

"Professor is concertmaster; he wants I study music. But I say radio is best way for people to hear my music."

"True," agreed Willis.

"I tell him I want radio training, not music," continued the Russian. "I see professor is disappointed, but he has old engineering friend at Queen's. So he writes."

"And . . .?" Willis prompted.

"And they agree to take me . . . on probation. Is called 'without records'."

"So here you are." Willis shook his head in amazement at the determination George had displayed.

"Here I am," agreed the Russian.

"That's one helluva story," said Willis. They had arrived outside his rooming house.

"Is different," George admitted with a shrug.

"I'll get the oil," said Willis. "Be right back."

Returning a minute later with a rag and an oilcan, he applied a few drops of oil to the rusty chain and rubbed it down with the rag. Then he squirted oil into the bearings.

"Can you lift the bike a little?" he asked George.

"Sure," said the Russian.

Willis the spun the wheels and noted to his satisfaction that they now ran silently. Then he pushed on the pedals and added more oil to the chain as it travelled smoothly over the cogs. "That should do it," he said.

"Is ready?" asked George.

"Yup," said Willis, wiping his hands.

George mounted the bike and pedalled onto the street. He turned in a wide circle, his face creased in a broad smile. "Is perfect," he said, and braked by the sidewalk. He started to reach into his pocket, but Willis shook his head.

"Not necessary," he said, "really."

"I buy you a beer," George countered.

Willis laughed. "It's a deal."

During that first year it was noted that Willis never seemed to walk anywhere; he cycled. Even during the winter months, when the temperature dropped, Willis could be seen pedalling from class to class. Only a major snowfall would ground him.

His roommate Hugh teased him about this. "You may not be a big wheel on campus, but you are definitely a wheeler." Thus was born the nickname that Willis would carry throughout his undergraduate days at Queen's, and beyond. To everyone he knew he became "Wheeler." His ability to maintain and repair bicycles added a certain cachet to this new name.

All the boys got packages from home that included various edibles, but Willis's mother's "fried cakes"—essentially doughnuts without the holes—were the clear favourite. Once a month these treats would be delivered by mail. Their arrival was always an occasion for celebration, and they were shared with Hugh, Don, and Mike. Anna Maria also liked to knit, and for years the only socks Willis wore were the heavy wool variety produced by his mother. He particularly appreciated their comfort when he cycled into downtown Kingston during the winter.

Willis struggled academically in his first year. Part of the problem was his social life. Many evenings when he should have been studying he was with his friends, playing pool at Ernie Cain's Billiard Parlor, bowling at the Frontenac Alleys, taking in movies at the Capital or the Grand, or watching hockey and basketball games. Instead of using the money that he earned repairing bicycles to pay his tuition, he used it to buy equipment from the Canada Radio Store on Princess Street and built his own wireless radio. His room soon became a gathering place for students who liked to listen to the broadcasts of games, particularly when the Queen's teams were playing away from Kingston. He dated a number of girls that first year, but none seriously. In short, he was having a good time—but at the expense of his studies.

In the spring, when he wrote his final exams, he knew he'd done poorly. He wasn't sure how poorly, though. He remained on campus, waiting for his marks. When they came he ripped open the envelope and, hands shaking, read the results. His mark in Mathematics II jumped out at him. It was 30, a failure. When he scanned the other subjects, he realized, with a surge of relief, that he'd managed to scrape through. He knew he'd have to take a supplemental exam in the Math course get into second year, but he was sure that with a summer to study he could bring it off.

Chapter 13

Leamington, June 1926

When Willis returned home he decided not to tell his parents about his academic shortcomings. He had a more pressing problem. The money he'd earned the previous summers working for his father wasn't enough to cover his living expenses in Kingston. His parents were paying his tuition, though, and he didn't want to ask them for anything more.

"Dad," said Willis on the first Sunday after his arrival back in Leamington, "I'd like your thoughts on something."

Clyde, who was reading the *Sporting News*, put the paper down. "Shoot," he said. He leaned back in his rocker expectantly.

"Well," Willis continued, avoiding his father's eyes, "I really enjoy working for you in the pool hall, but I thought this summer I'd like to try something different."

"What did you have in mind?" Clyde asked, and there was no reproach in his tone.

"I'd like to do something that relates to my studies," Willis continued, and looked up at his father.

"You could always try your hand at radio repairs," said Clyde. Willis had installed a rudimentary radio in the pool hall the previous summer. "You must have learned a few new tricks at Queen's."

"I could," Willis said, "but I don't think there's enough work to keep me busy full-time."

"I have a friend," said Clyde, "a regular at the pool hall, who works for Leamington Hydro. He might be able to suggest something."

A few days later Willis began work as a hydro repairman. His first week was spent under the supervision of an older man, Jack Wells, a First World War veteran who had learned his craft in the army. When Jack learned that

Willis planned to study electrical engineering he took him under his wing. Willis loved the work. No longer confined to the interior of a pool hall, he happily donned his spikes and tool belt every morning and scampered up telephone poles. In the beginning he was assigned to simple repair work, and he was in very little real danger as long as he made sure his safety harness was firmly in place. It was the perfect fit for a physically fit young man. In time, under Jack's direction, he took on more difficult tasks. By the summer's end he was repairing transformers.

Clyde had facilitated Willis's introduction to radio through his acquaintance with Sparks, and now his connections with Leamington Hydro had provided his son with a chance to gain practical experience that would prove invaluable in his career. Anna Maria may have been the one who gave Willis his solid sense of self-worth, but Clyde, too, in his own way, was instrumental in ensuring that his son had a chance to move beyond the life he himself knew as a failed ballplayer and a small-town businessman.

In Leamington Willis was still known as Willis, but back at Queen's "Willis" did not exist. In the fall he became "Wheeler" again, except to his professors. In September he managed to pass his supplemental mathematics exam, squeaking through with a mark of 51, and resolved to spend more time studying in the future. But once back at Queen's for his sophomore year he quickly fell into his old habits. His roommate had passed, but Don Phillips was on probation. Willis's room, with its wireless, was once again a hub of activity.

At the campus radio station there had been a major change. Professor Bain was no longer running CFRC, and Douglas Geiger had been appointed to replace him. Geiger had graduated from Queen's in 1923, had worked for the university as a demonstrator for a year, and then took a job with Bell Telephone in Montreal. He returned to Queen's in the fall of 1926 with ambitious plans for the station.

Willis found Geiger much more approachable than Bain, but it was Harold Minter, Geiger's assistant, who became his mentor. Minter was from Ottawa but he'd been born in England, and he retained the trace of a British accent. For this reason he often served as announcer, particularly for classical concerts and lectures, programs for which his cultivated voice provided an air of sophistication.

Minter was close in age to Willis, having graduated from Queen's in 1925, the year Willis had enrolled, and he took a liking to the enthusiastic sophomore. In November Willis helped Harold set up equipment in Grant

Hall for a live broadcast of the annual Science Social Evening, the first time the station had featured popular music with a local band. This program was more to Willis's liking than the classical concerts. He particularly liked the rendition of "Breezin' Along with the Breeze" by Sid Fox and His Serenaders, and found himself tapping his foot in time to the music.

After the broadcast they learned there had been transmitter problems, and the antennae current had had to be reduced. "At least the program was heard by the students on campus," said Willis, trying to cheer up the disconsolate Minter.

"I know, I know," replied Harold, "but I'm afraid that now the administration will want to confine us to the studio."

This gloomy prophecy was borne out when the next five musical programs were broadcast from Fleming Hall, which meant less involvement for Willis. Although Minter was able to make use of him when they were on location, space was limited in Fleming Hall and undergraduate participation was discouraged.

The station did continue to broadcast intercollegiate football games from Richardson Stadium, and this provided Willis with a chance to meet Foster Hewitt, a young man who would later become one of the most famous voices on Canadian radio.

On November 13 *The Toronto Star* sent Hewitt, a promising reporter, to Kingston to cover the game between Queen's and Varsity. The newspaper booked a telephone line to Toronto so that people in that city could listen to the game on radio station CFCA, direct from Richardson Stadium. The play-by-play commentary was to be provided by Hewitt.

Willis helped the crew from Toronto set up for the broadcast. The best vantage point was the steeply sloping tin roof of the stadium, more than fifty feet above the field. To reach the roof there was an iron ladder, and all the equipment, including the amplifier and the fifty-pound auto batteries, had to be carried up by hand.

Willis hefted one of the batteries to his shoulder. "Might as well get started," he said, and began scaling the ladder.

It took five trips to get everything into place.

Willis was introduced to Hewitt when they assembled on the stadium roof. "I have to thank you," the announcer said, glancing down at the batteries. "Those must weigh a ton."

It turned out that this had been the easy part of the afternoon. Hewitt was determined to gain the best possible view of the field. He pointed to the edge of the sloping roof. "I need to be down there."

The people from *The Toronto Star* were horrified. They had visions of their announcer slipping and falling into the crowded stands below.

"I have a suggestion," said Willis, and he explained his idea. If they tethered Hewitt with ropes looped under his arms and around his stomach, he could slide down safely to the edge of the roof. Once he was there they could secure the ropes to hold him safely in place, and he could brace his feet against a flagpole.

Hewitt was all for it, so Willis and the others fashioned a makeshift harness and then let out the ropes, a few feet at a time, as the announcer slid down the roof, hugging a microphone to the chest of his chinchilla coat. When he reached the flagpole he gave the men a thumbs-up signal. They secured the ropes to an iron railing.

The game began and Hewitt breathlessly described the action. Whenever he got too excited and leaned over the precipice, his crew would shout to him and tug the ropes.

In the fourth quarter it began to rain, which eventually met the bitterly cold wind from the lake and turned to hail. When the game was over Hewitt found himself frozen fast to the roof of Richardson Stadium. Willis and the others on the crest of the roof had to yank on the lines to tear him from the spot before they were able to drag him up the icy incline.

"Dammit," said a breathless Hewitt when he reached safety, "my coat is ruined."

Willis could make out what remained of Hewitt's chinchilla coat on the edge of the roof of Richardson Stadium. It stayed there for days, like the skin of an animal bagged by a big-game hunter.

Doug Geiger was at the controls that day, and Queen's prevailed. *The Kingston Whig* described the contest as the greatest game of football ever played. The story didn't mention Foster Hewitt.

A return match, this time for the league championship, was played two weeks later in Toronto. Hewitt's play-by-play was fed on a Bell line from Toronto to Kingston and carried over CFRC—the first time the station had ever arranged, and paid, to receive a program by a long-distance feed. Unfortunately for Queen's, Toronto's Varsity team emerged as winners.

The following day *The Toronto Star* reported, "Many Kingston people tuned in to Saturday's Queen's-Varsity struggle and say that that the results obtained were most realistic." The article added, without a trace of modesty, that "Foster Hewitt, who broadcasts for *The Toronto Star*, was said by the Kingston people who heard him on Saturday to be in a class by himself."

Another significant development in 1926 was the broadcast of the provincial election results on CFRC. Before that year, people seeking the latest election news would gather outside the offices of the town's two newspapers, *The Kingston Daily Standard* and *The Daily British Whig*, which merged in 1926 to form *The Kingston Whig-Standard*. As the returns came in on December 16, *The Whig-Standard* provided summaries for the campus station to broadcast, and the people of Kingston were able to hear the results in their own homes.

It was the first time the station had been involved in anything outside the university.

Chapter 14

Queen's University, 1927

When Willis returned to Queen's after the Christmas holidays he had an agenda he hadn't shared with anyone. He was twenty-one years old, and he was a virgin: that was something he desperately wanted to change. Although he was a sophomore and had gained a measure of self-confidence, in matters of sex he was still a novice. In fact, the idea of changing that status made him decidedly nervous. That was when George Ketiladze came to his rescue.

"What you doing Saturday night?" the Russian asked over a beer one evening.

"I don't know," said Willis with a shrug.

"You don't have girlfriend?"

"No," said Willis, then quickly added, "the curfew rules in Ban Righ Hall make dating difficult." It was a lame excuse, but he was reluctant to admit his lack of experience to George.

"Maybe you looking in wrong place."

"What do you mean?"

"No curfew at hospital."

"The hospital?"

"Nurses."

"Nurses?" Willis shook his head. "I don't know any nurses."

"Me," said George, pointing to himself with obvious pride, "I know nurses. I fix you up."

"How?"

"I play piano Saturday nights at local dance hall." He smiled. "Lots of nurses."

So at George's invitation, Willis turned up at the hall the following Saturday evening. He stood by the piano as his friend started to play. Soon a cluster of nurses joined them, requesting songs. When it was time for a break the women insisted George sit with them at their table, and Willis tagged along. George made the introductions. To Willis's surprise he seemed to know everyone.

Betty Murdoch wasn't the best-looking nurse on the staff of the Kingston General Hospital, but she was by no means plain. A little on the plump side, with curly brown hair, hazel eyes, and an infectious laugh, she was in her mid-twenties. After coming to Kingston from Toronto she had breezed through the training requirements and finished at the top of her class. Now, as a registered nurse, she was determined to get the most she could out of life. The "roaring twenties" had opened things up for women, and Betty took full advantage of this new freedom, refusing to be tied down to any one man. There would be time for that later, she reasoned, but for now she liked playing the field.

When it was time for George to play again, Willis asked Betty to dance. Sensing his nervousness, she initiated the conversation, asking about his studies. It was a wise strategy. If there was one subject Willis was comfortable talking about it was radio. He quickly relaxed as he explained his fascination with the new technology.

"I love listening to the radio," said Betty. "I have no idea how it works. It's magical."

During the evening Willis danced with other nurses, but whenever possible he sought out Betty. When he learned she roomed with another nurse on Stuart Street he worked up the courage to ask if he might accompany her home.

"I'd be delighted," she said.

As they approached her rooming house Willis took a deep breath. "How would you like to take in a movie with me . . . sometime this coming week?"

"That depends . . ."

"On . . .?"

"On what's playing," she said with a smile.

Willis found himself momentarily tongue-tied.

"I mean," she continued, sensing his confusion, "I may have already seen the movie. Which one did you have in mind?"

Willis found his voice. "The latest Douglas Fairbanks movie, *The Black Pirate*, is at the Capital." Willis hoped it was the right choice. A Laurel and Hardy Comedy was the feature at Kingston's other movie house. He was about to suggest this alternative when Betty spoke.

"I'm working evenings this week, but I'm free next Saturday night."

Willis relaxed. "I can call for you about seven," he suggested.

"Okay," she replied breezily. "I'll see you then."

The date went smoothly. He called for her promptly at seven and she was ready. She took his arm as they walked downtown, and any guilt Willis felt about his hidden agenda vanished. Betty had a way of putting men instantly at ease, and Willis was grateful. They both enjoyed the movie, and stopped in at the Queen's Café for a soda on the way home. There were a few students in the café, and one of the engineering students recognized Willis.

"Wheeler," he called across the room.

Willis waved an acknowledgement.

After they left the café Betty turned to Willis. "Wheeler?" she asked.

"It's my nickname," said Willis, a little embarrassed.

"How'd you come to be called Wheeler?"

"Well, I bike just about everywhere and I'm pretty handy at fixing bicycles."

"'Wheeler'," she said softly. "Yes, I like the sound of that."

Willis was relieved. He was pleased she liked his nickname, because he'd grown fond of it himself.

She took his arm and they walked back to her rooming house. Outside the front door she turned to face him. "Thanks," she said simply. "That was fun." She leaned forward and closed her eyes. Willis seized the moment. Their lips met and their tongues touched briefly. Then she broke the embrace, stood facing him, and smiled. Clearly the next move was up to Willis.

"Yes, it was fun," he agreed. "Are you free next Saturday?"

"Yes, I am," she said, then turned. "You have my number," she said as she started up the steps, and a moment later she was gone, leaving Willis alone on the street.

Walking back to his room, Willis found himself whistling a popular tune he'd recently heard on the radio. Although not musically educated he knew he was decidedly off key, but tonight that didn't seem to matter.

On their next date they went to the movies again, this time to see Victor McLaglen and Dolores Del Rio in a war film called *What Price Glory?* In the darkened theatre Betty permitted Willis to take certain liberties. It was all the encouragement he needed. He suggested dinner for the following Saturday, and when she agreed he made a reservation for two at the Venetian Gardens. It proved to be an expensive meal. With wine and a tip, the bill came to slightly more than ten dollars.

At the piano was his friend George Ketiladze, who gave him a knowing wink and played softly as they ate. When they had finished their coffee, Betty led Willis onto a small dance floor. He held Betty close and swayed to the music. It might have been the wine, but for the first time Willis actually felt comfortable dancing. It was almost midnight when they left. In the cab Betty told Willis that her roommate was working the night shift, and if they were very quiet he could come in for a drink.

Willis may have led on the dance floor, but in her room Betty was the one who led. Within minutes they had shed their clothes and were in bed.

"You came prepared, I hope," she whispered in the dark.

Willis was glad he'd managed to overcome his embarrassment at the Rexall Drug Store. "I'm a good Boy Scout," he said, and tore open the foil on the condom.

Betty must have sensed it was the first time for Willis. She took the condom and expertly slid it on. "You don't feel like a Boy Scout," she said with a giggle. "Maybe a Scout Master?"

It was all over in a matter of minutes, and Willis lay on his back, panting and flushed. He wondered what came next. He'd been so preoccupied with getting laid he hadn't given any thought to the post-coital protocol. Fortunately this was familiar territory for Betty.

"I think a drink is in order," she said, and rose. He could just make out her silhouette as she stood by the window, pouring. When she returned he sat on the edge of the bed and took the glass. The air held the scent of their lovemaking and a faint trace of her perfume.

"Cheers," she said, and they clinked glasses.

Willis sipped. It was rye whiskey, his favourite.

On the way back to his own room he stopped to light a cigarette, and as he drew in the smoke Willis wondered about the mysteries of sex.

This initiation had been pleasant, and he had certainly enjoyed it, but it hadn't quite measured up to his expectations. The experience hadn't really changed him. True, he was no longer a virgin, but in the end did it really matter that much? Maybe it would have been different if he'd been in love with Betty. He walked on through the night, vaguely troubled. Was it guilt he was feeling? Had he taken advantage of Betty? Or had it been the other way around—had she taken advantage of him? He took a last drag and flipped his cigarette onto the road, sparks flaring briefly as it hit the pavement.

It was a busy year for Willis. His social life improved, and he also spent a good deal of time helping out at the station. The spring graduation ceremonies were beamed live from Convocation Hall with no transmission problems. In fact, the program came in for praise in an article in *The Daily Standard*. It described the broadcast quality as excellent.

The word "excellent" was missing from Willis's exam results that spring. He had tried to cram, had stayed up nights studying, but there had been too much material to cover adequately. When he looked at his marks he felt his world crash around him. He had failed more than half of his courses. A letter was enclosed, asking him to report to the Dean's office the following morning. Don Phillips had already seen the Dean; he'd been told to pack his bags, because his days at Queen's were over.

As he sat outside the Dean's office fearing the worst, his palms sweating, he wondered what he would tell his parents. They'd had to borrow money for his tuition and now he'd let them down. When the Dean's secretary indicated it was his turn, Willis rose from his chair and walked into the office.

Dean Clark looked up from his desk. He was a rotund little man with a thatch of unruly red hair. He was holding a manila file folder, and when he opened it Willis saw that it contained copies of his exam results from both this and his freshman year.

"Well, young man," said the Dean as he removed a pair of wire-rimmed spectacles, "I've been going over your marks. Have you anything to say for yourself?"

Willis fought back a wave of nausea. "I'm just sorry I didn't do better, sir."

"A little too much partying?"

Willis shrugged. "I guess that was part of it."

"Just part?" The Dean arched his eyebrows.

"Yes . . ." said Willis, hesitating. "But I found some of the classes pretty tough."

The Dean smiled a thin smile. "Not quite what you expected?"

"No, sir."

The Dean sighed. "Didn't you learn anything last year?" He replaced his glasses and looked down at the papers on his desk. "Those marks . . . the supplemental math exam. They were a warning, a warning you obviously failed to heed."

"Yes, sir."

"Frankly, I'm surprised. It's not uncommon for us to lose students in their freshman year, but by second year most of the survivors have wised up."

Willis waited.

"There is only one possibility, as I see it."

Willis stared at the Dean. Was there hope, after all?

"You have a supporter, apparently, in Professor Minter. He thinks quite highly of you. So if you are serious and want another chance, you'll simply have to repeat your second year."

Willis felt a wave of gratitude. The work he'd done with Minter had paid off. But he was also filled with shame. Repeating a year was something he hadn't considered. Some of his friends had had to repeat a year, but that was in high school, not university.

"Well?" the Dean prompted.

"I'm not sure," Willis stammered. "It would mean an extra year."

The Dean closed the file. "Yes, it would, and it is not something we routinely suggest. However, in your case we are willing to make an exception."

Willis thought of Don Phillips, and realized at that moment that he was being offered something rare—a second chance. He wasn't going the pass it up. "I accept," he said; "and I appreciate the opportunity."

"Good," said the Dean, smiling. "Don't let us down."

"I won't," Willis said with conviction.

The Dean took a pipe from an ashtray on his desk, struck a match, and lit up. "Good luck, then, Willis," he said, and exhaled a plume of smoke.

"Thank you, sir." Willis stood, turned, and walked out of the Dean's office. As he did, he thought he recognized the smell of the Dean's tobacco. He was pretty sure it was the same locally grown, rough-cut mixture that his father favoured.

Chapter 15

Leamington, 1927

When Willis returned for the summer to Leamington and his job with Leamington Hydro, his mother told him in confidence that she was worried about Clyde's health, so he decided to withhold the news that he would be repeating his year at Queen's. It was unlike his father to complain about anything, but he seemed to lack energy, and had trouble sleeping. Anna Maria had observed an uncharacteristic irritability in him, and had tried to convince him to see a doctor on a number of occasions, but without much success. Late in July, though, he was overcome with such a profound sense of dizziness on his way home from work that he had to sit down on the front steps of a neighbour's home. He was glad that it was late and there was no one around to see him. Eventually his head cleared, and after mopping the sweat from his brow he managed to make it home.

The next morning at breakfast he told Anna Maria about it. "I don't know what it was," he said, shaking his head. "Never felt anything like it. It was as if I had just got off one of the rides at the country fair. Dizzy."

"I'm going to make an appointment for you with Dr. Brown," said Anna Maria in a tone that declared the discussion closed."

"I guess it wouldn't hurt," Clyde mumbled.

Anna Maria was filled with foreboding. She hadn't felt this way since Willis had come down with influenza. But she pushed the feeling away and poured Clyde a second cup of coffee.

When the results of Clyde's blood tests came in the doctor asked to see both Anna Maria and Clyde in his office. Anna Maria's sense of foreboding returned as the couple made their way downtown. It was unusual for a doctor to ask to see a wife when it was the husband who was the patient.

When they arrived they were immediately ushered into Dr. Brown's inner office. He motioned to the chairs. "Have a seat." He opened a manila file folder and looked up. "I'm afraid the news isn't the best," he said, tapping the papers before him. He looked directly at Clyde."There is simply no good way to put this. You have diabetes." He paused to let the word sink in.

Anna Maria broke the silence. "Diabetes?"

"Yes, diabetes. It happens when the body has a problem absorbing sugar properly."

Anna Maria had heard the word, and knew there had been some recent medical development in its treatment, but she hadn't paid much attention to the details. She remembered that some deaths at her father's tannery had been due to diabetes.

"What can you do?" asked Clyde, his voice barely rising above a whisper.

"Well," replied Dr. Brown, smiling for the first time, "diabetes used to be fatal, but a few years ago a team of Canadian scientists discovered an effective way of treating it."

"What did they discover?" asked Anna Maria, trying to keep the fear from her voice.

"It is something called insulin. In most people the body makes its own, but in diabetics it falls short. These scientists I mentioned have found a way of making insulin in the lab."

"Is it a cure?" asked Clyde, hopefully.

"In the short term, yes," replied Dr. Brown. "It only provides a temporary solution, though. In order to be effective it has to be repeated."

"How often?" Anna Maria asked.

Dr. Brown consulted his notes. "In Clyde's case it would be daily."

"Every day?" said Clyde in disbelief.

"I'm afraid so."

"You mean I'll have to take this medicine, this insulin, for the rest of my life?"

"Yes, you will," the doctor replied, and, after a moment's hesitation, continued. "The insulin has to be injected. That's why I asked to see your wife. I can teach her how to administer it."

Anna Maria didn't much like the idea, but she knew that she would overcome this obstacle, as she had so many others in her life. She would learn how to administer the insulin. It was clear that Clyde's life depended upon it.

Chapter 16

Kingston, 1927

Willis returned to Queen's in the fall with a new sense of resolve. His father's illness had put things in a new perspective. He was determined to do better. He realized that part of the problem was that his room in MacKinnon's boarding house, with its wireless, had become too popular. It was hard to get any serious work done there.

"I'm sorry," he said as he explained his decision to Hugh, "but I nearly flunked out last year. If I'm going hit the books a little harder I'll need a single room."

"I understand," said Hugh. "Maybe it's time for me to find a place of my own as well."

"Still friends?" asked Willis.

"Of course," Hugh assured him.

A week later Willis moved into a room in a house on Birch Avenue, closer to the campus. He was the only boarder.

He also sought out Harold Minter. He found him in the Fleming Hall studio. "I can't thank you enough," he said simply.

His mentor smiled. "You can thank me by spending more time on your studies this time around." He shook his head slightly. "In fact, I'm to blame for your some of your problems last year. All those hours you spent helping me . . ." His voice trailed off.

"But that's the stuff I love," Willis protested. "It's real, not theoretical."

"I know, Willis, but you're going to have to master the theory if you want to graduate."

"I know, I know," Willis echoed, "and I will, I promise." He told Minter about his decision to find a place of his own.

"Makes sense to me," the professor said.

Willis didn't abandon the radio station work entirely, but he rationed it out. Each day, after his lectures, he'd pore over his notes, making sure he understood the material. Only after that was done would he head over to Fleming Hall.

His relationship with Betty Murdoch had cooled. They were seeing less and less of each other. One night Willis dropped into the Queen's Café after an evening in the library and was surprised to see Betty with an older student, someone he didn't know. They exchanged polite greetings. At that moment it was clear that Betty had moved on. Willis found he wasn't jealous, simply relieved. It was time for him to move on as well.

More and more students were, like Willis, entering Queen's to study electrical engineering. In the summer of 1927, to accommodate this increase in enrolment, the basement studio was converted into one large classroom and the station was moved upstairs to the main floor. It was decided that this was a good opportunity to upgrade the equipment. New and larger vacuum tubes were now available, and these were ordered from Mullard in England. Doug Geiger was the driving force behind these changes. There was no time to test the new equipment before it was used for the first time, to broadcast the exhibition football game between Queen's and the Toronto Argonauts on October 8, but with Geiger at the controls it went off without a hitch. *The Kingston Whig-Standard* commented, "Reception was of high order and that station was in much better form."

CFRC was expanding its scope, and venturing into more ambitious programming. A play produced by the Queen's University Drama Club was broadcast live from Convocation Hall in the Old Arts Building, and jazz made its first appearance on CFRC in a program from Fleming Hall. In February, the Kingston Choral Society presented a ninety-minute recital entitled *Part Songs and a Cantata.*

There were also plans to produce a lecture series, and *The Queen's Journal* speculated that "at some time in the future, one may be able to obtain a degree over the air. One drawback would be that professors wouldn't be able to pick out the sleeping listeners who commonly frequent classrooms."

When the final exam results came out that spring it was clear that Willis hadn't been one of those dozing students. His hard work had paid off. There were no failures, and his overall average had jumped twenty percent from the previous year. Of course he was repeating many of the subjects, so some improvement could be expected; the real test would come in third year. But Willis was confident he had found a way to balance his studies, his involvement with the radio station, and his social activities.

When he shared the good news of his results with Harold Minter, the professor smiled. "I knew you could do it," he said.

Willis waited, sensing that Minter had more to say.

"I'm sorry I won't be around to see you finish up your degree."

"What?" said Willis, confused.

"I'm leaving Queen's," said Minter. "Bell has offered me an interesting job. Doug Geiger is leaving too. We're both venturing out into the real world. We want more experience." He smiled. "And we want a chance to make a little more money."

Willis felt a wave of sadness. "I'm going to miss you," he said. "It won't be the same around here without you."

"Stan Morgan is taking over the department. He's a Queen's man, class of 1916, and he's been teaching at the University of Alberta. So you'll be in good hands." Minter paused, momentarily lost in thought. "And Bill Richardson is going to fill in for me," he continued. "You must know Bill? He graduated in '26, your first year."

Willis vaguely remembered Richardson, but knew that he could never replace Minter, the man who'd demonstrated his faith in Willis when it mattered most.

"Well," said Willis, remembering his manners, "good luck in the real world."

"Thanks," said Minter with a laugh. "And who knows? In a few years maybe you'll be joining us." The two men shook hands and Minter turned and walked away, heading into town. Willis watched him depart, and the sadness returned.

Chapter 17

Queen's University, 1928–29

When Willis returned in the fall to enter his junior year, he realized that he would have to concentrate on his studies even harder if he was to maintain the level he had reached. This time he would enjoy none of the advantages he'd enjoyed the previous year, when he was repeating subjects he'd already taken.

The new men in charge at the radio station, Stan Morgan and Bill Richardson, were caught up in what was, for them, a new experience. They took on most of the interesting tasks themselves, with the result that there was less chance for the undergraduates to enjoy hands-on work at the station. Willis wasn't that disappointed. It wasn't the same with Minter gone. He deliberately cut back his time in Fleming Hall and applied himself to his course work.

He did suffer a physical setback in the first week when he sprained his right ankle in an intramural soccer game. He was barely able to walk off the field, but he made it to his room with the help of some friends. The next day he took a cab to the Kingston hospital.

"It's a bad sprain," said the examining doctor. "Too bad you didn't break it."

Willis was puzzled. "Break it?"

"With a break we could put you in a cast," the doctor explained. "That way you'd actually be better off. When a broken bone heals it's as good as new. A sprain can be more complicated."

"No cast?" Willis was secretly relieved.

"No, but you'll be on crutches. When the swelling goes down I'll bind it. From the look of things you'll have to stay off it for several weeks."

The next month was difficult for Willis. He hated the crutches; they made navigating stairs particularly challenging. His arms ached, and he developed blisters on his palms. During this time he became an even more devoted listener to the Queen's radio station.

When Willis finally got back on his feet he was able to attend the Queen's football games in Richardson Stadium as a fan rather than a technician, so he thoroughly enjoyed himself. George Ketiladze was often his companion. George had by then become something of a legend at Queen's. As a freshman he'd quickly won over his professors with his brilliant mind. Then in his sophomore year he took up wrestling for the first time, and promptly won the intercollegiate heavyweight title. George had never played football, so Willis found himself explaining the rules of the game to him.

It was a good year to be a Queen's fan. The team was exciting, and it capped off a great season with a win over Varsity for the league championship. The game was played in the stadium on a chilly Saturday afternoon in November, and Willis shared a flask of rye whisky with the big Russian. As a result the two were less than sober by game's end, a state they shared with a good many enthusiastic fans.

George didn't care for hockey, and it was of even less interest to Willis. Nonetheless, he did tune in when the season started in January, and was surprised to hear that the station decided to curtail these broadcasts after just two games. The February 26 edition of *The Queen's Journal* carried this explanation:

> The reason that hockey is not broadcast is because the cheering of the crowds drowned out the announcer's voice, and until a special box is made in the arena, broadcasting hockey will be abandoned. This year they are thinking of building a compartment on top of the stadium to facilitate broadcasting of rugby games. If this is done no longer will Toronto scribes complain of being held up by ropes nor will they graphically describe hair-raising perils of broadcasting on the roof of the stadium.

Willis knew there had been a good deal of talk about the need for a special broadcast facility in Richardson Stadium to shelter the play-by-play commentator from the elements, and it made sense to him. Harty Arena, though, was an enclosed space where weather wasn't a factor. To Willis it seemed the sort of challenge technology could solve. But no one had asked him.

Willis shared this opinion with the irrepressible George Ketiladze one evening when he dropped by at the Venetian Garden Restaurant, where George was now playing.

"There has to be some way of filtering out the background noise electronically," he said during one of George's breaks.

"If they let us at equipment," said George, sipping his rum and Coke, "I bet we find solution."

Willis shook his head. "That isn't going to happen," he said, lighting a cigarette.

"We could . . . how you say . . . sneak in?" George suggested with a grin. "Take look around. After dark. Everyone sleeping."

"Sure," said Willis with a sigh, "and if we did come up with the answer, imagine trying to explain how we managed it."

"True," the Russian mused.

Willis knew that with a little encouragement this was just the sort of thing George would undertake. "Nope," said Willis, "let's let the profs work it out. That's what they're paid to do. As for me, I have to get back to the books." He ground out his cigarette in an ashtray and rose from the table.

"And I get back to piano," said George, and drained his drink.

After the hockey broadcasts were cancelled that winter CFRC fell silent. When Willis dropped by he found that Morgan and Richardson were devoting most of their time to upgrading the outdated Mark IV transmitter. They were also lobbying the administration for additional funds for new equipment. It had been suggested that the station should rent out its broadcast facilities to commercial interests, but this provoked a heated debate. It had long been the policy of the university to insist that the station should be maintained for purely educational purposes.

New radio stations were going on the air weekly, and they were no longer confined to large metropolitan centres. It will only be a matter of time, Willis thought, before some entrepreneur will target Kingston. Why not allow limited commercial activity at CFRC? That way the station can finance itself, and it will be able to use some of its revenues to make sure that students have the most up-to-date equipment to work with. It might even discourage private radio from seeking a foothold in Kingston.

In the final months of the school year, though, Willis stayed clear of the station and its politics. He studied hard, in order to make sure he was ready for his final exams. He resolved not to have to repeat the fiasco of his sophomore year.

When the results were posted, his marks reflected all his hard work. He not only passed everything, he excelled. In Electrical Engineering he recorded 86, a personal best. His average was five points higher than it had been the previous year, when he'd been repeating subjects. Willis had learned his lesson.

Chapter 18

Kingston, Montreal, 1929

In his junior year Willis concentrated on his studies, but in the spring, when it came time to seek summer employment, he wanted a new challenge. He had spent three summers working for the Leamington hydro company, reading meters and climbing poles to make repairs. He lived at home, so there was no room and board to pay, but he wasn't learning anything new, and now, about to enter his final year at Queen's, he began searching for something more challenging.

One evening, after spending a couple of hours in the library studying for his final exams, Willis was joined by Ralph Curtis, a fellow engineering student. Willis liked Ralph, a short, stocky lad with a quick wit and an easygoing personality. They'd become friends when he'd helped restore Ralph's bike after it had been left out one summer, exposed to the elements. Ralph was from Montreal. His parents lived in the Notre Dame de Grâce section of the city. He'd told Willis the area was known to English-speaking Montrealers as NDG.

"NDG," he'd said, laughing. "They say it really stands for No Damn Good."

As they left the library together Ralph asked, "Feel like a beer?"

"Sure," said Willis, "we deserve a reward after all the hard work we've put in." He took out a package of cigarettes and offered one to Ralph. They lit up on the steps of the library, then headed downtown.

The night was mild; the last of the snow had disappeared and there was a hint of spring in the air, the faint scent of lawns and gardens awakening from their long hibernation.

"Speaking of work," said Willis, "have you got anything lined up for the summer?"

"Probably the same as last year," Ralph replied. "My dad works for Northern Electric. They hire students to fill in for workers who are on vacation. He can usually find something for me."

"What sort of work?"

Ralph shrugged. "Nothing very exciting. Assembly line mostly. The company makes telephones for Bell. What about you?"

"I can always go back home to Leamington and live with my folks," said Willis. "The local hydro company has work for me. I'm not sure, though. If I have to climb one more pole I may go nuts."

Ralph laughed, "Well you wouldn't find me up there. I'm afraid of heights."

"Not a job for the faint of heart," Willis agreed.

"You know," Ralph countered, "my dad might be able to find something for you if you'd be willing to come to Montreal."

"You're kidding!" Willis stopped and turned to face his friend. "Do you really think so?" He'd never been to Montreal, but the stories he'd heard about Canada's largest city, with its wide-open nightlife and its cosmopolitan mixture of French and English, had always intrigued him.

"I'll write him tomorrow," Ralph said, "and see what he can do for you."

Willis didn't consider himself the least bit superstitious, but he crossed his fingers anyway. In the distance he could see the moon reflected on the waters of Lake Ontario, and the lonely sound of a train whistle broke the silence of the night.

A week later Ralph gave Willis the news. "Dad came through," he said excitedly. "He's lined up something for you at the same Northern Electric plant where I'll be working."

"I can't believe it," said Willis, shaking his head. "You're not putting me on?" Ralph did have a mischievous sense of humour.

"No," scoffed Ralph. "I wouldn't joke about something that important."

"Well, make sure you thank your dad for me. And when I get to Montreal I'll thank him again myself, in person."

Privately, Willis wondered how his folks would take this news. He knew how much they looked forward to the summers, when he was back with them. He'd have liked to break it to them in person, but there wouldn't be time. Instead, he phoned.

When his mother answered they chatted briefly about his studies, and she brought him up to date on the latest Leamington gossip."

"How's Dad?" Willis asked.

"He's managing."

"Good," said Willis. He paused, then pushed on. "I've been thinking . . ."

"I don't like the sound of that."

Willis marvelled once more at his mother's uncanny instinct for picking up on the hidden meaning in the simplest phrase. "You know how much I like being home for the summers . . ."

"Yes, and how much we like having you home." Then it was Anna Maria's turn to pause. "But . . .?" she asked.

"But this summer I've been offered a chance to work in Montreal. Something exciting . . . but I wanted to check with you first, particularly considering Dad's condition."

When Anna Maria spoke, she didn't try to hide the note of disappointment in her voice. At the same time she surprised Willis. "You said the job would be exciting?"

"Yes, I did. It would be at Northern Electric in Montreal. They make telephones, public address systems, all sorts of equipment. Right up my alley, in fact."

"Well," said Anna Maria resignedly, "if that's the case, we certainly won't stand in your way. I'll explain to your father."

"Thanks, Mom," said Willis with a surge of relief, tinged with sadness.

"But Willis," his mother continued, "promise me that you'll call home every weekend."

"I will."

After they exchanged goodbyes, Willis gently replaced the phone on its cradle.

To save money, Willis decided to hitchhike from Kingston. He'd only been on the highway a few minutes when a car pulled over. "Where you heading, son?" the driver asked.

"Montreal."

"Well, you're in luck. I'm on my way there. Hop in."

Willis couldn't believe his good fortune—a lift all the way to Montreal. His benefactor was a travelling salesman, Terry Bennett, who was on his way home after a series of meetings in Toronto. The two engaged in small talk, Willis providing details of college life and Terry complaining about the amount of time he had to spend on the road, away from his family.

Willis's first view of the city through the windshield of Terry's Ford Model A coupe didn't live up to his expectations. In the distance, through a haze of smoke, he could see the land rise like the dark hump of a giant beached whale.

"There she is," Terry announced, then adopted a French accent. "Zee beeg citee of sin."

"Is that Mount Royal?" asked Willis, trying to keep the disappointment from his voice. He didn't want to offend Terry.

"Nope, that's Westmount," said the salesman. "Mount Royal is beyond it, to the east." As the car drew closer he pointed out a large domed structure. "That's Brother André's Shrine on the left."

"Must be big," Willis said. "It sure stands out."

"It's big all right," the salesman agreed. "Catholics come from all over the world to see it, climb the steps on their knees, hoping for a miracle." He glanced over at Willis. "You're not a Catholic, are you?"

"No," said Willis, shaking his head in amusement, "not a Catholic. What about you?"

"A Protestant, but not a very good one. The wife goes to church every Sunday, but I try to avoid it when I can." He smiled. "She drags me there on special occasions of course—you know, Christmas, Easter, weddings."

The car was closer to the city now, and through the open window Willis could smell a mixture of smoke, exhaust fumes, manure, asphalt, and something sour he couldn't identify.

As they drove along, Willis noticed a sign indicating an exit for Lachine, and he recalled that the early French explorers, expecting to find an easy route to the Far East, had mistakenly named the area where rapids blocked their progress "La Chine"—China—the country they were seeking.

The salesman turned off the highway, drove through an area of factories, and eventually came to a row of residential semi-detached homes. "This is far as we go, young man." He braked and the car came to a stop.

"You live here?" asked Willis.

"Not here, exactly. Couple of blocks from here. But this is a streetcar stop and the service is pretty regular." He looked at his watch. "There's one every fifteen or twenty minutes. You said your friend lives in NDG?"

"Yes," said Willis. Ralph's parents had offered to put Willis up until he found a place of his own.

"Well, you'll have to transfer, but the conductor will let you know."

Willis reached across and shook the man's hand. "I can't thank you enough," he said. "I could have been out there for hours, hitching. But you turn up within minutes and I land a ride all the way to Montreal."

"No problem," the salesman countered, "always willing to help a student. Wasn't so long ago I was hitching rides myself."

Willis opened the door and got out, then retrieved his suitcase from the back seat. He closed the door and leaned down to the window. "Thanks again," he said, and watched as the salesman put the car in gear and drove off.

Willis lit a cigarette and waited. When an eastbound streetcar appeared, he ground out the butt and climbed aboard. "I'm going to NDG," he told the conductor.

"NDG?" The conductor looked at him quizzically.

"My friend lives there, on Cavendish Boulevard."

The conductor smiled. "Ah, Cavendish . . . Notre Dame de Grâce. OK, but I can only take you so far." He spoke English with a French accent, but it was somehow different from the accent of Willis's high school French teacher, a Parisian who had regaled his students with tales of life in France.

Willis bought tickets and deposited one, and the conductor handed him a transfer. "You must change in Ville St. Pierre."

As they headed east out of Lachine, they rocked along into an area with a mixture of shops, small houses set close to the street, and the occasional factory. The conductor called out the stops in French. Eventually he turned to Willis and indicated that it was time to disembark. "You wait there," he said, pointing to a streetcar stop across the street. "Take Number 27."

Willis got off, hoping he wasn't lost. He had Ralph's home telephone number in his pocket, but decided he would wait for the Number 27 and see if it brought him closer to his destination. He noticed that most of the stores on the street bore French names. When his streetcar arrived he again asked the conductor directions, this time using the street name, Cavendish Boulevard.

"Cavendish?" said the conductor with a shrug. "Oui, d'accord." He took the transfer and Willis sat down. The ride this time took them up a long slope, through an area of open fields. He caught an occasional glimpse of water to his right. Probably the St. Lawrence River, he thought. Eventually the streetcar levelled off and moved east, entering a residential district with brick apartment buildings, the occasional grocery store, and a few single-family homes. Willis caught sight of a sign that said Sherbrooke

Street. This was good. He knew from the directions Ralph had given him that Sherbrooke would eventually cross Cavendish.

Willis listened as the conductor announced the stops, and he noticed that now most of the street names were English. They were also called out twice. It was "Aven-noo Westminster . . . Westminster Avenue." Eventually he heard the words "Boo-lee-var Cavendeesh . . . Cavendish Boulevard," and pulled the cord to signal his wish to get off. When the streetcar stopped the conductor turned and glanced back at Willis, who had already risen from his seat. He gave him a reassuring smile and the side doors opened. Willis stepped down, hefting his suitcase.

He found himself on the corner of the busy intersection of Sherbrooke and Cavendish. He crossed the street and made his way north on Cavendish, seeking the number of Ralph's home. It turned out to be a short walk—which was fortunate, because Willis's suitcase seemed to grow heavier with each step. The Curtis home was a red-brick two-storey house set well back from the street. A flagstone path curved through a recently cut lawn to the main entrance. Willis mounted the steps, rang the bell, and checked his watch. It was just after six. The door opened, and Ralph greeted him with a broad smile.

"Wheeler, you made it! Welcome to Montreal! Let me take that," he said, reaching for the suitcase, and ushered Willis inside. "Your timing is perfect. We'll be eating shortly. I hope you like pork chops."

Willis was too tired to protest that it wasn't really necessary, that he hadn't expected to be fed. But he was hungry, and pork chops sounded great. Ralph took the suitcase into the spare bedroom, then introduced Willis to his parents.

CHAPTER 19

Montreal, 1929

Willis woke to the sound of church bells and the smell of freshly brewed coffee. It took him a moment to get his bearings, but then he remembered that he was in Montreal, not Kingston, and felt the excitement that always came with the promise of a new experience. Tomorrow he would be starting his summer job at Northern Electric, and he wondered what sort of work he'd be doing. Ralph had been a little vague about it, but Willis decided he'd worry about that when the time came.

Today was Sunday, and because Willis was new to the city Ralph had promised to show him around. Willis looked at his watch. He was surprised to see that it was after nine o'clock. He was usually up long before this. He'd slept soundly. Not that sleep was ever a problem for Willis, but the trip from Kingston had taken more out of him than he realized. He stretched and pushed back the covers. There was a gentle tap at the door and Ralph poked his head in.

"Ready for some breakfast?" he asked.

"Sure." Willis rose from the bed and rummaged in his suitcase for his shaving kit. "Let me wash up first and I'll be down in a few minutes."

After bacon and eggs, toast, marmalade, and two cups of coffee, the boys were ready to set out. The family car, a dark green Chevrolet sedan, was in the driveway. Ralph climbed behind the wheel and Willis slid into the passenger seat. He whistled softly. "I can see we're going to be travelling in style."

Ralph pushed the starter and the engine caught. "She's a 1927 Chevy National. Four-cylinder engine. But Dad is going to trade her next year. Chevrolet's new line is more powerful, with six cylinders." He backed the car out onto the street and they set off.

To Willis, who didn't know a great deal about cars, the sedan seemed to have plenty of power.

They drove east, and Willis noticed that the modest houses of Notre Dame de Grâce soon gave way to larger, more expensive residences. "We're in Westmount now," Ralph explained, "where the real money resides." He laughed. "When you live here you've got it made."

Willis looked around him as the Chevrolet climbed steadily, and it seemed that the higher they went the larger the homes became. Eventually the road levelled off and the homes gave way to a wooded area. Ralph pulled into a space opposite the woods and turned off the ignition.

"This is the Westmount lookout," he said, getting out of the car. "It has a great view of the city."

There were dozens of people milling about, many with cameras. Some were leaning on a long concrete wall, peering south. When they reached the wall Willis looked down and saw a vast panorama of city spread out before him—mansions of Westmount in the foreground, then apartments, and then, farther down the slope, where the land levelled off, larger, more substantial office buildings and department stores. Beyond these lay a jumble of factories, identified by their tall smokestacks. Tracing down this sprawl of brick, stone, and concrete, streets stretched south like veins to the waters of the St. Lawrence River.

"Impressive?" Ralph asked.

"Amazing!" Willis knew that Montreal was the country's largest city, but the fact hadn't registered fully until now. And he'd never seen a river as wide as the St. Lawrence.

Ralph pointed into the distance. "If you look carefully, to the right of where the Victoria Bridge meets the land, you can just make out the Northern Electric Building."

When they got back in the car, Ralph headed east from the lookout. They drove down a road that wound through more expensive homes until they came to a stop at what was clearly a main street.

"This is Côte des Neiges," said Ralph. "It's the main road between Westmount and Mount Royal." When there was a break in the traffic he gunned the engine and cut across the street onto a dirt road. The car fishtailed slightly, kicking up dust, but Ralph quickly had it under control, and they drove on, up another slope. Now, for the first time, Willis saw before him the rise of land that had given the city its name. It was clearly higher than Westmount, but to Willis it didn't look like a mountain at all, but a huge hump-backed hill. There were no homes on this hill, and to the left, as far as he could see, were row after row of tombstones. To the right, as they rounded a curve in the gravel road, was an expanse of water. Ralph nosed the Chevrolet into a parking area, turned off the ignition, pulled on the emergency brake, and looked at Willis. "What do you think?"

Willis glanced through the windshield. "That's Mount Royal?" He tried to keep the disappointment from his voice.

"Yep, that's it, although I have to admit it looks a lot more impressive when you're downtown. Here, we're actually half way up the mountain."

Willis gestured to the water. "What's that?" He could see people strolling, kids playing with toy sailboats.

"That, my friend, is Beaver Lake. It's man-made, part of the park."

Looking into the distance beyond the lake, Willis watched a pair of mounted policemen making their way along a path that wound its way up the slope and through the trees. "I don't see any big homes here," he said, "nothing like the ones in Westmount."

"Well, there are a few up there," said Ralph, pointing to his right, "but most of the area is parkland, open to the public. It's our own version of New York's Central Park. In fact, it was Olmstead, the man who designed Central Park, who did the planning for Mount Royal."

Willis shrugged. "I guess when I heard people talk about Mount Royal—the mountain, I mean—I had visions of something bigger, more majestic."

Ralph laughed. "Well, it ain't the Rockies, that's for sure."

Ralph got out of the car and Willis followed him. They walked along the path that bordered the lake. The first thing Willis noticed were the women. They chattered animatedly to each other as they promenaded languidly along the water's edge in twos and threes. These women seemed to him to have stepped out of a fashion magazine. He'd seen some pretty ladies on brief visits to Detroit, but none like these. He gave a low whistle under his breath.

Ralph punched him playfully on the shoulder. "I told you Montreal was special, Wheeler."

Two of the women approached and gazed directly at the young men without a hint of shyness; it was simply a frank appraisal. They were, Willis noticed, perfectly made up, their hair was carefully styled, and their dresses showed a good deal of leg. Embarrassed, he glanced down at his shoes as they passed.

A faint fragrance lingered and Ralph sniffed their air.

"Ah," he said, savouring the scent, "that's what I missed in Kingston."

Willis shook his head in wonder. "I can see why." He searched in his mind for the right word to describe these women. Stylish came close, but it didn't seem complete. But then words were never his forte. His thoughts were interrupted when a bright red ball rolled across the path at his feet. He picked it up and heard a small voice calling, "Ici, Monsieur, s'il vous plaît."

Willis turned and saw a small boy standing on a slope of grass to his right. Several other children were clustered there, watching him. Smiling, he lofted the ball back and watched as they scrambled for it. Beyond the children groups of people were enjoying the day, their picnic blankets spread out around them. Some of the men dozed, fedoras shading their eyes, vests unbuttoned. Others smoked while the women fussed with food. These were not the glamorous ladies of the lake, but mothers who looked more like the women he knew back home.

"Families," said Ralph with a sweep of his hand. "They don't have much in the way of yards or lawns so they use the mountain whenever they can."

Willis nodded in the direction of the children who'd been playing with the ball. They'd been called to a blanket and were now squatting there, taking sandwiches and drinks from a stout woman in a print dress, her grey hair tied back in a bun.

"Must be a reunion," Willis commented. "At least a dozen or more kids."

Ralph shook his head. "I doubt it. I think what you have there is a typical French-Canadian family. Only two adults," he noted. "Mom in charge, and Dad . . . supervising." He smiled as the man, on cue, took a sip of beer from a bottle in a brown paper bag.

They continued around the lake, and Willis scanned the mountainside. There were many such family groups spread out on the grass. Some were eating, others playing games, and a few were flying kites. Wherever he looked people were enjoying the day, a tapestry of families assembled on the gentle slopes of Mount Royal, in the very heart of the city. This was completely at odds with the hustle and drive he'd expected to find in Canada's largest metropolis. He was glad Ralph had brought him here on his first day in the city. It was as if he'd been given a glimpse into something that was very public yet at the same time very private. Montreal had begun to captivate Willis—but on this Sunday, standing by Beaver Lake, he had no idea what an important role it would play in his future.

When they'd completed the circuit of the lake, Ralph suggested it was time for lunch. They returned to the car and drove back to Côte des Neiges.

"I'm going to treat you to one of Montreal's most famous culinary delights," said Ralph as they drove south toward the city centre.

Ralph made a left turn at a major intersection and Willis noticed a familiar sign. "Sherbrooke Street?" he asked.

"Yep, it's one of the longest streets in the city. All the way from NDG to the east end, where all the oil refineries are."

As they drove east along Sherbrooke they saw ornate apartments lining the left side of the street and, on the right, lower buildings that housed galleries and small stores that, to Willis, looked exclusive.

"This stretch used to be known as the Golden Mile," said Ralph. "It was where the wealthy congregated in the early days."

"It still looks pretty fancy to me," Willis commented.

"Well, some of the old families still live here, in expensive apartments, but most of them have moved to bigger homes in Westmount." He pointed to a building on his left. "That's the Montreal Museum of Fine Art, if you're into that sort of thing."

Willis shrugged. "Not my cup of tea, I'm afraid." He'd never had much time for art; he didn't know what to make of the abstract works that were currently in vogue. Ralph drove past the Ritz Carleton Hotel and then pulled over and parked opposite an imposing set of stone gates.

"That's the main entrance to McGill," he said. "A couple of my old high school friends are studying medicine here."

Willis could see a stretch of green beyond the gates, and a road that led up a gentle incline to a domed building in the distance. It wasn't what he expected, here in the heart of the city. Queen's and McGill were fierce rivals on the football field, the hockey rink, and the basketball court. He wasn't sure why, but he'd always pictured the campus as newer, with modern buildings. Instead, it had the look of a stately manor, its buildings nestled among mature oak and maple trees, set well back from the busy street. Something inside Willis had shifted. It was going to be harder to dislike the visiting teams from McGill when he was back in Kingston—harder, knowing that while the teams did indeed come from a modern industrial city, the McGill campus itself was . . . well, almost old-fashioned. Not that different from Queen's.

Ralph checked traffic in the rear-view mirror and pulled out, continuing east. To their left Willis noticed an imposing statue, set in front of steps that led to a large building.

"That's Queen Victoria," Ralph said with a wink. "Her Imperial Majesty guards a particularly valuable treasure."

Willis looked back questioningly. "Treasure?"

"The building behind her is the residence for girls training to be nurses," he explained. "Love those nurses," he said with a laugh. "They sure know how to take care of a guy."

Ralph began softly whistling a tune that Willis couldn't place. Eventually they stopped for a red light at another busy intersection.

"This is St. Lawrence Boulevard," said Ralph. "We call it 'The Main,' and it actually divides Montreal. Further east is where most of the French live. We English are on the west." He turned left and headed up a gentle slope. Willis noticed that the signs on the stores and restaurants were in a variety of foreign languages. "A lot of immigrants have established businesses here," Ralph explained. A few minutes later he pulled the car to the curb and parked.

"Come on," he said. "We're going to get us some lunch." He smiled a mysterious smile. "It's a new place, opened a year ago." Above the door were the words "Schwartz's—Montreal's Hebrew Delicatessen."

Inside, the dozen or so tables were crowded with people. The air was thick with smoke, and Willis suddenly realized that he was hungry. They sat down at the only seats available, at a counter that ran the length of the restaurant. "What'll it be?" asked a burly man behind the counter. He wore a greasy apron and was chewing on a toothpick.

"Two smoked meats, with dills and cherry Cokes," said Ralph.

"Lean?"

Ralph shook his head. "No, medium."

As the man retreated behind a curtain at the rear of the restaurant, Willis surveyed the crowd. They were a mixed lot, but from the snatches of conversation he overheard he couldn't be sure what language they were speaking.

He looked at Ralph. "Not exactly a French restaurant."

"Nope, but wait till you taste the food. I promise you it's right up there with the best French cuisine."

The man with the toothpick returned and set down two sandwiches, piled thick with smoked meat. He was back a minute later with an order of dill pickles and two glasses of Coke.

Willis bit hesitantly into his sandwich and chewed thoughtfully. The taste was like nothing he'd ever experienced. The meat was rich and savoury, and the touch of mustard a perfect complement to the thick rye bread. There were other flavours present, but Willis couldn't make them out. He swallowed his first bite.

Ralph watched him, expectantly. "Well?" he asked.

"Amazing!" said Willis, then shook his head and took another bite.

"Don't forget the pickles." Ralph reached for a dill and bit into his own sandwich. "You can't properly appreciate the meat without them."

The two men fell silent, concentrating on their food. When they'd finished Willis wiped his face with a paper napkin and drained his Coke.

"Dessert?" Willis wondered.

"Not here," said Ralph. "No desserts at Schwartz's."

When they returned to the car Ralph turned around and drove back down St. Lawrence, heading south. They crossed Sherbrooke and then stopped for a traffic light at another major intersection.

"This is St. Catherine Street," Ralph explained. He pointed to his right. "All the big department stores are here—Eaton's, Hudson's Bay, Simpsons, Ogilvy's."

When the light changed Ralph drove another block, then once more he pulled over and parked. "Wait here," he said. "I'm going to get us a little something else." He got out and quickly disappeared into a small restaurant, its windows so steamed over that it was hard to make out the interior.

Willis lit a cigarette and watched the people passing by. Some were clearly working-class men, and there was an occasional family. From time to time he noticed women who seemed out of place, a little over-dressed to Willis's eye.

Soon Ralph returned bearing a large paper bag. As he climbed behind the wheel the car was filled with the pungent tang of vinegar. This clearly wasn't a dessert.

"What's in the bag?" Willis asked.

Ralph smiled. "It's a secret."

They drove on south, and eventually the small shops and restaurants gave way to larger buildings. Ralph negotiated a number of turns and suddenly a large church appeared before them. It looked vaguely familiar to Willis, but he couldn't place it.

"That's Notre Dame Cathedral," said Ralph. "It's a smaller version of the one in Paris." He parked the car. "We can eat here, in Place d'Armes." They got out and made their way to a nearby bench. With a flourish, Ralph opened the bag and handed Willis a small cardboard container. "We can't have you spending your first day in Montreal without the taste of something truly French."

Willis looked down. "French fries?" he asked.

Ralph laughed. "Mais non," he said, "patates frites."

"I don't think these are exactly the 'fine French cuisine' that was mentioned in the articles I read about Montreal."

"No," agreed Ralph, "but you won't find any better in France."

Conversation ceased as the two men concentrated on their food. Willis noticed pigeons gathering at their feet, hoping to be fed. When they finished eating, they tossed the few remaining fries to the birds, rose, and headed back to the car. The pigeons jostled for the scraps.

As they drove on, Ralph resumed his tour guide role. "This is the city's business district," he said: "the stock exchange, the big banks, the court house, city hall." Eventually they reached the port itself, and Willis could make out large ships berthed at the wharves. He noted the many warehouses that lined the north side of the street. Eventually they reached a major intersection, turned right, and soon passed another park.

"Youville Square," said Ralph, then pointed to a steep rise in the distance. "And that is Beaver Hall Hill."

He dropped the car into a lower gear, and soon they were effortlessly climbing the slope. When they reached the top they headed west again.

"This is another main drag," said Ralph, "Dorchester Boulevard." He made a right turn and at the next intersection once again pulled to the curb, beside another large park.

"Dominion Square," he said. "But take a look up the street."

Willis did, and there, in the distance, was Mount Royal. "I see what you mean," he said. "It is a lot more impressive from down here." He could make out a large metal tower on the eastern edge of the mountain. "What's that?" he asked, pointing.

"That's our famous cross. The first explorers planted a wooden cross here three centuries ago, and this is the modern equivalent. But you have to see it at night. It's illuminated with thousands of light bulbs."

Willis made a mental note. Montreal was proving to be an impressive city. But they were seeing it by day—he was sure that another, equally interesting side to this vast metropolis appeared once the sun had set.

By mid-afternoon Willis was beginning to lose track of all the things they'd seen, and he was relieved when Ralph suggested they call it a day.

Over supper, Willis peppered the family with questions about Montreal. The topics were wide-ranging, from the prowess of the city's professional hockey teams to the widespread corruption in the provincial police force. They discussed some of the early history of the city, the dominant role of the Catholic Church among the French-Canadian population, the strange public school system, in which the English-speaking students were divided along religious lines—Protestant and Catholic—and the French children were taught by nuns and priests in their own language in separate Catholic schools. This didn't sit well with Ralph's mother. "I think their school system is why so many French Canadians don't go on to university. And, of course, the Catholic Church encourages large families. With so many children to feed they can't afford to educate them properly. In fact, most of their kids don't even finish high school."

Ralph's father changed the subject. "All set for tomorrow?" he asked Willis.

"I can't wait to get started."

It was rare for Willis to have trouble falling asleep, but this night was different. As he lay in bed his mind was awash with information. Today he knew they'd only scratched the surface. He'd been a little surprised at the hostility in Ralph's mother's voice as she talked about the French Canadians. He was determined to see for himself if this opinion was shared by others. There was so much more he wanted to see, to explore. He'd come to the city seeking a summer job that would provide him with a new challenge and new professional experience, but he now realized that Montreal itself had much to teach him, and he was determined learn everything he could in the weeks that lay ahead. He finally drifted off to sleep trying to picture how the city might look after dark, when viewed from the Westmount Lookout.

Chapter 20

Montreal, 1929

On Monday morning Ralph's father gave the boys a lift to the Northern Electric plant on St. Patrick Street, in the industrial sector of the city bordering the Lachine Canal. Willis was surprised at the size of the factory. It had been a mere blip on the horizon when seen from the Westmount Lookout, but the red-brick building actually occupied a full city block, towered eight stories high, and had large windows.

Mr. Curtis took Willis to the personnel office on the main floor and introduced him to Mr. Chamberlain, who was in charge of summer employment. The man rose from behind his desk, reached across it, and shook Willis's hand. Tall and somewhat stooped, he wore a vest over a white shirt and tie. His long, narrow face with its prominent jaw seemed an extension of the stoop. He smiled and handed Willis some forms. "Fill these out," he said, his voice deep and resonant, "and we will see what we can find for you." He indicated a small table by the office door. Willis sat down to begin working his way through the application forms and Mr. Chamberlain left the room.

When he returned a few minutes later, Willis handed him the forms. He spent a few minutes reading through the pages, then looked up. "I'm afraid we haven't much for you this week," he said. "It's early, so not many of our assembly-line people are taking their vacations yet. But there is some outside work I'd like to see taken care of." He beckoned Willis to follow, and the two men made their way along a corridor to the back of the building.

In a courtyard dozens of empty wooden crates were stacked in rows. "We get our electrical parts shipped here," the manager said. "In fact they come right into the factory on rail lines. We don't like to waste anything, so we take the crates apart, salvage the wood, and use it again when we have something ready to ship out." He pointed to some tools, including a crowbar and a sledge hammer, on a ledge by the door. "I know it isn't exactly the sort of work you expected to do, but if you strip these crates down and stack the wood we'll see if we can find something a little more interesting for you later in the week."

"Sure," said Willis as he reached for the crowbar. Mr. Chamberlain went back into the building. Willis was glad he'd taken Ralph's advice and worn work clothes. He set about prying the crates apart and stacking the wood. The work wasn't hard, but the courtyard was exposed to the sun, and it wasn't long before perspiration was dripping from his forehead. At mid-morning the door opened and Ralph emerged. He smiled sheepishly and glanced at the pile of wooden slats. "Not the sort of work you expected, I guess."

Willis shrugged. "It's not so bad. At least I'm outside—plenty of fresh air." He wiped his brow.

Ralph offered him a cigarette and they lit up. "Chamberlain told Dad he'd try to find you something more suitable. I mean, they don't need an electrical engineer for this." He pointed to the crates.

Willis laughed. "It beats climbing telephone poles, but not by much. That's what I did the past three summers."

The two young men sat down on the ground and leaned their backs against the brick wall. "I do have some good news for you," said Ralph. "At least I hope you'll think it's good news."

"Oh?" said Willis, turning to look at his friend.

"Well, I spoke to Mom this morning before we left. Since my brother won't be back this summer and his room isn't being used, she said it would be all right with her if you boarded with us. Of course she'd like you to contribute something for the meals, but the room would be free."

Willis couldn't believe his luck. "Are you sure?" He had thought of the arrangement with the Curtis family as something temporary, a place to stay until he found a boarding house.

"I'm sure," said Ralph, exhaling smoke. "Of course, we'll have to make our way here by streetcar. Our shifts start earlier, a good hour before Dad gets in."

"That's no problem. It was kind of your dad to drive us this morning."

"He's not a bad guy, as fathers go," said Ralph with a chuckle. He ground out his cigarette and rose. "I've got to get back inside. I'm soldering wires on a new line of telephones."

"Now that is something suited to your talents," Willis teased. "I knew all those hours in the classroom would come in handy."

"Well, it is a little closer to engineering, but I wouldn't want it to be a life sentence. See you at lunch." He went back into the plant.

At noon the whistle blew and the workers filed out of the plant. Some had lunch pails and lounged in the sun on a stretch of grass beside the building. A few drifted over to the Lachine Canal and ate there. But since Ralph and Willis hadn't brought their lunch, they joined a group of men who made their way to a nearby tavern. The waiter brought them draft beer and asked if they wanted to eat. He spoke in English, but with a pronounced French-Canadian accent.

Ralph glanced at a blackboard on the wall behind the waiter. "I'll have the special, the pigs' knuckles and sauerkraut."

Willis looked up. Written on the board in chalk were the words, "Knuckles 'n' Kraut—$1.00." Was this another Montreal specialty? He was familiar with sauerkraut, but not pigs' knuckles.

Sensing his uncertainty, Ralph spoke up. "Try them," he said. "They have a German cook here and they're his specialty. With the outside work you're doing you need a hearty lunch."

Willis nodded his agreement and the waiter left. He lit a cigarette and listened to the conversation of the men at the other tables. From what he could make out most of them were speaking French, but it didn't sound like the language he'd learned in high school.

"Are most of the guys working here French?" he asked Ralph.

"Most of the factory workers are, but the foremen and the bosses are English."

"Why's that?"

"I'm not sure," said Ralph with a shrug. "Maybe it's because as a rule French Canadians don't finish high school. Funny thing is, though, they all speak English."

The waiter arrived with their food and Willis tentatively tasted his pigs' knuckles.

Ralph took a sip of his beer. "How are they?"

To Willis's surprise the meat was tender and succulent. "Good," he said enthusiastically, and forked some sauerkraut into his mouth.

Willis and Ralph ate in silence. When their glasses were empty the waiter appeared, unbeckoned, with a tray of draft beer. Ralph nodded, and the empty glasses were removed and fresh drafts put down in front of them.

"He seems to be on the ball," Willis commented.

"Well, we only have an hour for lunch and the waiters know it. They try to keep things moving."

On the short walk back to the plant Ralph explained that the area they were in was known as Point St. Charles, a distinct part of the city. "But no one calls it that," he added. "To Montrealers it's simply known as 'The Point'." He pointed across St. Patrick Street: "Over there by the St. Lawrence is 'Goose Village.'"

Willis shook his head. "'Goose Village?' That's an odd name."

"I'm not sure why it's called that," said Ralph. "The Village is probably best known for its prizefighters. Most of the people there are of Irish descent."

"Any of them work for us? I mean at Northern Electric?"

"A few, but more of them work at the big meat packing plant in the Village," said Ralph with a smile. "But that's another story," he said as they approached the factory entrance. "You'll get the drift in time." With that he pushed through the door. "I'll meet you here later, when our shift is over."

By mid-afternoon Willis had worked his way through about half the crates. With sunlight no longer flooding the courtyard, Willis felt a chill, and he reached for his jacket, which he'd hung on one of the crates that morning. He was suddenly aware of a sharp, unpleasant odour. He checked the jacket, thinking that it might be the source. But it was relatively new and smelled fine. He sniffed the air, trying to identify the new and unwelcome scent. It reminded him a little of something he'd encountered during a university-sponsored visit to a pulp and paper mill. But that wasn't it. He would have to ask Ralph. In the meantime he went back to work on the crates. The sooner he finished, the sooner he'd be doing something more in keeping with his engineering background.

When the whistle sounded to end the shift, Willis made his way back through the building and joined the line of men punching out at the exit. Ralph was there, waiting, with that mysterious smile once more on his face. Outside, the smell was still in the air.

"Whew," said Willis, sniffing. "What in God's name is that?"

Ralph laughed. "I told you you'd get the drift of it. There's a slaughterhouse at the meat-packing plant in Goose Village. In French it's called an abattoir. From time to time, particularly if the wind is blowing in the wrong direction, we're treated to its pungent perfume."

"How often?" asked Willis as they walked to a nearby streetcar stop.

"Too often, if you ask me. Just be thankful we don't live down here. Can you imagine sitting down to supper with that smell in the air?"

Willis's mind flashed back to his lunch of pig's knuckles, and he felt a vague queasiness. Fortunately their streetcar arrived just then, and the two young men climbed on board, deposited their tickets, and found seats near the rear of the car. It pulled away from the plant and began its trek up the lower slope of Mount Royal. The smell lingered for a time, but grew fainter as they moved higher.

"Dad thinks they'll have something for you inside in another couple of days," said Ralph.

"That's good. Today wasn't so bad, though." In fact, Willis thought to himself, he'd picked up a good deal in his first day on the job. He was learning—not about the work, but about the city and its people.

True to his word, Mr. Chamberlain moved Willis inside two days later when he'd finished dismantling the crates. Over the next few weeks Willis filled in at a number of different jobs, usually on one of the company's many assembly lines. He learned that Northern Electric was actually owned jointly by Western Electric, a subsidiary of AT&T in the United States, and the Bell Telephone Company of Canada. It made most of the telephones sold in Canada, but the company was also beginning to branch out into other areas, such as public address systems and sound equipment for movie theatres. These fields were not open to summer students, but Willis eagerly sought out the engineers working in these departments to learn all he could from them.

He'd developed woodworking skills while in high school, and this paid off in the plant. He was often called upon to fashion tables and benches in areas where new projects were being undertaken. It was an entrée into the world beyond the assembly lines, and it also broke up the tedium of the more routine work.

Willis befriended a number of the French Canadians in the plant. One man in particular, Jean-Paul Trépanier, took a liking to Willis, and when Ralph wasn't around for lunch Jean-Paul and Willis often spent the hour together.

Trépanier was short and wiry, and wore his dark hair slicked back without a part. He had a pencil-thin moustache and deep-set brown eyes.

Willis could imagine him in an earlier era as a coureur de bois, paddling a canoe loaded with furs down the St. Lawrence. His English carried the French-Canadian accent but was surprisingly good. Willis eventually asked him about this.

"I was lucky," Jean-Paul said. "My grandmother on my father's side was English and she taught me."

"But the other men on the assembly lines seem to speak English too. Maybe not as well as you, but pretty good."

Jean-Paul shrugged. "We have no choice. If we want this sort of work we need to know English."

Willis nodded. It was true, he seldom heard French spoken within the plant. "Do any of the foremen speak French?"

Jean-Paul shrugged. "A few words, maybe, but nothing more."

"I wonder why they don't learn French?"

Jean-Paul smiled. "No need, I guess. And all of the bosses are English."

Willis was surprised that the words carried no hint of resentment. Just acceptance. It was, Jean-Paul seemed to be saying, simply the way things were. A fact of life.

"Not many of us French Canadians finish high school," Jean-Paul continued. "We have large families, and most of us boys are working by the time we're sixteen. The girls, too, usually in clothing factories. Or as maids, or housekeepers."

"How did you do in school . . . I mean before you left?"

"Oh, I did okay, I guess. But the nuns could make life difficult. There was a lot of emphasis on religion. Some of the brighter boys stayed on, and went off to seminaries to become priests." He laughed. "But that wasn't the life for me."

Willis decided not to press the matter. Jean-Paul seemed bright, and Willis wondered what might have become of him if he'd been given the chance for more education. The following weekend he sought out a bookstore and bought a French-English dictionary. He envied Jean-Paul's bilingualism. He wondered whether he could he pick up some French while he was in Montreal. If the blue-collar French Canadians on the assembly line could speak two languages, the least he could do was try to learn a little French.

Ralph came upon him one Saturday afternoon when he was looking up some words in the dictionary. "What are you up to?"

Willis explained.

"Well, good luck, but I think you're wasting your time."

"How so?"

"We studied French in high school here in Montreal, but when we tried it on the street no one seemed to understand us."

"Was it that difficult?"

"Nope, but our teachers were teaching us French the way they speak it in France, not here in Quebec, and there's quite a difference. So whenever we tried using what we learned on French Canadians, in stores, for instance, they'd answer in English."

"So you gave up?" Willis asked.

"It didn't seem worth the trouble," his friend replied.

"Most of the French Canadians at the plant speak pretty good English," said Willis.

"Exactly. I guess that's one reason why we never learned French. And I think English is easier to learn—I'm not sure about that, but I know French was always my worst subject in school."

Willis thought back to the limited French he'd studied in his high school days in Leamington, and recalled that without exception his French-language teachers had all been English. And they probably used textbooks from France. Perhaps this explained why he understood so little of the French he overheard in the tavern.

"Well," he said to Ralph, "I'm going to keep at it. See if I can pick up at least some French this summer."

"Good luck, chum," said Ralph. "I'm on my way to a Saturday afternoon matinee at the Palace. There's a new Douglas Fairbanks movie playing. Sure I can't entice you?"

"Nope," said Willis, although the idea was tempting. "I'm going to stick to this." He held up the dictionary.

"Bonne chance," said Ralph, and headed out the door.

French proved a difficult challenge. By the end of the summer he could, with the aid of his dictionary, manage to work his way through the weekend edition of *La Presse*, the daily newspaper. But speaking the language was an entirely different matter. When he tried his rudimentary French, at work or while shopping, the replies were usually in English. He wasn't sure—was his French that bad? Or was it because most of the French Canadians he encountered were bilingual, and, sensing his difficulty with the language, tried to help him out by switching to English.

Work was a different story. Willis was more successful at work than he was at mastering a new language. When he picked up his final cheque from the pay office at the end of the summer there was a note from Mr. Chamberlain. He wanted to see Willis before he returned to Kingston. Willis phoned from the shop floor and spoke to a secretary, who told him the personnel manager was free and he should drop by on his way out.

"Ah, Willis," said Mr. Chamberlain, looking up from his desk. "On your way back to Queen's I expect?"

"Soon," Willis responded, a little unsure of where this conversation was leading.

"Well, young man, I suppose you are doing the right thing. One more year and you'll have your degree?"

"That's right, sir. I'll be a senior."

The personnel manager looked down at some papers on his desk. "I've been going through the reports from the foremen. It seems you've made quite an impression." He looked up at Willis with a smile. "A very favourable impression."

"I enjoyed the work, sir," he said, a little embarrassed at the compliment. "And I learned a good deal."

"I can see that," said Mr. Chamberlain. "You've made yourself useful in a number of ways this summer. Quite unusual for a student."

Willis remained silent, sensing that the personnel manager had more to say. He was right.

"In fact, we'd be prepared to take you on permanently, right away, if you decide to forgo your final year at Queen's."

The offer came as a complete surprise to Willis. He had to admit it was tempting. But he knew his parents would be disappointed if he didn't finish his degree. "Thank you," he said. "I'm flattered, of course, but I've put in four years at Queen's and I like to finish what I start."

"Admirable," the personnel manager said, shuffling the file of papers on his desk. "I like that quality in a man." He rose, and as the two shook hands he added, "Keep us in mind when you graduate. We can always find work for someone like you."

"Thanks again, sir," said Willis, retreating to the door. "I'll certainly keep that in mind."

"Good luck, young man," said Mr. Chamberlain.

Chapter 21

Queen's University, 1929–30

When he returned to Queen's for his final year Willis discovered that the technical tinkering that had silenced CFRC for much of the previous term seemed to be over. Modifications to the Mark IV transmitter had been completed and the signal strength had been improved. Morgan and Richardson were now familiar with the workings of the station and took a more relaxed approach to undergraduates who were seeking hands-on experience. As a senior, Willis was in a privileged position. He became more actively involved, but was careful to keep up with his academic work.

Two additions to the Queen's staff that year made an important difference to Willis. Harold Stewart, a 1926 graduate, was appointed as a lecturer in electrical engineering and he quickly became involved in station work. He seemed open and friendly. But the big news was the appointment of Willis's old friend, George Ketiladze, who had graduated the previous fall, as Stanley Morgan's assistant. These two filled a void Willis had felt since the departure of Harold Minter.

The two men quickly breathed new life into the station. CFRC was no longer silent. The fall convocation was broadcast from Grant Hall, as was the annual meeting of the Queen's Theological Alumnae. Football, hockey, and basketball games were back on the air. In fact, George, with Willis's help, managed to solve several technical problems that had plagued the hockey transmissions. This improvement was noted in *The Kingston Whig-Standard*:

> The sounds of the whistle, bell and the crowd at the Harty Arena, could all be heard and the ones sitting at home had half the fun

> of being at the game itself. The announcer of the game deserves praise for the remarkable play-by-play report given, no action being missed at any time, which is something to be said when broadcasting a fast game like hockey.

In February another broadcast drew praise from the newspaper. The Kingston Choral Society was featured in a ninety-minute program, and the paper described the reception as "well-nigh perfect."

This outside recognition did not go unnoticed by businessmen in Kingston. They had begun to lobby the university, seeking use of the station for advertising purposes. They had already gained a measure of goodwill by subsidizing the costs involved when CFRC broadcast out-of-town football games. Gradually, Queen's authorities softened their stance against commercial involvement in CFRC policy. The Kingston Chamber of Commerce applied to the executive committee of the Queen's board of trustees for the occasional use of Queen's broadcasting facilities for advertising purposes. It was agreed to grant the request on condition that the relevant programs be submitted in advance for university approval. The stock market crash of 1929 reverberated through financial circles, and while it hadn't yet hit the universities, it did drive home the need for Queen's to explore ways of generating additional revenue.

Meanwhile, commercial radio was making huge technical leaps. In 1927, CFRB in Toronto became the first Canadian station to move from battery-powered direct current transmissions to alternating current, the more readily available and convenient power source. The shift itself was costly, but inevitable. It was clear to the men running the Queen's station that they were going to need more revenue if CFRC was to keep up with the times.

The facilities were still primitive. Half of the available space was walled off for the transmitter, while the other half served as a studio. All of the programming was live, most of it from Fleming Hall. Drapes were hung on the walls to absorb sound. The operator doubled as announcer for most programs. Despite these drawbacks, Willis still filled with pride when the station signed on and he heard the words, "This is CFRC Queen's University in Kingston, Ontario," usually spoken by Stanley Morgan.

But it was clear to everyone that the station needed a major overhaul. With the door now open to commercial revenue, the university trustees authorized the station to explore plans for expansion. The biggest single expense would be the acquisition of a new transmitter. The Mark IV, purchased in 1923, had to be replaced, and estimates were sought from manufacturers and suppliers.

George kept Willis apprised of developments. "Is finally happening!" he reported, excitedly. "We emerge from dark ages!"

Willis smiled at his friend's enthusiasm, but he knew that it would take time before the changes were implemented. It would be another year before the project was complete—too late for Willis. In his final months at Queen's, with the finish line in sight, he applied himself to his studies.

In the spring his efforts were rewarded. Clyde and Anna Maria travelled from Leamington so that they could be present to watch proudly as their son received his degree in electrical engineering. Fittingly, the convocation ceremony was broadcast live on CFRC.

IV
Career Begins

Chapter 22

Schenectady, New York, 1930

Willis first heard the word "television" mentioned in the fall of 1928, during one of Professor Harold Stewart's lectures at Queen's.

"Some interesting research has been carried out recently in the United States," the professor told his students. "You've all heard about the contributions that Canada's Reginald Fessenden made in the early days of radio. Well, there's another name I'm sure you'll be hearing in the future. It's Ernst Alexanderson. Alexanderson has been working at the General Electric plant in Schenectady, and it seems that he's succeeded in transmitting images—moving images with sound—wirelessly. He's calling his invention television."

The classroom suddenly became very still. Most of the engineering students had heard the rumours about a new form of communication involving images, but until now it had seemed the stuff of science fiction. There were murmurs, and soon hands were raised.

Professor Stewart waved off the questions. "I know, I know," he said, "it sounds pretty far-fetched, but my source, *The Scientific American*, is reliable." He held up the magazine. "You can read for yourself."

This led Willis to wonder about the future of broadcasting as he knew it. Radio might not be a final destination after all. Intrigued, he began to research the subject. He learned that Alexanderson was just one of many inventors working on systems to transmit moving pictures wirelessly. In England, John Logie Baird had achieved some modest results as early as 1923 with an invention he called "radio vision." Similar research was being carried out in Europe and Japan. In 1926 Philo Taylor Farnsworth, an American, applied for a patent for his system in California. On the commercial front, RCA Victor, Westinghouse, and AT&T were beginning to show interest.

But it was Alexanderson's progress that particularly interested Willis. He worked for General Electric, a company that offered summer internships to electrical engineers. Willis applied for one of these internships and was accepted, so within days of his graduation he was on a train heading for Schenectady, his trusty CCM bicycle safely stored in the baggage car.

The company had provided a handbook that included information about accommodation. Soon after Willis arrived he rented a room close to the GE plant, then set about exploring the town on his bike. He learned that Schenectady was about the same size as Windsor, with a population of close to 100,000.

General Electric had been established by Thomas Edison, the American best known to the public as the inventor of the first practical light bulb, and the phonograph. Willis knew from his studies that many of Edison's numerous other, less publicized inventions, some of which had been essential in the evolution of radio, had been equally important. To Willis, Edison, like Alexander Graham Bell and Marconi, was as historically important as the early explorers. So it was with something approaching reverence that he biked to the plant for his first day at General Electric.

He dismounted and gazed in wonder at the vast sprawling complex of buildings on the banks of the Erie Canal.

"Can I help you?" called a guard, breaking Willis's reverie.

Willis wheeled his bicycle across the street to the gate. "I'm an intern," he said.

"First day?" asked the guard.

"Yes," said Willis with a nod, taking in the buildings. He'd read about the plant, but nothing had prepared him for its massive size.

The guard smiled. "Pretty impressive, isn't it?"

"I'll say," Willis answered.

"Well, you can leave your bicycle in that big shed over there," said the guard, pointing to his left, "and then go right on in."

Willis and a dozen or so other interns were taken on a tour of the plant. This in itself took two full days, and even then it didn't cover everything. Parts of the research facilities were off-limits, but the rest of the factory, where the company manufactured everything from household appliances to industrial equipment, was proudly displayed. By the end of the second day Willis had lost count of the buildings they'd toured and the products they'd seen.

On his third day at GE he was put to work. His previous summer job, at Northern Electric, had included some time on a generator assembly line, so he was assigned a similar task under the supervision of a gruff, taciturn foreman. It was clear that this man viewed his university-educated intern with a mixture of curiosity and contempt. The foreman had come up "the hard way," through the ranks, having joined GE to serve an apprenticeship right out of trade school. Each summer he'd been saddled with a new group of university interns, and had found that as soon as they had been trained and were of some real use, they usually left. Willis sensed some resistance in the man, and set about winning him over. But he kept his real agenda to himself. He wanted to see a television transmission, and perhaps even meet the legendary Ernst Alexanderson.

As it happened, he didn't have long to wait, at least as far as seeing television was concerned. Proud of its progress, in 1928 GE had established WGY, the country's first television station. By the time Willis arrived it was transmitting three programs a week. Of course only a small number of residents of Schenectady could view these programs, because receivers weren't yet available to the public. There were a few exceptions, though, and one afternoon Willis and three other interns were taken to the home of a senior research scientist near the plant.

They were seated in the living room in front of what appeared to be a standard radio receiver housed in a wooden cabinet. But where the dial was usually located there was, instead, a small glass screen, some three inches by four inches. When the scientist turned on the set the screen emitted a ghostly glow, and then a stationary picture slowly appeared.

"That's a test pattern," explained their host. It was accompanied by a sound that Willis recognized as a radio signal.

The scientist checked his watch. After a few seconds the test pattern vanished and in its place appeared a flickering but distinct image of a man standing by a microphone. He glanced down at some sheets of paper in his hand and began to read.

"Good afternoon, ladies and gentlemen. Welcome to station WGY in Schenectady. Our television broadcast today originates from the studios of General Electric."

Willis and the others looked on in amazement. He'd heard about it and read about it, but nothing had prepared him for the actual moment when he saw his first television images. It was a memory he would retain for the rest of his life.

Willis had to be patient in fulfilling his second objective, the chance to talk to Ernst Alexanderson. Thomas Edison was the man most of the other interns wanted to meet, but unfortunately Edison lived in New Jersey and spent most of his time in Menlo Park, the company's research facility there. Willis knew that Alexanderson, unlike Edison, worked in Schenectady, but as time passed he was rotated through a series of jobs at the plant, and he began to worry that he might not meet the inventor.

Then, late in August, the company held a small informal reception for the interns. Most would soon be departing, some back to universities, others to work at companies throughout the United States. A few of the more gifted would be staying on at General Electric. It was at this reception, held outdoors in one of the parks GE provided for its employees, that Willis finally met Ernst Alexanderson.

Although Willis was determined, he was also shy, so it took him some time to approach the inventor. When he did, he found him in conversation with two of the interns. As he listened he was surprised hear the man speaking with a familiar accent—one that Willis had heard all his life. Alexanderson sounded just like his Swedish mother. This gave Willis the courage to join the discussion, and soon he was discussing the technical intricacies of television transmission with Alexanderson.

Alexanderson was a tall, rather imposing figure with a square jaw, dark hair, and a full moustache. He was dressed formally for the occasion, with a three-piece suit, a crisp, white, high-collared shirt, and a striped tie, but to Willis he had the rugged appearance of cavalry officer, a person who belonged outdoors, not cooped up in a laboratory.

Eventually there was a lull in the conversation when one of the interns left to get Alexanderson a glass of lemonade. "Where in Sweden are you from?" Willis ventured.

"Uppsala," answered Alexanderson, a little surprised at the question. "It's about seventy kilometres north of Stockholm. Why do you ask?"

"My mother is Swedish," Willis explained. "She emigrated to the United States in 1887."

"I'll be damned," said Alexanderson. "That's the year I was born. Where is your mother from?"

"Uppsala," replied Willis, and with that the ice was broken. He described his mother's harrowing journey to Wilcox, Pennsylvania, her eventual marriage, and her move to Canada.

"That's quite a story," said Alexanderson. "But tell me about yourself. You're a Canadian, I take it."

Willis filled in the details—his love for radio, his time at Queen's, and his hopes for the future.

"These are difficult times," Alexanderson mused. "The Depression has taken its toll. Have you lined up work for the fall?"

"I think I might be able to get something in Montreal, at a company called Northern Electric. I worked there last summer. It isn't radio, but it's a start."

Alexanderson shook his head. "I don't know much about Canada, but if you're offered a job, young man, I'd strongly advise you to take it. We are in for some very trying times in the coming years."

As Willis cycled home after the reception he considered Alexanderson's advice. Perhaps the inventor was right. It was time for him to put personal preferences aside and take whatever he could get, time to begin earning his own living. He wanted to repay something to his parents, who had gone heavily into debt to finance his tuition at Queen's. The Depression had hurt the pool hall business, and now his father's diabetes was taking its toll. His eyesight was failing.

Willis parked his bicycle by the side of the boarding house and hurried up the steps to his room. He'd made up his mind. He sat down at a small table beside the window, took out his fountain pen, and began a letter to Mr. Chamberlain, the personnel director at Northern Electric. When he finished, he folded the letter carefully, slid it into an envelope, addressed and sealed it, then applied a stamp and hurried back outside to the nearest mailbox.

Chapter 23

Montreal, 1930–31

Willis was incredibly lucky. Although Northern Electric was laying off employees, the demand for the company's public address systems remained strong, and there was a shortage of engineers familiar with that technology. Willis not only understood the theory, but he had picked up valuable practical experience when he was sorting through some of the problems with the PA systems used in Richardson Stadium and Harty Arena at Queen's. He was an ideal fit.

He arrived in Montreal at Windsor Station on a warm September evening. Ralph Curtis was there to pick him up. Willis noticed he was still driving his father's 1927 Chevrolet National, but he decided not to ask what had happened to the plan to trade the car in on a newer model. It was, he assumed, a reflection of the financial stress imposed by the Depression.

"You're looking great, Wheeler," said Ralph as he drove west along Dorchester Boulevard. "I can't believe we'll be working together again."

"I'm damn fortunate to have a job," said Willis. "If it wasn't for you and your dad I'd never have connected with Northern Electric in the first place."

"If it wasn't for Dad, I'd probably be out of work too," said Ralph. "Times are tough."

"What've they got you doing these days?" Willis asked.

"I'm in the office, pushing paper," said Ralph, disappointment in his voice. "And my brother is back home. There were big cutbacks at the pulp and paper mill where he worked."

"Your folks okay?" asked Willis.

"I think Dad's job is pretty secure. But in these crazy times, who knows?" Ralph turned the car into the driveway, cut the engine, and pulled on the emergency brake.

Willis stayed the night, sleeping on the sofa in the Curtis living room, but the next morning he took the streetcar downtown and began to search for a room of his own. There was no shortage of places to rent. Many people were offering spare bedrooms in an effort to make ends meet. He had considered an apartment, but decided it would be too expensive.

He found a room to his liking on Dorchester Boulevard, just west of Atwater. It was in an old grey stone residence that had, at one time, been home to a single wealthy family. Now, although converted into a boarding house, it still featured high ceilings and ornate wooden panelling. Willis took a fancy to a corner room at the back of the building overlooking a small park. The price, six dollars a week, was well within his budget.

The landlady was Mrs. Whelan, a short, plump woman with curly red hair and an Irish accent. There appeared to be only one rule. "No women friends overnight," she warned. "We have a reputation to uphold."

Willis smiled. "No women overnight," he agreed, and paid a month's rent in advance.

In the three days that remained before Willis had to report for work, he took the opportunity to reacquaint himself with Montreal.

A good deal had changed in the time he'd been away. Montreal was the banking capital of Canada, and the crash of the stock market in 1929 had cut deeply into the business community. Willis had read about the suicides—the executives facing financial ruin who had jumped from office buildings—but it was on the streets that Willis came face-to-face with the reality of the Depression. On every corner men were asking for "spare change." The more enterprising of the unemployed were selling apples, pencils, or shoelaces. As he passed the Salvation Army Willis saw his first soup kitchen—men lined up along an entire city block, desperate for something to eat. Willis didn't notice any women in these line-ups. What were they doing for food? He didn't like to think about it.

The *joie de vivre* had clearly gone out of Canada's largest city. And yet, according to the newspapers, illegal enterprises seemed unaffected. The after-hours drinking establishments, the bookies, and the brothels were apparently immune to the hard times. Willis wondered whether the crime rate had risen.

These thoughts were put aside on Monday morning when he stood before the Northern Electric plant in Ville St. Pierre. At eight o'clock sharp he was ushered into the familiar office of the personnel manager. There were papers to fill out—more this time, because he was becoming a permanent employee. By nine he was finished, and Mr. Chamberlain escorted him to a part of the plant where the company's public address systems were manufactured.

"I'd like you to meet Mr. Knobloch," he said, "the man in charge here."

They entered an office, and a large beefy man looked up from a set of blueprints. "It's about time," he said, addressing his remarks to Mr. Chamberlain. "We can certainly use some help around here."

The personnel manager introduced Willis and then departed with a wave of his hand.

Mr. Knobloch looked over the frames of thick spectacles. "A Queen's man?" His eyes were robin's-egg blue. Inquisitive.

"Yes, sir."

"Well, you have one helluva football team. I went to McGill, myself, but don't hold that against me." He smiled. "Tell me what you know about PA systems," he said, then gave a resigned sigh. "I just hope you know your stuff. We've got lots of orders, but we're decidedly short on expertise."

Fortunately Willis did know his stuff, and by the end of the day this was clear to Mr. Knobloch.

Over the next few weeks Willis worked on a variety of systems. Northern Electric's customers included municipalities, professional sporting teams, and owners of roller-skating rinks, carnivals, and bingo parlours—in short, any business that wanted its announcements or its music amplified, indoors or out.

It was a challenge, and Willis liked the diversity of tasks. He had to select and assemble the equipment that was best suited to a particular venue. His biggest challenge was to meet the budgetary constraints imposed by the customers, which could be frustrating. There were times when the loudspeakers, amplifiers, and microphones were, for financial reasons, less than ideal for the location. But even in such cases Willis found original ways of mixing and matching, of stringing components together to gain improved quality at no extra cost.

On weekends he hung out with Ralph. When the family car was available they'd go for drives in the country. One Saturday in November they crossed the St. Lawrence River on the Victoria Bridge and headed for an apple-growing region southeast of the city. They were in search of hard cider. It was illegal to sell the alcoholic version of this drink, but it was available if you knew the "right" farmer and were willing to pay a premium price. Ralph had an address in Abbottsford.

Willis learned two valuable lessons on this trip.

On the outskirts of the town the two young men were pulled over by a motorcycle cop, an officer of the Quebec Provincial Police.

"What seems to be the problem?" Ralph asked the policeman innocently.

"You were going a little too fast," replied the cop, in heavily accented English.

"I'm sorry," said Ralph. "I didn't realize I was speeding."

"Your driver's licence and registration?"

Ralph reached across to the glove compartment and took out a leather case. He handed it to the officer, then dug out his wallet and opened it to display his driver's permit. The officer peered at the licence and then moved to his motorcycle. A moment later he returned, and handed the registration back to Ralph.

"Only a warning this time, Monsieur Curtis," he said with a wink, "but in future please be more careful."

Ralph put the registration back in the glove compartment and watched in the rear-view mirror as the officer mounted his motorcycle and kick-started it into life.

"That was lucky," said Willis as the policeman drove off.

"Lucky, my ass," laughed Ralph. "Give me the registration."

Willis reached into the glove compartment and handed it over. He watched as Ralph extracted a ten-dollar bill from his wallet, folded it, slid it neatly behind the registration form, and handed the case back to Willis.

"Insurance," Ralph explained.

"I don't get it," said Willis as he put the case back in the glove compartment.

"These provincial cops don't make much money," said Ralph. "A little bribe goes a long way."

"You mean there was money with the registration?"

"Yep, and guess what: it wasn't there when he gave it back."

A few minutes later they stopped at a farmer's stand by side of the highway.

"This is the place," said Ralph, and they got out of the car. "Let me doing the talking," he cautioned Willis.

The farmer appeared, wearing a straw hat, a scuffed leather jacket, and faded overalls. "What can I do for you?" he asked, his French accent even more pronounced than the cop's.

"We'd like a little cider," said Ralph.

"Take your pick," said the farmer, indicating a row of bottles at the back of the stand.

Ralph continued to choose his words carefully. "The thing is," he said, "what we'd really like is some . . ."—and he paused as if searching for the right word—"some old cider . . . very old."

The farmer's weathered faced creased in a knowing smile. "One minute," he said. "I see what I can find." He retreated down the lane to a nearby shed. A moment later he was back with a brown paper bag. "Good stuff," he said, and handed the bag to Ralph. "Only two bucks."

Ralph handed the bag to Willis, took two dollars from his wallet, and paid the man.

As they drove off Willis extracted a dusty gallon jug from the bag. Ralph asked him to hold it to the sunlight that was streaming in the window. "Looks pretty cloudy," Willis observed.

"Good," said Ralph. "The non-alcoholic cider they sell to the public is clear, like apple juice. It's the cloudy stuff that has the kick."

Back in Montreal they stopped at Willis's room to sample their purchase. By the time they finished the bottle Willis had learned the second lesson of the day. The cider they paid the premium for was nothing more than cloudy apple juice. It had no kick at all.

"That old frog," Ralph said in disgust, "selling us swill!"

"What's that Latin saying . . .?" asked Willis innocently.

"*Caveat emptor*," Ralph responded. "Let the buyer beware. Especially when you're dealing with pea-soupers."

Willis didn't like the reference, but decided to let it pass. "Well, I know one thing for sure," he said with a laugh. "This buyer will be up all night pissing out that swill."

Although Willis's previous stint with Northern Electric had given him a sampling of life in the big city, it had been a summer job, so he hadn't been exposed to the hardships of a Montreal winter. Nothing in Leamington, or Kingston for that matter, had prepared him for the combination of below-freezing temperatures, heavy snowfalls, and driving winds that the city regularly endured.

The winter clothing he'd brought with him simply wasn't warm enough. Although he was trying to economize so that he could send part of his earnings back to his folks, he realized by January that he'd have to dip into his savings for a warmer overcoat, decent gloves, proper footwear, and a woollen scarf. He even bought himself a toque, because the fedora he usually wore didn't keep his ears warm. Shivering in the cold, waiting for a streetcar to arrive, he marvelled at the variety of ways Montrealers bundled up against the elements. Style no longer ruled. It didn't matter how you looked as long as you stayed warm.

Willis was also impressed at how efficiently city crews managed to keep the streets clear, even during the worst storms. A blizzard hit the city one Saturday in February, and he watched from his window as snow accumulated on the tree branches, turning the park into a frosted fairyland. Winter could transform the city into a thing of beauty, he thought, as long as you didn't have to venture out. When he did leave the house for something to eat he was struck by the silence. The snow had stopped falling, and it was as if Montreal was sleeping, tucked in beneath its mantle of white.

One Sunday Willis rented skis for a trial run down the gentle slopes of Mount Royal. It looked simple, and he thought he could handle it. At the crest of the hill he fastened the straps of the skis to his boots, then used his poles to push off. At first he concentrated on balance, determined to stay upright. But the ankle he had sprained at Queen's had never properly healed, and as gravity took over and his speed picked up he felt the skis go out from under him. Soon he was tumbling head over heels. It wasn't until he was halfway down that he was able to stop. He felt embarrassed as he lay panting in a heap, but as near as he could determine he had escaped serious injury. After wiping the snow from his face he removed his skis and rose unsteadily to his feet. He limped the rest of the way down the hill.

Willis returned the skis to a chalet near Beaver Lake. The old man serving him smiled knowingly. "Want to try de skates?" he asked. "Easier dan de skies."

"I don't think so," he said, humbled by his skiing misadventure.

The lake had been cleared of snow, and Willis watched as skaters glided effortlessly across the frozen surface in time to music. The sound came from speakers mounted on the chalet roof. Suddenly Willis felt a little better. He recognized those speakers. They'd come from the Northern Electric plant where he worked.

By March the snow had begun to melt, and there were days when it was apparent that winter was relaxing its icy grip on the city. Willis had weathered the worst the city had to offer. He'd also gone to the Forum to see his first professional hockey game, taken in a vaudeville show at the Gayety Theatre and an opera at His Majesty's Theatre, and climbed to the top of Notre Dame Cathedral. He was beginning to feel like a true Montrealer. But there was one thing left to do. During their fall trip in search of hard cider, Ralph had driven them through the south-shore community of St. Lambert, and Willis had regretted that he hadn't brought his camera with him. He wanted a picture of the city from across the river. So, on an unseasonably warm Sunday morning in April, he boarded a Montreal and Southern Counties streetcar with his folding Kodak in his hand for a trip across the Victoria Bridge to St. Lambert.

He got off at the corner of Victoria and Green Avenue, in the centre of the town, and walked the few blocks down to the water. The view was spectacular. The St. Lawrence River was more than a mile wide at that point, and in the distance Montreal was spread out around the foot of Mount Royal.

He set the camera exposure for 1/25th of a second at f8 and cocked the shutter. He peered down through the tiny prism of the viewfinder. A few clouds had formed on the horizon. Perfect. He pressed the release. Then, just to be sure, he advanced the film, cocked the shutter again, and took a second picture. He remained there, taking in the view, until a cloud obscured the sun. Shivering, he folded the camera and headed back the way he'd come.

It was lunchtime and he was hungry. Although it was Sunday and the businesses along the main street were closed, he did find a variety store that was open. He pushed through the door and a small bell announced his arrival. A petite middle-aged woman emerged from a room at the back of the store.

"Can I help you?" she asked, looking across a counter at Willis.

"I was looking for a restaurant."

She smiled apologetically. "I'm afraid you're out of luck. It's Sunday and we're about the only place open today."

Willis looked down at what could be seen beneath the glass top of the counter. Candy, gum, chips, and boxes of Cracker Jacks. Not his idea of a suitable lunch.

The woman apparently shared this sentiment. "I could probably put together a sandwich for you," she said hesitantly, "if you don't mind waiting. We don't normally serve lunch here."

"That would be great," Willis replied, surprised at the woman's generosity.

"Ham and cheese be okay?" she asked.

"Ham and cheese is fine."

"We do have soft drinks." She pointed to a cooler across from the counter. "Help yourself while I fix the sandwich."

Willis took a Coke from the cooler, uncapped it, and drank from the bottle. He glanced around the sparsely lit interior of the store. The scent in the air was vaguely familiar. It took him a minute to place it. Then he remembered. It was the same musty smell that had greeted him when he used to open his father's pool hall in the morning.

The woman returned with the sandwich. "Here," she said as she handed him the plate. "It's odd how things work out. We're usually closed Sunday, but I came in to catch up on the books."

"Well, I'm certainly glad you did," he said with a smile. "I'm Wheeler Little, by the way."

"And I'm Helen O'Hearne. My husband and I inherited the store from Jim's folks."

Willis chewed thoughtfully. "Tell me a little about St. Lambert," he said.

Over the course of the next quarter-hour Willis learned a good deal about this south-shore community. Most of the five thousand residents were English-speaking. It was largely Protestant, though there were enough Catholics to support their own school, which was staffed by nuns. Protestant children were served by two elementary schools and a high school. A good many residents worked in Montreal, commuting daily by streetcar or train. The only major industry was the Waterman Pen Factory, but the town had its own police force, post office, and train station. There were several parks, one of which had a bandstand. There was also a movie theatre, a bowling alley, and a pool hall. Willis reflected that the town was not that different from Leamington.

He thanked Helen O'Hearne for the lunch, but when he tried to pay she told him she would only accept money for the Coke.

"Are you sure?"

"I am. The sandwich is on the house." She looked around and laughed. "Or should I say, on the store?"

Chapter 24

Montreal, 1931

As spring gave way to summer, Northern Electric received fewer orders for new public address systems. Willis was aware that a lot of businesses cut back in the summer, the time when employees tended to take their annual vacations, and he thought that might explain the lull. One afternoon in late July he was summoned to Mr. Knobloch's office.

"Close the door, Willis."

Willis did so. When he turned, he knew instantly, from the look on his supervisor's face, that his theory about a seasonal slowdown was not the whole story.

"I hate this part of the job," said Mr. Knobloch as he shuffled some papers. "I'm sure you've noticed we haven't been very busy lately."

Willis nodded.

"I've been ordered to cut staff. And, as the last man hired, I'm afraid you'll have to be the first to go."

Willis felt a weakness in his knees.

"It's no reflection on your work, Willis. God knows you're the best engineer I have. But it's policy, company policy."

"I see," Willis managed to say.

"Normally we give people just two weeks notice. But in your case I think we can extend that a little.'

"Extend it?"

"Yes. I can carry you until the end of August. But not a day beyond that." Mr. Knobloch looked up. "I just hope the extra time will give you a chance to find something else."

"I appreciate that," mumbled Willis.

"And if you need a recommendation I'll be more than happy to provide it." He rose, reached across the desk, and shook Willis's hand. "Good luck, my boy."

When his shift ended Willis found Ralph waiting for him.

"I just heard," his friend said. "I'm so sorry."

Willis tried to keep it light. "Last in, first out."

"Shitty policy if you ask me," said Ralph. "What'll you do?"

"Dunno." Willis shrugged.

"Feel like a beer?"

"No. Thanks anyway."

"Sure?"

"Yep. I've got some serious thinking to do."

"Well, drop by the house if you change your mind."

"Maybe this weekend."

When Willis returned to his room, he was depressed and more than a little worried. Times were tough everywhere and jobs were scarce. Should he look for work in Montreal? He wasn't sure. He knew that back home in Leamington his father's business was suffering. His parents had come to depend on the money he was sending them.

On his way out for dinner he saw that he had a letter in his mailbox. He'd been so distracted on the way in he'd failed to notice it. He took the envelope from the slot and recognized the handwriting of his old university friend, George Ketiladze. The two had stayed in touch. George, who was still teaching at Queen's, had been particularly interested in Willis's experiences at General Electric. He was intrigued by television and curious about Ernst Alexanderson. Willis had also written him about his Northern Electric work. George, for his part, had kept Willis up to date on developments at Queen's, and provided the latest campus gossip.

Willis put off reading this latest letter until he had finished dinner. After the waitress had cleared away his plates and refilled his coffee, he tore open the envelope. George wrote that the Depression was beginning to take its toll on the university. Enrolment was down and outside funding was drying up. George had lobbied for a new transmitter at the radio station but was told that it was out of the question, although some money had been provided to upgrade the old one. Harold Stewart, the man in charge, had decided that if the university couldn't afford to buy a new transmitter outright, he might be able use the available funds to build one, from scratch, himself. George described the work in detail:

> We built crystal controlled set, right on breadboard layout, just nailed stuff down to a tabletop. In old set, modulator and modulated tube operate at same voltage. We change circuits so modulator operate at 500 volts, and tube at 250–300 volts. Enough power, this way, to run amplifier. We carry R.F., feed it through crystal oscillator.

George mentioned with pride that the first signal from the new transmitter was heard in Oklahoma City, more than a thousand miles away. Willis envied the role his friend was playing. This was exactly the sort of challenge he was seeking.

George also wrote about a notable embarrassment the station had suffered. During the broadcast of a football game, the play-by-play announcer, Henry Mungoban, had become so caught up in the action that when a Queen's halfback fumbled he forgot himself and announced to his listeners, "Jesus Christ, he dropped the ball!" Those were the last words broadcast from the stadium that afternoon. Harold Stewart, who was in the control room, had panicked and simply terminated the broadcast.

The rest of the letter was personal. George had enrolled in the master's program, part-time. He was still playing piano professionally to "make ends meet," and had run through a string of new girlfriends.

It was his final paragraph that grabbed Willis's attention. "Too bad you in Montreal," George wrote. "Our best demonstrator leaving this month . . . going to try luck in England. We could use someone like you."

Willis quickly put the letter into its envelope, paid for his dinner, and hurried back to his rooming house. He sat down at his desk and wrote a quick reply to George, explaining his predicament. He knew that the university would need a replacement for the departing demonstrator by September, when the new term would begin. He'd be available, and told George he would jump at the chance to work at Queen's. As he mailed the letter he wondered whether he was too late; he hoped they hadn't already found someone.

Back in his room, he sat down at the desk once more and wrote a second, more measured letter, this one to Dean Clark, formally applying for the position.

Willis rarely had trouble sleeping. No matter the circumstances, he could usually compartmentalize his worries and put them aside until morning. But this night was the exception. His mind was filled with contradictory thoughts. Unemployment, or a return to Queen's? The two possibilities fought for his attention. One minute he was back at university, earning money, the next minute he was jobless, on the dole, one of the many thousands seeking work.

He had barely drifted off when the alarm clock awoke him. He dragged himself out of bed, shaved, dressed, and left the house. He bought a coffee while waiting for his streetcar, then lit a cigarette and replayed the previous evening's events in his head. He realized that the matter was out of his hands. He'd done everything he could. Now he'd simply have to wait. He boarded the streetcar and sipped his coffee. It was a bright summer morning and all around him the city was coming to life.

When he got to the Northern Electric plant he decided to say nothing to Mr. Knobloch. There would be no slacking off and no worrying or daydreaming. He had a job to do and he would do it.

No matter how hard he tried, however, he couldn't help wondering, as he mounted the steps to his rooming house each evening, whether there would be a letter waiting for him. He knew it would be at least another week before he could reasonably expect an answer, but he checked his mail slot every time he passed it, just in case.

As it turned out Willis did not have long to wait. In the evening of the third day after he sent his letter, his landlady knocked on his door.

"There is a telegram for you," she called.

Willis hurried to the door.

"It came this afternoon," said Mrs. Whelan, excitement and curiosity evident in her voice. "I told the telegraph boy you were at work." She held the telegram to her bosom. "He made me sign for it," she continued. "I hope that is okay with you?"

"It's fine, sure . . ." Willis tried to keep the impatience from his voice. "May I have it?"

"Oh, of course," said Mrs. Whelan, flustered, and thrust it forward.

"Thank you." Willis took the telegram and turned, anxious to read it, but wanting the privacy of his room.

The landlady's curiosity overcame her embarrassment. "Must be important?" she asked. "Hope it isn't bad news."

But by then Willis had closed the door. He ripped open the flimsy yellow envelope.

> SPOKE TO CLARK STOP JOB IS YOURS STOP BEGIN SEPTEMBER SEVEN STOP REPLY ACCEPTANCE BY TELEGRAM

Beneath the message was a single name. GEORGE. Willis threw himself on his bed with a whoop. His old friend George, last of the big-time spenders—a telegram, no less! George had understood the need for urgency. Now there would be no more nervous waiting. He decided this called for a celebration.

On his way out he encountered Mrs. Whelan, still hovering in the hall. "Everything all right?" she asked.

"Everything is fine," said Willis, smiling. "Better than fine, in fact."

The next day Willis broke the news to Ralph during their lunch break.

"Sounds like you've landed on your feet," said Ralph, and Willis thought he detected a touch of envy in his friend's tone. It was no secret that he wasn't happy doing office work.

"I don't think the money is going be nearly as good at Queen's as it is here."

"Well, at least you'll be doing interesting work."

"There is that," Willis acknowledged, "and I'm looking forward to getting more hands-on experience with the radio station. Now that I'm no longer an undergraduate, they may even let me play with some of the more expensive toys."

"I'm curious," said Ralph, "what exactly will your duties be?"

Willis hadn't had a chance to ask. "I'm not quite sure. I'll have a better idea when I hear from the dean."

As it happened, the dean's letter didn't spell out his responsibilities. He would be filling the position of demonstrator in the Department of Electrical Engineering. The letter did confirm the dates of employment. He would begin, as George had said, on September 7, and would work the fall and winter terms, finishing on May 7. His salary was $500 a term, and he would be paid in monthly instalments.

Ralph threw a party for Willis on the Saturday of his final week at Northern Electric. Ralph's parents had thoughtfully picked that weekend to visit friends in the country, so the house was free. Among the guests were friends, a few employees from the company, wives, and girlfriends. Willis had dated occasionally during his year in Montreal, but tonight he was on his own.

Sitting on a sofa in the living room, listening as a Russ Columbo record played softly in the background, Willis mused about how different this party was from some of the wilder undergraduate bashes he'd attended at Queen's. The conversation here was muted; the guests rarely raised their voices. He sipped his rye and ginger ale and pondered the reasons for the subdued nature of the occasion. It was a farewell party, of course, but that didn't completely explain the lack of excitement. It seemed to him that the Depression was at the root of the lassitude. None of the men in the room could say with complete certainty that their jobs were secure. Willis was a reminder of what could happen to anyone. If a man with an engineering degree could be let go, anything was possible.

Willis lit a cigarette and thought about a phrase he'd overheard that day on the streetcar. "The Roaring Twenties have given way to the Dirty Thirties," a nondescript man had muttered, to no one in particular.

Chapter 25

Kingston, September 1931

Ralph turned the wheel of the Chevrolet and eased onto Bagot Street while Willis scanned the houses, looking for number 112.

"There it is," he said, pointing to a two-storey limestone house set back from the street, with a sloping roof and twin dormers.

"You sure?" Ralph asked. "Doesn't look like a boarding house to me." He pulled over and parked.

"It's the right address," said Willis. "The owner's a widow and she has two rooms." He smiled. "She doesn't take students, just faculty."

"Excuse me, sir," Ralph teased. "Good thing you're not a lowly undergrad."

"Nope, those days are behind me."

Ralph helped Willis unload his two suitcases and a steamer trunk. He had offered to drive Willis to Kingston when news of the teaching job came through.

"Want some help getting the stuff inside?" he asked.

"Sure," said Willis, "and you need a break before that long drive back to Montreal."

Willis and Ralph carried the suitcases up the walkway, then returned to the sidewalk and hefted the trunk, putting it down beside the front door. It opened before they could knock. An attractive middle-aged lady wearing a tailored jacket and matching skirt stood sizing them up.

"I'm Muriel Abbott," she said.

They deposited the suitcases on the porch. "And I'm Wheeler Little." He extended a hand. "I wrote about the room. And this is my friend, Ralph." He glanced back at the car. "He was kind enough to give me a lift."

"Well," said Muriel Abbott, "you two look like you could use a cup of tea. Why don't you leave your things here on the porch for now."

They followed her into the house and along the hall to a kitchen. Mrs. Abbott filled a kettle and put it on the stove.

"While we're waiting for this to boil," she said, turning to Willis, "I'll show you your room. Your friend can wait in the parlour."

His room, at the rear of the house, was larger than the one he'd occupied as a student. The furnishings were simple but tasteful. There was a single bed, a nightstand with a reading lamp, and a matching bureau. At the foot of the bed was an upholstered easy chair and a floor lamp. A desk by the window was positioned to offer a view of the backyard, in which Willis could see a small vegetable garden.

"The closet is here," said Mrs. Abbott, indicating a door to her right, "and the bathroom is down the hall."

"This is great," said Willis. "Perfect."

Downstairs as they sipped tea Willis filled in some additional details about the nature of his work. She knew from his letter that he'd be teaching, but was interested to learn about his involvement with the Queen's radio station.

"I have a radio," she said, "but I'm afraid the reception isn't very good."

"I'd be happy to have a look," said Willis. "I'm sure it's something I can fix."

Ralph glanced at his watch. "I'd better get a move on," he said. "It's getting late and I want to be home in time for supper. But you'll need a hand with that trunk."

"I have a suggestion," said Mrs. Abbott. "Why don't you take what you need up to your room. Then you can store the trunk out of the way in the basement."

"Good idea," said Willis, and he and Ralph carried his books and a few clothes upstairs. He decided to leave his overcoat in the trunk.

"I'll show you the way to the basement," said Mrs. Abbott, and led them back into the kitchen. "It's down there," she said, opening a narrow door beside a cupboard. They manoeuvred the trunk carefully down the stairs, and Willis called back, "Where would you like me to put it?"

"Oh, against a wall would be fine. Can you see okay?"

"It's fine," he said. The cellar was illuminated by a single bulb that hung from the ceiling.

Back upstairs, Willis turned to Ralph. "Can I buy you dinner? It's the least I can do."

"Thanks, but I should be getting back to Montreal."

"You sure?"

"I'm sure. I've got a date tonight."

"Well," Willis teased, "I hope she's worth it."

Ralph smiled. "She is."

They went out to the car. Ralph got in and pressed the starter.

Willis, standing on the sidewalk, bent down and peered through the passenger-side window. "I can't thank you enough," he said.

Ralph shrugged. "That's what friends are for." He put the car in gear. "Good luck."

"You too," said Willis, and watched his friend drive off.

The following day Willis went to Queen's and reported to Dean Clark, who outlined some of his responsibilities as a demonstrator. There would be occasions when the professors would expect him to set things up to illustrate how a particular piece of equipment worked. Some of this was familiar territory that he'd seen as a student. He would also be expected to assist undergraduates with projects and evaluate their performances.

"Any questions?" the dean asked.

"No, sir." Willis shook his head.

"Well, Willis, welcome once again to Queen's." He glanced down at some papers on his desk, papers that Willis recognized as transcripts. "You've come a long way. We took a chance on you after your second year, and it's clear you made the most of it."

Dean Clark rose from his desk and the two men shook hands.

Willis then went looking for George Ketiladze. He was pretty sure he knew where he'd find him. When he pushed through the doors of the studio in Fleming Hall he wasn't disappointed. The big Russian was bent over, fiddling with the wires of a new microphone.

He glanced up. "Wheeler!" he bellowed, then hurried across the room and enveloped Willis in a bear hug. "So good to see you, old friend." He held Willis at arm's length. "You looking good."

"And you too," said Willis. "George, I don't know how I can begin to thank you."

George waved his hand dismissively. "No problem."

"Your timing was uncanny. And that telegram! I had no time to fret or worry."

George, clearly embarrassed by the praise, changed the subject. "Let me show you around." He gestured to the control room. "Many changes since last time you here."

The first thing Willis noticed was the Mark V transmitter. It looked brand new, as if it had only recently arrived from the manufacturer, but Willis realized that it was the piece of equipment that had been built painstakingly from scratch, piece by piece, by Harold Stewart and Stanley Morgan.

"Amazing," he said, running his hand appreciatively over the shiny black casing.

"In many ways is big improvement," said George. "Better than new. We buy parts we need, special for us. We got generator from Lancashire Dynamo in Montreal. A lot of money—five hundred bucks—but worth every penny." George sounded for all the world like a parent extolling the virtues of a particularly bright child. "Valves standard, from Mullard in England, but we shop around for other stuff—condensers, resisters, frequency transformer. And this," he said, moving from the control room to the studio and picking up the microphone, "is latest, an R600A. Can't even fart in studio. Picks up every sound."

Willis laughed. His friend could always find a way to make technology a source of humour. "I'll try to remember that," he said. "But I doubt that you'll ever find me speaking into a microphone. I'm a behind-the-scenes guy, happiest to be twiddling dials in the control room. What about you?"

"Well . . . my accent . . . they won't let me announce," George admitted. "But play piano."

"Are you still at it outside, professionally?"

"Sure," said George with a smile. "I told you. Big hit with ladies."

Chapter 26

Kingston, 1931

The next few weeks went by quickly for Willis. He had no trouble satisfying the needs of the professors, even those who were the most demanding. He also found, to his surprise, that he liked working with the students, and they seemed equally taken with him. Willis wondered if his own academic struggles as an undergraduate were what made the difference. Willis was aware that most faculty members had been brilliant students who had breezed through their courses, so many of them found it hard to relate to the less talented students. Willis, though, could sympathize with those who were struggling.

Bill MacKenzie, a freshman from Toronto, came to him one afternoon in state of despair, with an assignment he was finding particularly difficult. "Maybe engineering isn't my cup of tea," he said, shaking his mop of unruly hair from side to side.

Willis smiled. "I wouldn't give up on myself this early in the game," he said. "Let's have a look." He took the young freshman through the various stages of the assignment patiently, step by step.

Eventually the penny dropped. "I get it!" said Bill excitedly. "Why didn't I see that in the first place? You've made it all seem so logical."

"If you keep at it, the logic will become clear to you, too." Willis could see potential in Bill MacKenzie, and decided that this was the right time to confide in him. "You know," he said, "it wasn't easy for me either, in the beginning."

"You?" Bill looked surprised.

"Me," said Willis. "In fact, I did so poorly I had to repeat my second year."

"You're kidding!" Bill found it hard to imagine that a teacher might have been anything less than brilliant at course work.

"Nope," said Willis. "I was a mess in my second year. Fortunately I found someone who believed in me and I was given a second chance."

When the student left the room at the end of their session together, Willis noted a marked bounce in his stride.

Willis was at the station on October 17 when the new Mark V transmitter went into regular use for the first time for the broadcast of a football game from Richardson stadium. The equipment functioned flawlessly. In the control room Willis watched carefully as a nervous Harold Stewart rode the levels to make sure that the signal strength was constant. He was also present when the new transmitter was used for the first time in a studio broadcast, a series of lectures on the League of Nations. And he finally learned from George why undergraduates were rarely asked to fill demanding roles in the running of CFRC.

"Is a matter of money," his friend explained. "Radio work just a sideline for them."

"A sideline?"

George nodded.

"A sideline for who?"

"For the professors. Stan Morgan, he gets an extra $300 a year for it. Not sure about Harold Stewart. Maybe half that."

Willis was shocked. "I thought it was all part of their job as professors, something they were expected to do."

"No," said George, "they teach, do research. Radio work is extra."

"I'd be willing to do it for nothing," said Willis. "Just for experience."

"Exactly. But you start that and bosses ask why they should pay people to run station. Especially now, money so scarce."

"But none of us could have built the new transmitter," Willis countered.

"True," said George, "so I take master's degree. Next time, I can build."

"You scheming SOB," laughed Willis. "Always thinking ahead."

Word soon spread among the undergraduates that Willis could usually be counted upon to lend a sympathetic ear when problems arose, and he found that he was spending more and more of his time working with students. This did not escape the notice of George, who asked him one day if he'd given any thought to teaching as a career.

"You a natural," the Russian observed.

Willis shrugged. "I have to admit, I do enjoy it."

"They come to me, too," said George, "but I don't have patience like you."

"I know, but I still have a hankering to get into radio."

"You should keep options open. Work is scarce. As professor, you always sure of job."

"My folks would flip," said Willis. "I can just imagine them telling all their friends that their son is a college professor."

"'Professor Little,'" said George. "It has nice ring to it."

"Never, in my wildest dreams!" said Willis with a laugh Then his friend brought him up short. "But you need master's degree."

"That does it," Willis said. "I don't think I could stand any more exams. Not to mention the cost."

Chapter 27

Leamington, 1931 Kingston, 1932

When Willis returned to Leamington for the Christmas break he discovered that his father's health was failing; his eyesight in particular was deteriorating. Anna Maria hadn't wanted to worry her son, so she hadn't mentioned it in her letters. Willis also learned that his dad, always a soft touch when it came to money, had extended unlimited credit to a number of his customers in the pool hall. Unfortunately, most of these men were now out of work and in no position to pay off their debts. Unpaid bills were piling up. Clyde was depressed and didn't want to discuss the problem, but one evening, when he was at the pool hall, Anna Maria provided Willis with the details. They were sitting in the living room.

"We may have to declare bankruptcy if things don't pick up."

Willis was stunned. The idea that Clyde Little's Billiard Parlor might go under had never occurred to him. It was a Leamington institution. It was only then that Willis realized the importance of the money he'd been sending home.

"What about the bank?" he asked.

Anna Maria merely laughed.

"The Masons?"

"Oh, they'd like to help," Anna Maria sighed, "but their books aren't in the best of shape either."

"What will you do?" Willis could not bring himself to use the word "bankruptcy."

"I'm not sure," said his mother, shaking her head. He had rarely seen her this way. Anna Maria had always seemed ready to take on whatever the world threw at her.

"Would it help if I was here?"

Anna Maria's head snapped up at these words. "No," she said sharply, the steel in her voice once again. "You stay on at Queen's. There'd be no work for you here even if you did come home."

"I could help out at the pool hall," offered Willis.

"Don't even suggest it," said Anna Maria. "That's not an option."

"I'll try to send a little more money," Willis promised. "I'm afraid I haven't saved as much as I should have. I didn't realize how tough things had become for you."

Anna Maria smiled at him through tears. "That would be nice," she said, "but only if you can." She rose from her chair, wiping her eyes. "I don't want you skimping on anything. We'll get by somehow." She moved into the kitchen. "Right now I think a cup of tea would hit the spot."

As Willis lit a cigarette he looked down at the package it had come from. He could always roll his own, he thought, or quit entirely; that was one way to save money. But he'd tried to kick the habit in the past without success. It had been tough. There had to be an easier way to economize.

Mrs. Abbott could tell that something was bothering Willis as soon as he returned to Kingston. His usual cheerful disposition had disappeared. He was polite, of course, but distant. One evening in January when he came in from work she invited him to join her for coffee and dessert. She had baked an apple pie that afternoon.

"What's the matter, Wheeler?" she asked. "You're not yourself these days."

Willis looked across the kitchen table. It was not his nature to share bad news. He hadn't even confided his family's difficulties to George. But there was something in Mrs. Abbott's voice, something in her look, that unlocked his customary reserve.

"It's my folks," he said, hesitantly. "Dad's business is failing, and so is his health."

Mrs. Abbott sipped her coffee and remained silent.

"They borrowed money to get me through university and now I can't begin to pay back all they've given me. I'd like to help them out, but I don't make enough at Queen's."

Mrs. Abbott reached across the table and patted Willis's hand. "These times are trying for everyone," she said soothingly. "Your folks aren't alone. What about your brothers, your sisters? Can't they do something?"

Willis shook his head. "I'm an only child, I'm afraid."

"What does your dad do?"

"He owns a billiard parlour, but these days no one has the money to play. And then there's his diabetes . . ."

"Oh, my," said Mrs. Abbott, shaking her head sadly. "That does complicate matters."

"I do send them something every month, but it isn't enough."

Mrs. Abbott liked Willis, and felt a surge of sympathy for the young man. "Is there anything I can do?"

"I don't think so," replied Willis. "Thanks for listening. It's helped. I haven't told anyone at the university."

"A sympathetic ear is the least thing I can provide." Mrs. Abbott was no stranger to life's challenges. She'd lost a husband in the war and it had taken her years to come to terms with that.

The following evening she once again invited Willis into the kitchen for coffee. "I think I might be able to help out in a small way," she said.

"A small way?" asked Willis.

"My other boarder, the retired officer, won't be back with us."

"Oh, I'm sorry," said Willis. "Can't you find someone else for the room?

"I doubt it," said Mrs. Abbott. "At this time of the year it would be hard."

"What will you do?" It was Willis's turn to be sympathetic.

"Oh, I'll get by," she said with a smile. "As you probably know, Major Steel's arrangement called for room and board. He made his own breakfast and I prepared supper for him."

Willis wasn't sure where this conversation was leading.

"I came to quite enjoy his company at dinner," said Mrs. Abbott. "Eating alone can be depressing. I thought you might like to take his place, keep me company."

"But I can't afford . . ."

"I know. Your rate needn't change. It would be my way of helping. If you can save on dinner expenses you'll be able send a little more money home."

Willis didn't know what to say. This act of generosity was unexpected. But Mrs. Abbott was right. Eliminating the cost of eating out would be a big saving. "I'm touched," he managed, a little overcome by her offer. "But are you sure? Times must be tough for you too."

"Well, my late husband's family is pretty well off," she said. "They paid off the mortgage on the house and still help out from time to time. But the decision to take in boarders was my own. I didn't want to be dependent on them."

Willis looked across the table at Mrs. Abbott. He had begun to see her in a different light. She was no longer just his landlady, she'd become a friend.

"You're sure?" he asked again.

"I'm sure," she said with resolve. "And I hope you like liver and bacon, because that's on the menu tomorrow evening."

At Queen's the next day Willis finally opened up to George. Willis was surprised when his account of his family's difficulties brought tears to the eyes of his friend.

"Is rough," said George, not the least bit embarrassed by his show of emotion. "My folks are dead, God rest their souls."

"You must miss them," said Willis.

The silence that followed brought on more tears. "Yes, I do," said George. "Was long time ago but I still miss them."

Willis felt the need to move the conversation to safer ground, so he told George about Mrs. Abbott's offer.

His friend's face creased in a knowing smile, the sorrow gone as quickly as it had come. "She a good friend."

Chapter 28

Kingston, 1932

The New Year brought a flood of new projects to the radio station. An ambitious Extension Lecture Series was launched. In the first two months alone, CFRC broadcast seven programs. The topics were wide-ranging. They included a discussion of the gold standard, an appropriate subject for the times; a tribute to England's poet laureate John Masefield; and, on the medical front, a talk entitled "Bacteria and the Age of Man." The *Whig-Standard* printed the station's programming schedule each week to let the people of Kingston know what was on.

By far the most ambitious undertaking was the broadcast of a charity concert from the Masonic Temple in Kingston on the evening of February 3. It was a fundraising venture, one of the first of its kind, organized by Lorne Richardson, Master of the Lodge. The goal was to raise money for unemployment relief, to be distributed by the local Red Cross Society.

Willis and George surveyed the hall. "Place is huge," said George.

Willis, a Mason, was no stranger to the hall. "If you think this hall is big, wait until you see the choir. There'll be more than a hundred people singing."

"Going to be challenge," said George with a grin.

In the end it was decided that the technical demands of the project could not be met with the equipment available at CFRC. It was Stanley Morgan who came up with a solution.

"I know some people at CKNC in Toronto," he said. "I'll see if they'd be willing to help out."

His connections paid off. The Toronto station not only provided additional equipment, but it also sent along a technician.

In the days before the concert a team of technicians assembled under the direction of Stanley Morgan set about wiring the hall, and finding the appropriate positions for the microphones. It was decided that two professors would share the responsibility of announcing the program. The team carried out extensive testing, but none of them had had previous experience with a broadcast of this magnitude.

"We can test," George remarked to Willis after a rehearsal, "but acoustics change when we have audience."

On the evening of the concert the Masons of Kingston, along with their wives and their invited guests, packed the hall. The irony of their formal attire was not lost on Willis. "The cost of all those gowns and tuxedos could feed a small army," he mumbled.

To say that the concert was a success would be an understatement. From a technical standpoint everything worked to perfection. The following day, *The Kingston Whig-Standard* carried this account:

> The Masonic Temple was filled to capacity and it is estimated that several thousands listened in to the broadcast throughout the City and district. Messrs. W.H. Mills and Duncan McArthur were the radio announcers during the evening and they reported over the air on every subscription that was telephoned in to the 4 phones at the Temple, which were receiving donations. Some were phoned from Ottawa and other distant points.

Announcements about subscriptions were made after each number. They also received many telephone calls asking for encores, and although the concert was already very long, a soprano solo and a number by the Choir were added. The charity concert raised about $1200 for unemployment relief.

Among those in the audience that night was Mrs. Abbott. As a Mason, Willis could bring a guest, and he'd extended an invitation to his benefactor.

"I'd be delighted," she'd said.

When she arrived, outfitted in a burgundy evening dress, Willis pointed her out to George. The Russian gave a low whistle. "That's some landlady, Wheeler."

Slightly embarrassed, Willis realized for the first time that Mrs. Abbott was, in fact, quite a stunning woman.

There was a small informal party for the technical crew in the hall after the performance, but Willis passed it up, deciding instead to escort his landlady home.

"That was a wonderful performance," she said. "I don't think I've ever heard such beautiful singing from so many voices."

"I have to confess that I'm not very musical," said Willis, "but that choir moved me in a way I hadn't expected."

"What do you mean?"

"I guess I was thinking of all the people who were unable to attend. At least they could hear it at home on their radios."

"And you were part of the team that made that happen."

"I was," said Willis. He spoke with modesty, but he welcomed the praise nonetheless.

When they reached the house, they mounted the steps and Willis unlocked the door, then held the door open for Mrs. Abbott. Inside, he helped her off with her coat and hung it, along with his own, in the hall closet.

"I think this calls for a celebration," said his landlady. "Something a little stronger than tea or coffee?"

Willis grinned. "Sure, I could use a drink. I've been worrying all week about broadcasting such a big production. Now that's all behind me and I can finally relax."

"Well," said Mrs. Abbott, "help yourself." She pointed to a collection of liquor bottles on a sideboard. "I'm going to change."

Willis selected a bottle of Seagram's Canadian Club rye whisky and poured himself a generous measure. He found some ginger ale in the icebox in the kitchen, and was in the living room sipping his drink when Mrs. Abbott returned.

She'd changed from her formal evening gown into something casual, yet elegant. He wasn't sure, but he thought it was Japanese. He rose when she entered the room.

"What can I get you?" he asked, moving to the sideboard.

"What are you having?" she asked.

"Oh, my old standby, rye and ginger."

"Rye and ginger, a man after my own heart. I'll have the same." She crossed the room and sat down on the sofa.

Willis mixed her drink. When she took it from him she patted the space beside her. "Here," she said, "sit next to me so that we can toast tonight's success together."

Willis joined her and she raised her glass. "To you, to Queen's, to the Masons, and to everyone who made the concert such a treat."

They touched glasses. "I thank you," he said, "on behalf of Queen's and the Masons. As for me, I was honoured to have you as my guest."

They both drank. Willis was aware that although his landlady had changed out of her formal gown, she had not removed her makeup. And he was certain she had added a touch of perfume.

He brought out his cigarettes. Until now he had smoked in his room, but never downstairs. He knew that Mrs. Abbott didn't smoke. "Do you mind?"

She laughed. "Go ahead," she said. At that moment Willis knew that the permission she was granting him went beyond smoking. When he tried to light his cigarette, he had trouble with his hands.

"Let me do that," she said, and took the matchbook from him. She struck the match, and when it flared he leaned forward to draw in the smoke. As he did, his hand rested on the soft fabric covering her thigh.

She pursed her lips and blew out the flame. "Wheeler Little," she said, her voice a whisper, "you are such a sweet boy." She kissed him lightly on the lips. "You must call me Muriel."

Chapter 29

Kingston, 1932

What followed with Muriel never became a love affair. It was quite simply an arrangement, one that met the needs of both of them, a lonely widow and a troubled young man. They comforted one another.

As Muriel quipped at breakfast the next morning, "I guess it gives new meaning to 'bed and board.'"

Willis, for his part, told no one of the new development, not even his best friend. He explained to George that the reason he no longer dated was so that he could save money, since he was sending his parents everything he could spare.

At Queen's, the success of the Masonic Temple charity broadcast gave rise to a second fundraiser. This event was organized by the Kiwanis Club, and was held in the club's auditorium. CFRC broadcast the concert, again with technical support from CKNC. A member of the Toronto contingent noted at the time that Queen's facilities were first-rate, and the station required outside help only because of the sheer size of the undertaking. With the Masonic experience behind them, George, Willis, and the rest of the team brought off the broadcast without a hitch.

Two other developments occurred at the station that spring. First, the university decided to stop broadcasting athletic events, because attendance at the games themselves was dwindling.

"They must be desperate," said Willis to George. "But I don't think radio is the reason for the poor turnout. Folks just don't have the money these days for anything but the basic necessities."

"I agree," said George. "It's stupid."

The Queen's Journal shared that view, commenting in an editorial, "The putting on the air of these contests not only enhances the university's prestige, but encourages good will in the community."

The second development had far greater significance for Willis. A financial representative from a company with radio interests in Windsor and Detroit visited Queen's to explore the possibility of expanding operations to Kingston.

Essex Broadcasters Limited was planning to launch a radio station in Windsor to serve that city and Detroit. George McLeod Coates met the Queen's principal, William Phyfe, with a proposal. He wanted to broadcast programs to Kingston listeners from his new station in Windsor, using CFRC facilities. This concept, the creation of what was later called a "phantom station," was not unique. It was spreading across Canada and the United States. It allowed established broadcasters to reach a wider audience without incurring the cost of building new stations. Phyfe was hesitant, and asked for time to consult other members of the administration.

Coates asked if he could check out CFRC's facilities. Willis happened to be in the studio at the time, so he was the one who showed the visiting executive around.

"This is first-rate equipment," Coates remarked, looking at the Mark V transmitter.

"On clear nights our signal can carry a long way," said Willis. "We've had letters from listeners all over the States."

"And it's a good microphone," Coates commented. "The studio is a little cramped, though. Is that because you do most of your work outside, on location?"

"We used to, but I'm not sure about the future. The administration has decided to cut sports programming."

"What? I can't believe it. Surely that's where you get your biggest audience?"

"You'd think so," said Willis, but he left it at that. He didn't want to seem disloyal to the university. "Are you going to be broadcasting games from your station in Windsor?"

"By all means," said Coates. "And don't forget that our signal will be heard in Detroit. Americans love sports."

"My dad will be pleased," said Willis. "He's a big baseball fan."

"Where are you from, young man?"

"Leamington, sir, not so far from Windsor."

"Well, I'll be damned," said Coates. "My own backyard. Look, I'm not leaving until the morning. Why don't you join me for dinner tonight?"

"I'd be delighted."

At the Venetian Gardens Coates again expressed his frustration with Queen's. "I don't get it," he said. "We'd be paying handsomely for the arrangement. Your university clearly needs the money. I thought they'd jump at the chance to join forces with us."

Willis agreed, but remained silent.

"Academics!" said Coates. "They don't live in the real world."

This mention of the "real world" prompted Willis to shift the conversation to his experience in Schenectady with General Electric. "It was amazing!" he said of the experience of watching his first television broadcast.

"It's where broadcasting is headed," Coates commented. "And if it wasn't for this damned Depression we could be there in the near future. The way things stand now, though, it's probably a long way off."

In the course of the evening Willis found that Coates had an insatiable curiosity. He was intrigued when he mentioned that he had spent some time at Northern Electric in Montreal, and questioned him in great detail about the plant, the work, and his general impressions of the company. He proved to be a good listener, and Willis opened up, even revealing personal concerns about the state of his father's business.

They finished eating, and when were drinking their coffee Coates asked the waiter to call him a cab. It soon arrived, and as they moved out onto the sidewalk he said to Willis, "Can I drop you somewhere?"

"No, sir, I live pretty close by and I need the exercise. Thank you for the dinner."

"It was my pleasure, Willis," said the executive, and the two men shook hands.

The taxi drove off and Willis walked home. He had no idea that this chance meeting with Coates would lead to a significant change in his life.

Most of the letters Willis received while he was working at Queen's were personal, and were sent to him at Muriel Abbott's address, so when word reached him that there was mail waiting at the porter's office, he was curious as to who would be writing him care of the university.

The porter at Fleming Hall handed him the envelope. It was from Essex Broadcasters Limited in Windsor. As he opened it, he remembered his dinner with George Coates.

He read the letter through quickly, and then again more carefully. He could not believe his luck. The new station in Windsor was set to begin broadcasting at the end of May, in just two months' time. They were looking for a Chief Control Engineer. Would Willis be interested? If so, he was to contact them immediately.

Willis sprinted across the campus to the office of George Ketiladze.

"Can I use your phone?" he gasped. "It's for a long-distance call to Windsor."

"Of course, of course," said the Russian. "But you must sit. Catch your breath first."

"You're right," said Willis, and collapsed in the chair across the desk from his friend.

"What so important in Windsor?"

Willis handed George the letter.

The Russian read it and gave a low whistle. "You one very lucky guy."

"I am," said Willis. "If this comes through it will be the answer to my prayers. A job in commercial radio. I can live with my folks, and really help them out financially."

George pushed the telephone across the desk. "Make call." He rose to leave the office. At the door he paused. "And good luck." He closed the door gently behind him.

A few minutes later, as Willis put the telephone back in its cradle, tears welled up in his eyes. His dream had come true. He would begin work at radio station CKOK in Windsor on May 31. His starting salary would be $200 a month, twice what he was making at Queen's.

George had not yet returned to his office, but Willis was sure he wouldn't mind if he made a second long-distance call, this time to Leamington. He couldn't wait to tell his parents the good news.

Willis sat silently for a few minutes after the call to his parents, savouring his good fortune. When George returned, he stood in the doorway with a questioning look. Willis rose from the desk and gave him a thumbs-up.

"You got it?" asked the Russian.

"I did."

George gave a whoop and crossed the room. Before he knew it Willis found himself embraced in a bear hug.

"We celebrate!" said George, stepping back.

"Tonight," Willis promised. "And the drinks are on me."

Later that day Willis stopped by the administration office to give his notice. Dean Clark was quick to congratulate him, admitting that the prospects for future work at Queen's were questionable.

"These are difficult times, Willis," said the dean, shaking his head. "We've been told to cut staff again."

"Well, I've enjoyed my time here, sir," said Willis, "and you gave me the chance when jobs were scarce. I appreciate that."

"I'm glad we were able to help," said the dean. "And keep in touch. Who knows? If this damn Depression ever lets up, we'll be hiring again. You made a very good impression this year, particularly with the undergraduates. We'll miss you, young man."

"Thank you, sir." The two shook hands.

That afternoon, when Willis returned to his rooming house, he wondered how he would break the news to Muriel. In the beginning their relationship had been a simple business arrangement. That had changed when she began helping him financially by providing meals. And now it had become even more complicated. So when he told her about the Windsor job, she surprised him.

"Why that's wonderful, Wheeler," she said, and kissed him lightly on the lips. "Now you'll be working in radio, where you've always wanted to be. And," she added, "you'll be back with your folks, able to help them out."

"I'm just sorry to be leaving you," said Willis. "You've been more than kind, and so . . ."—he paused, searching for the right word—". . . understanding."

"Oh Wheeler, I've always known Queen's was a temporary solution for you, something you did out of necessity. Of course I'll be sorry to see you go, but at least I'll know you have landed on your feet."

Willis admired the way Muriel took the news. He knew she was putting up a brave front. He would miss her, of that he was certain, but at least he was moving on. Her future wasn't nearly as exciting.

"When do you leave?" she asked.

"The end of the month."

"Well, that gives us almost three weeks," she smiled teasingly, and draped her arms around his neck. "Let's make the most of it, while we can."

V
Windsor Years

Chapter 30

Leamington and Windsor, 1932

Willis managed to get away from Queen's the weekend before he was due to report for his new job in Windsor. It was late afternoon when the taxi drew up in front of the house on Talbot Street in Leamington and Willis paid the driver. As he headed up the walk the front door opened, and his mother, in a floral housedress and a maroon cardigan, stood there smiling.

"Welcome home, stranger," she said.

Willis mounted the steps, put his bag down, and gave Anna Maria a long hug. "It's good to be home."

She held him at arm's length and gave him an appraising look. It had been some time since they'd seen each other. "Just in time for supper," she said.

Willis sniffed the air. "And if I'm not mistaken, a fresh batch of fried cakes."

Anna Maria smiled in acknowledgement. Willis went inside and climbed the stairs to his room. As he unpacked his bag he realized that very little had changed. His crystal set, now sadly out of date, sat on the desk where he had left it. A Queen's pennant was still tacked to the wall above his bed. Through the window he could see the leaves beginning to fill out on the maple tree. He could hear his mother moving about downstairs in the kitchen as she prepared the evening meal.

At supper Willis outlined what was expected of him in his new job. His earlier telephone conversation from Kingston had been long-distance call, so he had kept the details to a minimum.

"Chief Control Engineer," she said, with pride in her voice. "Sounds like a lot of responsibility."

"Oh, it's nothing I can't handle," said Willis. "And to be in at the beginning, when everything is new . . . well, I'm anxious to get started."

"You'll have to repeat all this for your father when he comes home later tonight," said Anna Maria.

"How is Dad?" Willis asked.

"He's managing," said Anna Maria. She had decided to keep things on a positive note. "And he can't wait to see you."

Sensing that the time wasn't right to press the matter, Willis changed the subject. He'd written his mother about his meeting with the Swedish inventor, Ernst Alexanderson, in Schenectady. Now he described it in greater detail.

"From Uppsala?" she said. "Who would have thought? Small world, isn't it?"

"And his latest invention is going to take the world by storm," said Willis, and then described his first glimpse of television.

"Land o' Goshen!" exclaimed Anna Maria, shaking her head in disbelief. "What's next? I'm still getting used to the wireless radio."

Willis laughed. "It will be a long time before we have television. But in the meantime radio is really taking off."

It was after midnight when Clyde returned from the pool hall, but Willis had waited up for him. When the front door opened he was shocked at the change in his father's appearance. Clyde had lost considerable weight, his hair was now completely grey, and he was wearing glasses.

"Dad!" said Willis, crossing the room.

Clyde put down his satchel. "Son!" He rested a hand on Willis's shoulder and gave it a squeeze. "It's good to have you home, my boy."

"And it's great to be home, Dad. I've missed you and Mom."

Clyde moved into the living room and settled in a rocking chair with a grateful sigh.

"Can I get you something, Dad? A cup of tea?"

"Tea would be nice."

"Tea coming up," said Willis, and he went into the kitchen.

A few minutes later he returned and sat on the sofa. "You look tired."

Clyde shrugged. "Oh, nothing a good night's sleep won't fix. It's been a long day." He removed his glasses and began polishing them with a handkerchief.

Willis decided this wasn't the time to probe, to ask about his father's health or the state of the business. Instead he outlined the work he would be doing at the radio station in Windsor.

As tired as Clyde was, he seemed to revive as he learned about his son's new responsibilities. "You know," he said, "we are very proud of you."

"I couldn't have done it without you," said Willis. "And I know my Queen's tuition put you in a hole."

"Worth every penny," said Clyde, managing a smile.

"Well, now, living at home, I'll be able to repay you."

Clyde waved his hand dismissively. "All in good time," he said.

Willis rose and took his father's empty teacup. "I don't know about you," he said, "but I'm ready for bed." He went into the kitchen and rinsed out the cup. When he returned, Clyde had dozed off in the rocking chair. Willis decided against waking his father. Instead, he carefully draped an afghan over Clyde's lap and went quietly upstairs to his own room.

Willis rose shortly after six on Monday morning, but when he went downstairs he found his mother already in the kitchen, preparing breakfast.

"Can't let you go off to work on your first day without a good breakfast," she said with a smile, and cracked two eggs into a frying pan where bacon was already sizzling. It was clear that Anna Maria had been up for a while, because the old wood stove in the kitchen was already radiating warmth.

"You didn't have to," said Willis.

"Pshaw!" Anna Maria waved away his protest. She placed slices of white bread flat on the stovetop, waited a few seconds, then flipped them over. She nodded to the pot percolating noisily. "Coffee is just about ready. Help yourself."

Willis took down a cup from the shelf above the sink and got a pitcher of cream from the icebox. He poured himself some coffee and sat back down at the kitchen table, where he added some cream and a couple of teaspoons of sugar. He stirred and sipped.

Anna Maria brought a plate to the table and put it down in front of him. The eggs were sunny side up, just the way he liked them. And the toast was Anna Maria's specialty, something he found only at home. Both sides were lightly charred, but the centre was still soft and doughy. Buttered, it was delicious. He dipped one of the slices into the soft centre of the egg yolk and began to eat. His mother poured herself a cup of coffee and sat down opposite him.

"Excited?" she asked.

"A little," Willis replied. "The only person I know is George Coates, one of the businessmen involved in the project."

"Well, that's a start," said Anna Maria.

Willis grinned. His mother was always upbeat. "It is," he admitted.

When he was ready to leave, Anna Maria handed him a brown paper bag. "Your lunch."

"Oh Mom, you're spoiling me." He shook his head, but took the bag.

"Good luck," she called as he turned and headed down the walk.

Willis gave a wave and Anna Maria watched from the door as he strode off along Talbot Street in the direction of the streetcar. It was a warm spring morning, and Willis found himself whistling as he walked.

The radio station was in an unassuming two-storey red brick building on the corner of Ouellette Avenue and Tecumseh Road. Above the main entrance the letters CKOK were painted in bright red. A sixty-foot steel frame antenna poked skyward from behind the building. It was shortly after eight o'clock when Willis pushed through the front door.

A middle-aged woman looked up from behind a counter. "Good morning," she said, her voice cheerful. "Can I help you?"

"I'm Wheeler Little," he said, and waited.

She looked puzzled. "Wheeler Little?" She glanced down at a list.

"Sorry," he said. "It's 'Willis Little'."

"Ah, there you are . . . our new Control Engineer." She smiled. "I'm Violet, and let me be the first to welcome you to CKOK."

"Thank you, and please call me Wheeler."

"All right, Wheeler. But you're early."

"I know, but I couldn't wait," Willis admitted.

"Well, George is already in, so I'll let him know you're here."

When Coates appeared he immediately spotted Willis. "Ah, there you are," he said, and the two shook hands. Coates was accompanied by a second, older man of medium height with closely cropped grey hair and the trace of a moustache.

"This is Major Andrew Cory," said Coates, "our station manager."

"Welcome," said Cory, and Willis noticed a hint of a Scottish accent in his voice. The major smiled. "George has told me all about you."

Willis decided to risk a little humour. "Not everything I hope."

"Just the good things," said Cory with a smile. "Now, let me show you around."

Willis noted with satisfaction that most of the equipment at CKOK was new. He was particularly impressed with the studios. Both of them were lined with sound-absorbing material and were furnished with expensive microphones. Glass panelling separated the studios from the control rooms.

"You've got a good set-up here," he said to Major Cory.

"Nothing but the best," Cory replied. "Our signal will be 500 watts when we launch, but there are plans to boost the power to 5,000 watts next year. We want to make sure we can be heard in Detroit as well as in Windsor."

The Major introduced Willis to two technicians who were working on the wiring for the Mullard transmitter. "You'll meet the rest of the staff later," he said. "But let me show you your office."

Willis wasn't sure what to expect, but when the manager led him to an office on the first floor, close to the studios, he was pleased. He wanted to be a hands-on engineer, and the office afforded him easy access to the control rooms. There was a window facing north, with a view of an expanse of lawn and, in the distance, Tecumseh Road. It wasn't a particularly large office, but there was enough room for a good-sized desk, a row of filing cabinets, and three wooden chairs.

"This is great," said Willis.

"I'll let you get settled in," said Cory as he turned to leave. "If you need anything, give Violet a shout."

Willis moved about the office to get a feel for the space. The desk drawers were empty, as were the filing cabinets, but he knew that that would change. He sat down and lit a cigarette. It was hard to believe, but here he was, about to begin work at a radio station. It was something he'd dreamed about since that afternoon years ago when for the first time he had heard a voice in the earphones of his primitive crystal wireless set.

The rest of the morning he spent being introduced to the rest of the staff. He met two staff announcers, Charles Hill and Terrence Odell, Gordon Fleming, a pianist, and Mary Hodge, who was in charge of children's programs. At about noon Coates poked his head into his office and invited Willis to join him for lunch.

"I want you to meet the boss," he said, and Willis followed him out into the lobby, where Coates told Violet they'd be back in an hour or so. Coates crossed the lawn to a black Nash Town Sedan in the parking lot.

Willis climbed in, and they drove onto Ouellette Avenue, then headed south into the downtown area.

"Malcolm Campbell is our top man," Coates explained. "He heads a group of local businessmen. They put up the money to establish Essex Broadcasters Limited and CKOK."

"It must be tough to find funds these days," Willis mused.

"True enough in general," said Coates, "but radio is an exception. People need entertainment, and radio provides it, free of charge."

He parked the car in a space opposite the Lasalle Hotel. "Mac is going to meet us inside," he said. "He was downtown this morning on business and suggested we have lunch in the hotel dining room. There aren't many places to eat near the station."

Willis smiled to himself at the memory of his mother's carefully prepared lunch he'd left in his desk drawer.

In the dining room Coates quickly spotted Campbell, and they made their way to his table. Coates introduced Willis, the men shook hands, and they sat down. Malcolm Campbell was tall, his complexion a ruddy contrast to his silver hair, and Willis guessed he was in his fifties. He wore an expensive three-piece suit, a high-collared shirt, and a silk tie.

"Welcome aboard," he said in a deep, resonant voice. "George has told me a little about you already, but I'd like to hear more, first-hand, from you." He looked across the table, his dark eyes assessing his new employee.

Willis was about to speak when a waiter appeared. After they had placed their orders, Willis gave an account of his early interest in radio, his studies at Queen's, his work in Montreal, and his brief time at General Electric. Campbell was particularly interested in television.

"That's the future," he said, looking at Coates, who nodded in agreement.

Their food arrived. Willis had ordered pot roast, the special, while his companions had opted for the lamb chops. As they ate, Campbell held forth on his plans for the station. "This launch, next week, is just the beginning," he said. "It's a foot in the door, so to speak. It will give us a chance to experiment with the equipment. But the real challenge is a year or so off, when we increase our signal strength. That's when the folks in Detroit will be able to hear our station clearly. And that's where the real potential lies, with all those listeners across the river. A chance for our advertisers to reach a vast audience." He wiped his mouth and sipped his ice water.

"The first voice I ever heard on the radio was a ham operator in Detroit," said Willis.

"Right," said Campbell. "The Americans are ahead of Canada when it comes to radio. But we're going to beat them at their own game."

Chapter 31

Windsor, 1932

The days before the official launch were busy ones for station personnel. Their families had all been given radios so that the signal strength of the station could be tested in various parts of the city. The results varied from day to day, and from evening to evening. Anna Maria and Clyde proudly reported that they were able to hear the station in Leamington, forty miles from the transmitter.

It soon became clear to the people at CKOK that Willis was the right man to guide them through the many technical hurdles they encountered. "I like the calm way you handle difficulties," Major Cory commented after Willis solved a particularly vexing frequency problem.

"When emergencies arise you've got to keep your emotions under control,'" said Willis. "Emotions can get in the way of judgment."

"My sentiments exactly."

The official launch was scheduled for noon on Tuesday, May 31. Most of the staff worked through the weekend leading up to the great event, and everything went smoothly. George Coates spoke at a meeting he had scheduled to follow Sunday's technical rehearsal.

"I want to thank all of you for the extra hours you've put in during these past few days. It's been a trying time, but now the finish line is in sight. Go home, get a good night's sleep, and take tomorrow morning off. We'll meet at noon and go over everything again, one last time."

Willis welcomed the chance for a break. In spite of his excitement he dozed off during the streetcar ride to Leamington, and the conductor had to waken him. When he arrived home he found his father sitting on the front porch, smoking a pipe. The poolroom was closed on Sunday, his only day off.

"All set for Tuesday?" Clyde asked.

"I think so," said Willis as he sat down on the steps. "Knock on wood." He rapped his knuckles on a railing.

Anna Maria came out from the kitchen. "You look exhausted," she said when she saw Willis.

"It's been a busy week," Willis admitted, "but things are under control, and we're ready . . . or as ready as we can be. I'm confident the equipment will work. Now it's up to the announcers, the musicians, and the producers."

"Well, dinner will be ready in a few minutes," said Anna Maria. "Why don't you come in and wash up?"

In spite of having been given the morning off, Willis found he was too keyed up to stay home. By ten o'clock he was back in Windsor. Major Cory was already at the station and the two men went over the schedule. The rest of the staff drifted in for the final rehearsal, which was set for one o'clock. Willis was handed a typed schedule of the day's programming:

12:00–3:00 p.m.	*Afternoons with Charlie* Charles Hill
3:00–5:00 p.m.	*Kiddies' Playtime* Mary Hodge, accompanied by Gordon Fleming on piano
5:00–5:05 p.m.	*News*
5:05–8:00 p.m.	*Beloved Vagabond* Bert McPhail
8:00–12:00 p.m.	*Evening Serenade* Tenor Tomas D'Hobbins and the Walkerside Singers

It was close to midnight when the rehearsal finished. Despite a few minor hitches, it had gone well, and Malcolm Campbell was impressed. After he thanked the staff and they prepared to leave, he called Willis over.

"It's a little late for you to be making the trip back to Leamington," he said. "Why don't we put you up for the night here in Windsor? I'll call the hotel, make a reservation. That way you'll be available first thing in the morning."

Willis was grateful. He wasn't looking forward to the long streetcar ride home.

"That would be great," he said.

Willis accepted Major Cory's offer to drive him to the Lasalle. When he got to his room he called home to his folks, then lit a cigarette and looked out the window. Across the St. Clair River he could make out the glittering lights of the Detroit skyline. He wondered how many of his American neighbours would be turning their radio dials to CKOK tomorrow evening. With this thought he stubbed out his cigarette, undressed, and pulled back the covers. Within minutes he was sleeping soundly.

Shortly after seven Willis awoke, showered, shaved with a razor borrowed from the hotel, and dressed. Downstairs he had breakfast in the dining room, then took a cab to the station. It had rained overnight, but the skies were clearing by the time they got there. As he made his way up the walk to the entrance the sun broke through. He hoped this was a good omen.

"Morning, Violet," he said to the receptionist.

"Good morning," she replied. "All set?"

"As set as I'll ever be."

During the morning Willis and Major Cory carried out a series of last-minute sound checks. Charles Hill rehearsed his lines in the studio while technicians rode the levels in the control room. Later, in a second studio, Mary Hodge rehearsed the introduction to *Kiddies' Playtime*, and Gordon Fleming played the piano. When troublesome feedback developed, Willis went into the studio and re-positioned the microphones. After a second test the Major gave Willis a thumbs-up from behind the glass.

"I'm still a little worried about the Walkerside Singers," said Willis when he returned to the control room. "There are five of them and we only have three mikes."

"They sounded fine during the rehearsal," the Major countered.

"Well, they're not in yet, but I'd like to test the balance again during the news break."

The countdown started just before noon.

"One minute to air," the producer announced from the control room.

In the studio Charles Hill cleared his throat.

Willis and the Major waited.

"Thirty seconds."

In the boardroom on the second floor George Coates and Malcolm Campbell paced nervously, eyeing the speakers.

"Ten, nine, eight, seven, six, five . . ."

The technician pressed a button and the turntable came to life. The needle was gently dropped into place, and at precisely twelve o'clock, music filled the control room and found its way to countless radio sets in the Windsor-Detroit area.

The red *ON AIR* light glowed in the studio.

"Fade the music . . ."—the producer brought his hand slowly down—"and cue Charles."

The announcer spoke. "Good afternoon, everyone, and welcome to the inaugural broadcast of CKOK—540 on your dial. I'm Charles Hill, and we are coming to you with 500 watts of power from our studios in Windsor, Ontario. Our programs are produced by Essex Broadcasters Limited, and we hope you'll make 540 a habit in the future."

As Hill continued to talk, outlining the program schedule for the day, the Major clapped Willis on the back. "Well done!" he said.

Upstairs, the station executives breathed a sigh of relief. "I think this calls for a drink," said Malcolm Campbell, and he moved to a liquor cabinet. "Scotch?"

"Scotch sounds fine," George Coates replied.

In the lobby Violet and an assorted group of guests broke into applause.

The day's programming went smoothly, but Willis remained in the control room until the final sign-off at midnight. A small informal staff party was held to mark the success of the launch. There were drinks all round, and when things broke up at about two, Willis was again put up overnight in the hotel.

He woke up the next day with a mild hangover. After breakfast in the hotel dining room he walked to a nearby barbershop for a shave, then visited a department store, where he bought a fresh change of clothes—underwear, a shirt, and, as a reward for his role in the station's successful launch, a four-dollar imported silk tie.

It was understood from the beginning that the 500-watt signal was merely the first step for CKOK. Willis recalled Malcolm Campbell's words; he'd called it "a foot in the broadcasting door." One of the main reasons Willis had been hired was to oversee the complex technical changes that would be involved when the station ramped up to 5,000 watts in the year ahead.

When Willis explained this to his father, Clyde said, "Sounds to me as if you're moving up from the minor to the major leagues."

"Exactly!" Willis smiled at his father's ability to bring baseball into any discussion.

The launch had not proved a serious challenge for Willis. True, he had uncovered and solved a number of problems during rehearsals, but the equipment was already in place and working when he arrived, so it had presented no major difficulties. What was more significant during those first few days was that he'd come to know and like the people he'd be working with.

In the months that followed he began to correspond with several major electronics manufacturers that would provide the new equipment required for the upgrade. Mullard, in England, still had the best reputation, but American firms were quickly closing the gap. It didn't sit well for American station owners to have to import equipment from England, so RCA and General Electric were determined to take the British firm's place at the top of the list. This worked in Willis's favour when it came time to order. He found that the mere mention of a competitor's name often resulted in a substantial price cut.

Another major change was the relocation of the station's studios. Space was leased on the tenth floor of the Guarantee Life Building in downtown Windsor, and workmen soon began gutting it. Willis could not believe his luck. He was not only overseeing the technical changes that would boost the station's signal strength, but he was also involved in designing the studios and control rooms. It was a far cry from his days at Queen's, where money was always a problem. Even with the Depression, the owners of CKOK spared no expense when it came to technical improvements.

Living in Leamington was posing a problem. Willis was working long hours, and while the station had generously provided hotel accommodation during the launch, it wasn't willing to make the arrangement permanent. Willis gently broached the subject to his mother during the Christmas holidays.

"I've been thinking about taking a room in Windsor," he said. He knew that his parents had come to depend on the income he provided, so he added, "I'm making good money at the station, and I can easily continue our arrangement here. I'll be home weekends, anyway."

"Are you sure?" said Anna Maria, trying to hide the disappointment in her voice.

"It will only be temporary—until we get the new station ready. Once the work is complete and we're on the air things will settle down. My hours should be more regular then, and I can come back home." Willis wasn't sure this would be the case, but he hoped it would make the idea more acceptable to his mother.

"Well," she said, "it does make sense, I guess. But are you sure you can afford it?"

"I'll find a place close to the station," said Willis. "That way I can walk to work and save the cost of commuting."

Anna Maria smiled, but she wondered if Willis would ever be back with them in Leamington. He was a young man, and a city the size of Windsor had much more to offer, not to mention having Detroit just across the river. But she kept these thoughts to herself.

Christmas at the Littles' was a subdued affair that year. It didn't feel right to be celebrating when so many people were out of work. Then there was the fact that Clyde's business was suffering, and that Willis wouldn't be around as much in the New Year. They did decide to have a tree, however, and Anna Maria cooked a small turkey.

The family always exchanged gifts on Christmas Eve—a Swedish tradition. Willis gave his father a Dunhill pipe, and he had found a fine silk shawl for his mother. Anna Maria, for her part, had been busy knitting—another tradition—and gave Willis several pairs of the woollen socks he favoured. Clyde surprised Willis when he proudly presented him with a gold Masonic ring.

There were hugs and thanks all around, and then Willis excused himself and went out onto the porch for a few moments. When he returned he was struggling with a large cardboard box festooned with a bright red bow. He placed it in the centre of the living room and invited his parents to open it.

Clyde took out his pocket knife and looked questioningly at his son.

"Go ahead, Dad—but be careful." Willis sat back and watched as his mother removed the bow and his father cut the cords that bound the box.

Soon a Stromberg Carlson radio emerged. It was a floor model, the latest, and Willis rose, plugged it in, and turned the dial. The room was filled with the full, rich sound of a church choir singing "Silent Night."

Anna Maria and Clyde stared in amazement at the radio. "It's beautiful," said his mother in a whisper. His father remained speechless.

"Wait until the summer, Dad," said Willis. "With this set you'll have your pick of baseball broadcasts—Detroit, of course, but on good days probably Boston, Chicago, New York, and even Cincinnati."

When the family went to bed later that evening the radio was still playing, tuned to CKOK.

Chapter 32

Windsor, 1933

In January Willis found a place on Donnelly Street within walking distance of the new station in the Guarantee Trust Building. It was in an old Victorian house that had seen better days. His room was on the third floor and it had a window facing south. His rent was six dollars a month and did not include meals. The two other boarders, a salesman and a bank teller, had both opted for breakfast and supper, which added two dollars to their rent. Willis reasoned that with the long hours he was putting in at the station he'd be better off eating out. He was making good money and could have afforded an apartment, but since he was still helping his parents financially he decided it would be prudent to keep his expenses to a minimum. He reasoned that all he really needed was a place to sleep and keep a few personal belongings.

Willis found he was dividing his time between the new studios, still under construction on Victoria Avenue, and the old CKOK site on Tecumseh Road. Getting from one location to the other was never a problem, because personnel were constantly driving back and forth. New equipment was arriving daily, and Willis had to check each item to make sure it was what had been ordered and that it hadn't been damaged in transit.

The price on one shipment of vacuum tubes seemed high to Willis, and he brought it to the attention of Major Cory. The station manager checked his files and found the original invoice.

"You're right, Wheeler." He looked up from his desk. "I'll call the company and see what's going on."

A day later he called Willis into his office. "You don't have to worry about this," he said, holding up the invoice. "The original manufacturer no longer makes the tubes we ordered, so the supplier substituted a more expensive brand." As Willis was about to leave, the Major spoke again. "Close the door; there's something I'd like to tell you." He lowered his voice. "Keep this under your hat for now. Our station is going to form a partnership with *The London Free Press*. When we increase our signal strength in the fall we'll have listeners in London, so the newspaper is sharing some of our expenses."

"Well, thanks for the warning, Major." He paused. "Will our programs still be under the control of Essex Broadcasting?"

"Sure, but they haven't worked out the details. There may be a change in the company name, but not in our plans."

Later, word leaked out that the new company would be called Western Ontario Broadcasting. The station had applied for permission to alter its frequency from 540 kHz to 840 kHz. This change would come about in the spring, when the signal strength was to be increased to 5000 watts. The call letters would also change, from CKOK to CKLW, to reflect London and Windsor.

Willis wondered about the possibility of conflicts when the time came to make programming decisions. The station was already serving two cities, Windsor and Detroit, and the prospect of adding London to the mix troubled him. However, he kept these thoughts to himself. He concentrated instead on making sure the equipment was properly installed and thoroughly tested, so that there would be no hiccups when CKLW was introduced to listeners for the first time, in May.

The one luxury Willis did allow himself that winter was a new brand of cigarettes. He had been introduced to them by chance when, after lunch with an electronics salesman in a Detroit restaurant, he'd been offered a smoke.

"No thanks," he said, "I prefer my own." He brought out his Buckinghams. "I like a strong tobacco. Like the first cigarettes I smoked as a kid."

The salesman smiled. "I used to roll my own. But try these, you may be surprised." He extended the pack to Willis.

Not wanting to offend, Willis took one. He lit it and inhaled. He felt the rough texture of the smoke as it passed into his throat, the satisfying click as it filled his lungs. "Wow," he said, "these are something!"

"Old Kentucky," the salesman said, and lit his own cigarette.

"Sounds more like a brand of bourbon." A few more puffs and Willis was hooked. "I wonder if I can get these in Canada," he mused.

"I don't know, but you're in Detroit often enough on business, you can always stock up when you're here." The salesman extended the pack to Willis. "For now, keep these. I have a whole carton back at the office."

When Willis returned to Windsor he sought out a smoke shop close to his rooming house. The owner didn't stock Old Kentucky, but assured Willis he could import the brand.

"They'll be a little more expensive," he said apologetically.

"How much more?"

"A nickle."

Willis considered this. He was smoking a pack a day now, but he was working hard, so he decided he deserved a reward.

One afternoon when Willis was back at the Tecumseh facility he was asked to replace a technician who had called in sick. A new singing group, Charmaine and the Chansonettes, had been booked for the evening program and were scheduled to rehearse.

Through the glass in the control room window he saw a pretty woman handing out sheet music to the other singers. He wondered about their French name. They didn't look French, nothing like the girls he'd seen in Montreal. Maybe they'd chosen it to set them apart from other quartets. Or perhaps the French name referred to the songs they'd be performing.

He pushed through the door into the studio and the producer introduced them. "This is Trudy Champion," he said, "and the Beckett sisters, Genevieve, Marion, and Pauline. They're from Wheatley."

Trudy smiled. "Hi, Wheeler." Her voice was playful, almost teasing.

Willis found himself speechless. Trudy was quite simply the most exquisite woman he'd ever met. And she radiated confidence.

Fortunately the producer filled the awkward silence. "Wheeler will be behind the glass in the control room. He's our Chief Control Engineer, so you'll be in good hands."

"I like a man with good hands," she said, and grinned. "We'll do our best. I hope you won't be disappointed."

Back in the control room Willis watched in fascination as they rehearsed, going over the songs again and again. He had no ear for music, but he knew from the first note that they were good. Eventually he turned off the studio microphones and just watched them talking, their lips moving wordlessly on the other side of the glass. In their colourful frocks they were like lovely tropical fish swimming silently before him.

Even without the sound it was clear that Trudy was the leader. She stood out from other people. It wasn't simply her looks—but God, she was beautiful—there was something else, an energy, maybe, or an aura. Whatever it was, Willis knew he wanted more of it. He just wasn't sure how to get it. There had to be other guys in her life. The very thought made him jealous. But that was silly, they'd just been introduced. What right did he have?

The door to the studio opened and Trudy poked her head in.

"Hi, Wheeler," she said with a smile. "I don't suppose you have a cigarette, do you?"

"Sure," he said, fumbling in his pocket for the package. He tapped the base of the pack and extended it to her.

Trudy crossed the control room, took a cigarette, and placed it to her lips. When Willis offered a light she leaned forward. The flame touched the tobacco, and she drew the smoke down into her lungs. She coughed and put her other hand to her mouth.

"These are strong," she said, tears welling up in her eyes.

"They're Old Kentucky," said Willis apologetically. "They're not everyone's cup of tea."

"I can see why," said Trudy, laughing. She took another drag and tentatively eased the smoke into her lungs.

"They probably go better with coffee," she said.

"Definitely better with coffee," he acknowledged. Then he asked, "Would you like a coffee?"

"Sure. We're taking a break. The piano is out of tune again. Jack says it will take twenty minutes or so to fix it."

Trudy watched as Willis shut off the switches on the console. She liked this tall, quiet man. Even when they were separated by the glass partition she had sensed his competence. And he'd been comfortable with it. He was one of the few men at the station who hadn't tried to impress her, hadn't forced an introduction.

He held the studio door open for her. "I can't promise the coffee will be fresh," he said as they headed down the hall to the cafeteria, "but it will be hot."

As they sipped their coffee Trudy looked at Willis quizzically. "I have to ask . . ."

"Ask away."

"It's about your name."

Willis laughed. "My nickname you mean?"

Trudy nodded.

"My real name is Willis—after the company that manufactures pianos. My mother chose it. Dad suggested William, but Mom remembered seeing a Willis piano at her wedding reception, so I became Willis, and that's what everyone called me until I arrived at Queen's University. While I was there I biked everywhere, even in winter, and word got around that I could fix bikes as well, so before I knew it everyone was calling me Wheeler."

Trudy smiled. "'Willis Little,'" she said thoughtfully.

In the days that followed Willis saw more of Trudy, but always at the station, after rehearsals. She began to appear more frequently on *The Evening Serenade*. When the program concluded she usually left with the Beckett sisters. Occasionally she left in the company of men. Willis noted that it was rarely the same man twice, and this emboldened him.

Over coffee one afternoon after a rehearsal, he steeled his nerve and suggested they have dinner together.

"Why, Wheeler," she teased, "I thought you'd never ask."

"I wasn't sure," he said hesitantly. "I thought you might . . ."

"Be spoken for?" she said, completing his thought. She smiled. "Nope, I like my independence too much to be tied down to any one man."

Willis had regained his composure. "So dinner with me wouldn't be tying you down?"

"Not at all," Trudy said. "It would take more than one dinner to tie me down, I can assure you."

Willis decided to press his luck. "And what about a movie after dinner?"

"Hmm . . ." Trudy pondered the question. "Only if I get to choose," she said with a laugh.

"It's a deal," said Willis.

They had dinner in the dining room of the Lasalle Hotel. Trudy wore a navy skirt and jacket with a pale blue blouse, open at the throat, exposing a simple strand of pearls and matching earrings. Willis was glad he'd changed into his only suit for the occasion. After they'd ordered, Trudy asked Willis about his family.

"I know your dad owns the poolroom in Leamington," she said, "but what about your mother?"

Willis explained Anna Maria's Swedish background and how his folks met for the first time when Clyde's barnstorming baseball team was visiting Wilcox, Pennsylvania.

"That's amazing," said Trudy, ". . . and very romantic."

Willis smiled. "It was a while before they were able to marry," he said. "Dad's future father-in-law wasn't exactly thrilled at the prospect of giving the hand of his only daughter to an itinerant ballplayer. So it wasn't until Dad took over the pool hall that Andrew Anderson relented."

"And they were married in Wilcox?" Trudy asked.

"Yep, at the Lutheran Church."

"Lutheran?" said Trudy. "That's interesting. My dad's a minister. He was Methodist, but became United when they and the Presbyterians joined forces a few years back. I don't think he liked the idea, but he didn't really have a choice."

Willis had heard about Trudy's father. J.B. Champion had made quite a name for himself in Essex County with his outspoken views on a good many controversial subjects. But what Trudy said next did surprise him.

"Dad's nothing but a blowhard," she said emphatically. "A big phony. He preaches one thing but practises another."

Willis was taken aback by this outburst. "So you're not on the best of terms with him?"

"Worst of terms would be more like it," said Trudy. "My eldest sister ran off with some guy when the family was living in the Maritimes. He swore at the time that in the future no daughter of his was going to embarrass him. Since that day my other sisters and I have been paying the price. He watches us like a hawk."

The waiter arrived with dinner. Trudy had opted for the chicken pot pie, while Willis was having a steak. They began to eat, and this gave Willis the chance to consider Trudy's feelings about her father. Her attitude was disconcerting, particularly since the man was a minister.

"You're still living at home?" he ventured.

"For now," said Trudy. "I teach piano at the parsonage in Wheatley. But when I graduated from high school I went off to Toronto to the Royal Conservatory of Music. I was there for two years. It was a breath of fresh air, being on my own. Now it looks as if Dad is going to be transferred to a parish in Ethel, up north. He doesn't know it yet, but I'm not going with them. For one thing I have students in Wheatley to consider, and for another, I like the radio work, and Ethel is long way from Windsor."

"Ethel?" Willis had heard of the town, but wasn't sure where it was.

"Up near Georgian Bay, more than a hundred miles north of here."

"So what's your plan?"

"I'll probably live with the Beckett sisters in Wheatley. And I'm pretty sure I can use the church there for my music students."

"And your father?"

"Oh, he'll hit the roof, I'm sure. But I'm twenty-two now and there's no way he can stop me."

Willis looked at the woman opposite him, and at that moment he was sure that J.B. would be no match for his daughter when the showdown came. As they were finishing dessert he decided he'd move the conversation to safer ground.

"Have you decided on a movie?"

"Yes, I have," said Trudy with enthusiasm. "I hope you like musicals, because the latest Ginger Rogers and Fred Astaire movie is at the Lux."

"What's it called?"

"*Flying Down to Rio.*" Trudy checked her watch. "The next show is in fifteen minutes."

"Well," said Willis, beckoning the waiter, "in that case we'd better take off."

Chapter 33

Windsor and Wheatley, 1933

In spite of the hard times the country was enduring, the mood within the broadcast community was decidedly upbeat. The Depression meant that people stayed home, and radio programs provided free entertainment. This optimism prompted Willis to do something he'd being dreaming about since his graduation from Queen's. Although he was sending money home to help his folks, his simple living arrangements in Windsor left him with a surplus at the end of each month. In April, to mark the completion of his first year's work at the station, he decided to treat himself.

He had read an ad for a 1930 Ford Model A coupe in the *Border Cities Star*, and called the owner. The asking price was $200, less than half the cost of a new automobile. The owner lived a short distance from his rooming house, so on a warm spring Saturday morning he walked over to have a look.

"She's a beauty," said the owner, with obvious pride. He was a man in his thirties, short and powerfully built. "Low mileage, too," he added.

Willis tried to hide his excitement. He didn't want to appear too interested. The car, a dark blue six-cylinder model with a rumble seat, was just what he was looking for. Sporty, but practical.

"Want to take it for a spin?" the owner asked.

"Sure," said Willis, "but you drive."

"Why are you selling it?" he asked as they climbed in.

"Been laid off," said the man, shaking his head. "I worked for Ford, on the assembly line. I'd been with the company for eight years, but when sales dropped they had to cut back."

"I'm sorry," said Willis, and decided wouldn't bargain for the car. He was sure he could get if for less than the asking price, but he didn't want to profit on someone else's misfortune.

Trudy's appearances on *The Evening Serenade* drew favourable responses from the listening audience, and soon she was appearing two or three times a week. She was seeing more and more of Willis, but most of the staff still saw them merely as friends, or colleagues. At first, true to her word, Trudy dated others, refusing to be tied down. Eventually, though, she came to realize that Willis offered her something unique.

One evening there was a party at the home of Major Cory. Willis was there, but Trudy had to finish some musical arrangements for her show, so it was late in the evening when the door opened and she walked in. Willis was sitting on a sofa, finishing off his rye and ginger ale.

"All work and no play . . ." Trudy announced, by way of explanation. There was a smattering of applause and she gave a mock curtsy. Trudy looked around, seeking somewhere to sit in the crowded room. One of the sales representatives got up from his chair and offered it to her. She was starting toward it when Willis called from across the room.

"Why don't you sit here, next to me?" He patted a place beside him. "This is where you belong."

That was the night Trudy decided she had played the field long enough. She felt completely at ease with this quiet, gentle man. He was everything her father wasn't. He was confident, but without arrogance. He seemed never to lose his temper, never even raised his voice. She particularly admired how he managed to remain calm when others were in a state of panic. He was, in a word, reassuring. And, of course, he was crazy about her.

She made the new arrangement conditional upon one fundamental change. She wanted Willis to drop his nickname. "I like the name 'Willis,'" she said, "and I think that while 'Wheeler' may have been appropriate for university life, you should phase it out. If you are going to meet my parents, I would like to introduce you as Willis, not Wheeler."

"But everyone knows me as Wheeler," Willis protested.

"They do now, but over time I think they'll come to think of you as Willis."

Willis felt a twinge of regret at the prospect of the name change. He'd come to think of himself as Wheeler. But it was a small sacrifice. If it would make Trudy happy, he would go along with it. "Okay," he said, "but how do we make the change? I can't very well make an announcement."

"No," agreed Trudy. "Do it gradually. Have new business cards printed and throw out your old ones. Get them to change the nameplate on your office door. Start signing your name Willis instead of Wheeler. And next year, when the new phone books are printed, ask them to list you as Willis Little, not Wheeler. It will take time. I'll make sure my friends use the name Willis, and you can encourage your friends to do the same. Don't make a big to-do about it."

Willis agreed, and he began using that name when new people at the station were introduced to him. Some of the older employees made the shift as well. Many still called him Wheeler, and Willis didn't object. He knew that if he did, it would only make them less accommodating. And Trudy let those references to "Wheeler" go unchallenged, because she too understood that the change could only be carried out gradually.

They'd been a couple for six months when, on a warm Saturday morning in May, Trudy decided they'd been putting off the inevitable long enough. It was time to introduce Willis to her parents.

"Wear a jacket and tie," Trudy advised. "We don't want to give him anything to nitpick about."

Willis and Trudy had already worked out their approach to the meeting. They agreed that she should remain silent. She had coached Willis about what to expect, what questions her father was likely to ask. She was afraid that if she took part in the conversation she might lose her temper.

Willis didn't know what to expect. He couldn't tell whether his future father-in-law was really the ogre Trudy described or merely a grumpy old man. He'd heard from a few church-going friends that the United Church minister was outspoken and opinionated, but he had been able to deal with difficult people in the past—particularly at Queen's, where some of his professors had had an inflated sense of their importance.

"How do I look?" Willis asked, standing by the only window in his room.

Trudy assessed him. "Good," she said, then moved closer and straightened his tie.

"And you look great, as usual," said Willis with a smile.

"Save your flattery for my father."

Trudy had also dressed for the occasion. She wore a new outfit, a simple beige skirt and a matching jacket. Her blouse, a delicate pink, added the necessary touch of colour. Willis knew that she'd had her hair done the previous day.

They were both as ready as they could be for the challenge that lay ahead.

In the car she felt the anxiety that welled up whenever she introduced her father to a new boyfriend. She swallowed hard and resolved that this time things would be different. For the first time her father would be meeting someone she hoped might become more than a mere boyfriend. She hadn't told her parents it was that serious, though, because she had been waiting for what she hoped would be the "right moment." She had already decided it wouldn't be during this visit. That would be rushing things.

As they drove into Wheatley Trudy pointed out the United Church, which was set back from the main street on a rise of ground. "That's Dad's," she said, then added with a touch of sarcasm, "It's where he holds forth every Sunday."

Willis didn't know much about churches, but he had to admit the building was impressive—almost too large, he thought, for such a small town. "Well, it certainly is big," he said.

"Wait till you see the parsonage," Trudy replied. "Remember, I warned you, there's nothing small about my father, except perhaps his mind."

Willis smiled, uncertain as to how to react to such judgments.

Following Trudy's directions, Willis extended his arm to signal the turn and swung left at the next intersection. As he shifted gears he noted a sign. "'Little Street?'" he said with a laugh. "Was it named in my honour?"

"I wondered if you'd notice," said Trudy, smiling. "Isn't that too much? I mean, Dad living on a street with the name of his future son-in-law."

"Is that a good omen or a bad one?" Willis asked.

"I hope it's a good omen, but I have my doubts. Dad has yet to meet the man he thinks suitable for the hand of his youngest daughter."

"Fingers crossed," said Willis.

"Indeed," said Trudy, holding both hands before her, fingers crossed. "Legs too," she said, with a laugh, "but it's a little late for that."

Willis chuckled.

"That's our place on the left." Trudy indicated a large three-storey red brick house on a corner lot. "You can park in the driveway."

Willis turned in, braked to a stop, and removed the keys from the ignition. As he stepped from the car he noticed the fragrance of flowers. There were lilac stands in full bloom on each side of the main entrance. Willis took in the size of the house. It was at least twice as large as his parents' place in Leamington. There was an expanse of lawn, and flowerbeds nestled neatly against the walls. He made out a row of beehives behind the house. Trudy had mentioned that beekeeping was her father's hobby. To Willis the hives had the look of gravestones, but he shook that thought from his head.

Trudy mounted the steps and pushed open the front door. Willis followed her in.

"Anyone home?" she called, keeping her voice light. Inside the house the scent of the lilacs gave way to the aroma of something sweet, recently baked.

Trudy's mother, Blanche Champion, answered from the kitchen. "I'll be right out," she called. "Sit yourselves down in the parlour. I'll be there in a minute."

Trudy ushered Willis into a formal room with an upholstered settee, matching chairs, and a window that looked out onto the front lawn. Heavy drapes bracketed the window, and a lace curtain filtered the light that fell on an oriental rug. Willis moved across the room to an open fireplace. Above the mantel hung a seascape set in an ornate gilt frame. He leaned closer and made out the name Anna Gertrude Champion written in the bottom right corner.

He turned to Trudy. "You did this?"

"Yes, I did," she said. "But it isn't original. I copied it from an English painter."

"Impressive," he said.

"My daughter is very talented." The words came from a tiny woman who'd come quietly into the room. She was carrying a silver tray on which was a tea set. She put it down on a table by the settee and rose to face Willis. "I'm Blanche, Trudy's mother."

Willis took her hand, and became aware how fine-boned it was, like the woman herself.

"Mother," Trudy said, "this is Willis, Willis Little." She kept her voice light, careful to hide any sign of nervousness. "He manages the technical side of things at the radio station."

"Well, young man," Blanche said, releasing her hand, "sit down and make yourself comfortable." She moved to the table, lifted the teapot, and began to pour.

Willis watched how effortlessly she managed. He found himself comparing this woman with his own mother, who was a good six inches taller and had none of the refinement that characterized Blanche.

"How do you take your tea?" she asked.

"A little milk, no sugar."

Blanche poured the milk and handed the cup and saucer to Willis. "Josiah will be down in a moment. He's upstairs in his study finishing up tomorrow's sermon."

Willis was thankful for the chance to size up Blanche and get to know a little about her before meeting Josiah. Trudy had warned him about her father, but she hadn't said much about her mother.

"Trudy tells us you're indispensable at that radio station," said Blanche.

Willis smiled. "I think she exaggerates my role," he said as he sipped his tea, wishing it was coffee. "It runs pretty well on its own. I just make sure things don't break down."

Blanche added sugar to her own tea and stirred it. When she looked up at Willis he saw in her blue eyes a strength he'd not noticed when he and she had been introduced. These eyes wouldn't miss much, he thought. Blanche was wearing a high-necked floral blouse with a cameo broach. Her hair was silver, cut short in a bob. She sat forward on the edge of her chair much as a tiny bird might perch on a branch.

"Trudy says you were a music teacher?" Willis continued.

"Oh, my, that was years ago. In New Brunswick. I gave it up when we moved from the Maritimes. Trudy is the music teacher now," she added. "Or at least she was until this radio business came along."

Willis noticed a trace of disappointment in this last remark. It didn't come as a surprise. People of Blanche's generation tended to dismiss radio, although Willis knew that they were often its most devoted listeners. He was sure Blanche wasn't one to miss her daughter's weekly program.

They heard a door close on the second floor, and then the sound of someone descending the stairs. "That will be Josiah," said Blanche, and Willis rose.

The man who entered the parlour was an inch or two taller than Willis and wore a dark three-piece suit, set off by a white clerical collar. He crossed the room and extended his hand. His grip was firm, and Willis found himself comparing this handshake to Blanche's. They were opposites, these two. Josiah was a big man and he dominated the room.

Trudy cleared her throat. "Dad, this is Willis Little," she said. With these words she sat down, and resolved to remain silent.

"Willis Little," Josiah said, "our Trudy's latest beau." His smile was devoid of humour. "Well, I'm sure you know by now I am Josiah Brooks

Champion . . . the Reverend Josiah Brooks Champion." He released Willis's hand and sat down opposite him.

The man who sat before Willis couldn't be called handsome, but he had strong, even features dominated by a nose that seemed a little too large for his face. His scalp was bare apart from a fringe of hair that remained above his ears. The smile had revealed even white teeth that Willis was sure were not Josiah's own. Josiah positioned himself so that the diffused light from the window reflected off his rimless spectacles.

"Where was it Trudy said your folks are from? Leamington?" said Josiah in a deep, resonant voice.

Willis was sure the man already knew the answer to this question. Next, he was sure, would be a query about his father's occupation. He decided to pre-empt this strategy.

"Yes," he said. Then he added, with a touch of pride, "My dad owns a billiard hall."

"You mean a poolroom?" Josiah asked, with feigned uncertainty, as if he was unfamiliar with the term billiard hall.

Willis was ready. "Yes, a poolroom," he said confidently. "But he sells tobacco, pipes, cigarettes, and smoking supplies as well." He would have given anything at that moment for one of his own cigarettes, but controlled the impulse.

Josiah picked an imaginary piece of lint from the knee of his trousers. "I see."

Trudy had told Willis that her father smoked cigarettes, but that he did so secretly. From the pulpit he railed against tobacco, lumping it with hard liquor as a vice to be avoided.

"And what faith do your parents follow?"

Again, Willis was ready for the question. "My mother's Swedish, and she was raised as a Lutheran."

"And now?"

"Now?" Willis understood the question but decided he would let Josiah lead.

After a pause, Josiah pushed on. "So your family is Lutheran?"

"I guess you could say my mother is a lapsed Lutheran. Dad and I are both Masons, so we're freethinkers. We don't attend church, except on special occasions." There, Willis thought: Trudy's concerns—his father's questionable occupation, the absence of religion in his home—had all been laid out before his future father-in-law.

Josiah shifted in his chair. "Masons," he mused. "Well, I suppose it could be worse. You could be Catholics."

At this point Blanche intervened. "More tea?" she asked. Willis heard a hint of strength in this simple query. There was more to this woman than met the eye, however frail she looked. It seemed to him that she had skilfully brought this part of the interrogation to an end.

"No, thank you," he replied. "I'm fine."

"Willis runs the radio station in Windsor," continued Blanche.

"Without him," added Trudy, breaking her vow of silence, "you wouldn't hear my program."

"Your program?" said Josiah, turning to Trudy. "Yes, I suppose that's true." He shifted his glance back to Willis. "It's a shame your station doesn't carry opera. Trudy's sister, Edith, has a fine soprano voice."

Again Willis was ready. Trudy had warned him to expect this comparison with her talented older sibling. "That's interesting," he said in a matter-of-fact tone. "Did you know that the first commercial radio transmission was a performance from the Metropolitan Opera in New York. I believe the year was 1911."

"Really?" For the first time Josiah seemed to be surprised. Then he rallied. "Yes, Edith will be off to Europe next year. The Europeans are far more appreciative of fine music."

Once again Blanche intervened. "Have a scone," she said as she removed a napkin from a plate on the table. "They're freshly baked."

"I thought something smelled awfully good when we came in," said Willis. For the next few minutes the group was largely silent as they sipped tea and ate. Again Willis found himself comparing Blanche to his own mother. Anna Maria would have served doughnuts, and probably offered coffee as well as tea. "These are excellent," he said to Blanche.

"Mother is a terrific cook," Trudy added. This was safe territory. She'd been uncomfortable as the silent partner, listening as her father questioned Willis but unable to come to his defence. It was a relief to speak up, even if her contribution wasn't part of the interrogation. "I wish I had Mom's talent in the kitchen," she said ruefully.

"You have the musical talent," said Blanche, deflecting the compliment. "That's far more important."

When they finished, Blanche rose. "I'll just take these," she said, rising to clear the tea service from the table.

"Let me help, mother," said Trudy, and the two women retreated to the kitchen.

Willis was left alone with Josiah. He felt as if he was in a prizefight; each competitor was jabbing, feeling out his opponent before trading serious punches. It was clear that Josiah was no lightweight.

"This radio business," said the minister, returning to the attack, "do you think it's more than just a fad?" He took a handkerchief from his jacket pocket, removed his glasses, and began polishing them.

"Oh, it's no fad, I can assure you," said Willis, resisting the urge to clean his own glasses.

"And the work you do, keeping the station running, what training did that require?" Willis was sure the man knew the answer, but played along anyway. "I studied at Queen's University in Kingston. I have a degree in electrical engineering."

"Queen's?" said Josiah. "A Presbyterian school, I seem to recall."

"Right," said Willis. "Founded in 1841 to train Presbyterian clergy. Still does, in fact."

"Presbyterian," said Josiah, almost to himself. "Well, at least it's a Protestant faith."

Willis smiled to himself as he recalled the divinity students he'd known. They were some of the wildest undergraduates on campus. "After I graduated I worked for a year in Montreal," he continued.

"In radio?" asked Josiah.

"No, for Northern Electric."

"And what did you think of Montreal?" Josiah leaned forward, and Willis was relieved that the conversation was moving to safer ground.

"It's a huge city, and . . ."

Josiah cut him off. "Roman Catholic—and French."

"Not all French," Willis countered. "Canada's first commercial radio station was started in Montreal, and it was an English-language station."

Josiah looked surprised.

Sensing he had the advantage in this verbal sparring, Willis pressed on. "Then, after my year in Montreal, Queen's invited me back to teach." Willis paused, wondering if any of this had impressed his future father-in-law.

"To teach?" This put Josiah off balance. "You were a professor?"

Willis chuckled. "No, nothing so grand. I worked with professors, demonstrating electrical equipment for undergraduates."

"And you left Queen's for a job with this radio station in Windsor?" The tone of Josiah's question left little doubt that he thought the move was foolish.

"Well," said Willis, "CKLW was a new station and I thought it offered the perfect chance for me to get in on the ground floor."

"I see," said Josiah, sounding unconvinced.

Willis was growing tired of this sparring. Most people he met were thrilled to know someone involved in radio. He'd come here at Trudy's insistence, and now he wanted nothing more than to be back in his car, returning to Windsor.

As if reading his mind, Josiah rose from his chair and buttoned his jacket. "Well," he said, "all this is very interesting, but I'm afraid you'll have to excuse me. I have a sermon to complete." He walked from the room and climbed the stairs.

As he stood alone in the parlour, Willis realized there had been no parting handshake.

Back in the car Willis lit a cigarette and inhaled deeply. "Well," he said, "I'm glad that's over with."

Beside him Trudy fumbled in her purse for her own cigarettes. "That bastard!" she fumed. "I warned you."

Willis blew smoke through the open window and turned to Trudy. "You did. And I can see now why he gets under your skin."

"Under my skin!" She put the cigarette to her lips, but her hands shook as she tried to light it. "Under my skin," she repeated. "That's putting it mildly."

Willis reached across and flicked his lighter. His hand was steady as he lit her cigarette. "He certainly is a piece of work."

She looked out of her window at the parsonage. "What are you waiting for?" she asked. "Let's get the hell out of here."

Willis pressed the starter button and put the car into gear.

"That son of a bitch," said Trudy, "grilling you like a criminal. He already knew about your dad, about your job."

"But not the stuff about the Masons, or about our family's religion—or the lack of it." Willis smiled at the memory. "That seemed to take him by surprise."

"I'm not so sure," said Trudy. "He has spies everywhere. I bet he had a buddy in Leamington check you out."

"Maybe," Willis mused. He wasn't as upset as Trudy. He felt he'd held his own with Josiah.

"You know what really gets me, what bothers me most about him? He is such a hypocrite. He lies, chases women, smokes—and yet he pretends to be this model of virtue. About the only thing he doesn't do is drink."

"Be thankful for small mercies."

As they drove past Josiah's church, Trudy flicked her cigarette from the window.

"I like your mother," said Willis, trying to change the subject.

Trudy sighed. "She's okay, I guess. But I don't know how she puts up with J.B."

"The scones were great," Willis added.

"Yes," she said, "I'll give you that. But they were the only good thing about today's visit."

"At least it's over with. Now we can get on with our lives."

"Oh," said Trudy with a sigh, "I wish it was that simple. I'd be happy if we never saw him again. But I think this was just the first round."

"We'll wear him down over time."

"I'd like to wear him out."

Trudy switched on the car radio, and when music from station CKLW—Dorothy Fields singing "Don't Blame Me"—filled the car, she leaned back against the seat and closed her eyes.

Chapter 34

Windsor, 1933

Things moved quickly at the station during the spring of 1933. The Canadian government, fearing the increasing influence of programming from American stations located close to the border, had established a parliamentary committee to study the problem. The result was the creation of the Canadian Radio Broadcasting Commission (CRBC), a network of stations in cities across the country. This ensured that the people of Canada would be able to hear original programs produced by Canadians and featuring Canadian performers. Although CKLW was already in a unique position, with listeners in both Canada and the United States, its owners still wanted to be part of the new Canadian network, so they applied for membership. They also needed government approval for the planned increase in signal strength and the shift of its position on the radio dial from 540 to 840 kilocycles. Optimistic that these requirements would pose no problems, CKLW announced that the change would happen on May 1.

The station was now on the verge of becoming a broadcasting powerhouse, and there were inevitable growing pains. The first casualty was George Coates, the man who'd hired Willis. He opposed the partnership with *The London Free Press*, and resigned when the new board of directors approved the merger. Willis was sorry to see him leave, but, in truth, he hadn't seen much of Coates during his time at the station. The man had been primarily involved in the business side of the operation, while Willis spent all his time dealing with technical problems.

A second change, however, had an immediate effect on Willis. The station hired William Carter as Chief Engineer. Carter had an impressive background. He had owned a telephone communications company and had worked as a marine transmission inspector for the federal government. When the appointment was announced, Willis realized that he would no longer be the station's most experienced technician. Any doubts he might have had about the pecking order were quickly laid to rest when he saw how the announcement of Carter's hiring was worded. He was the Chief Engineer, and Willis, as Chief Control Engineer, would be reporting to him.

On the day Carter took up his appointment he summoned Willis to his office on the second floor. "Come in, come in," he said. He rose from behind his desk and shook Willis's hand. "Have a seat," he said, motioning to a chair.

William Carter was a good deal older than Willis. He was tall, thin, and slightly stooped, and his voice had the rasp of a heavy smoker.

"I've been looking forward to our meeting," he said. "In fact, I asked about you when I signed on with the station. I like to know the calibre of people I'll be working with."

"I see," said Willis, and waited for Carter to continue.

"In fact, you are one of the reasons I took the job. You've an impressive background, Willis."

"Thank you, sir."

"As for me, I don't have your educational qualifications. In my day they weren't offering degrees in electrical engineering. But I do have other attributes," he said, with a smile that was almost apologetic. "I know my way around bureaucratic regulations—and that will be critical in the near future."

"I understand you worked for the government?" Willis ventured.

"I did, but frankly I found it tedious. I think the private sector is much more challenging."

"I agree," said Willis. "I liked teaching at Queen's, but it was all a little too theoretical."

"Exactly," said Carter. He leaned forward. "I'm going to be frank with you, Willis," he said. "I'm going to need someone with first-hand radio experience, someone I can trust. I want you to view me as a colleague, not as a boss. Is that clear?"

Willis felt a surge of relief. "Yes, sir," he said. Any concerns he'd had about the new chief engineer were dispelled. "I look forward to working with you."

Carter reached in his pocket and extracted a package of Buckingham cigarettes. He offered one to Willis. "Smoke?"

"Thank you," said Willis. He lit up and inhaled deeply.

"And please call me Bill."

Another major change occurred in March. When Willis arrived at the station on the first Monday of the month Violet beckoned him to the reception desk.

"Have you heard?" she asked in a whisper.

"Heard?"

"About the Major?"

"What about the Major?"

"He's been sacked. Let go."

"I don't believe it," said Willis. He shook his head. "Are you sure?"

Violet nodded. "Apparently he was told Friday afternoon. Came in over the weekend and packed up all his stuff."

In his office Willis mulled over the news. It wasn't a complete surprise to him; he'd overheard snatches of conversation from various members of the management team about the Major's shortcomings. He'd been the right choice in the beginning, when the station needed his military discipline to get things up and running, but when it came to programming he was out of his depth. But Willis was shocked by the abruptness of the dismissal. And it brought home to him again how quickly things could change.

The staff was introduced to the new station manager at a meeting after lunch. Garfield Packard's background was in advertising. About the same age as Willis, he had boyish good looks and a mop of unruly hair, and was dressed casually in a sports jacket and a brightly patterned tie. He seemed relaxed yet confident as he gave a brief speech outlining some of the plans he had for new programs, and then patiently answered questions.

One afternoon a couple of weeks after his appointment, Packard appeared in Studio A while the Chansonettes were rehearsing. The song was "You're Getting to Be a Habit with Me," made popular by Bing Crosby in a recent recording with Guy Lombardo's orchestra. Packard watched with fascination as the women went over the song again and again, refining it under Trudy's direction. He had studied music when he was a teenager, and appreciated the gradual improvements that were being made. He was also impressed with the patience Trudy showed as she worked with the Beckett sisters, rehearsing them individually, then as a group. Eventually she seemed satisfied, and Packard realized that the song had been transformed into something quite distinct from Crosby's rendition. It now played up the humorous aspect of the lyrics, changing the song from a lover's lament into something lighter. It was particularly noticeable on the lines,

> *Oh, I can't break away,*
> *I must have you every day*
> *As regularly as coffee or tea.*
> *You've got me in your clutches*
> *And I can't get free;*
> *You're getting to be a habit with me.*

As he prepared to leave, he spoke to the studio technician. "I notice Trudy accompanies the group on piano," he said. "Is this just for rehearsals?"

"Nope," he said, "it's the same deal for the program."

"Thank you," said Packard, and he closed the door quietly on his way out of the control room.

Two days later Packard asked to see Trudy. As she approached his office she tried to keep her fears under control. Always the perfectionist, she wondered if the new man had found something he didn't like in the group's performance.

The door was open, and his welcoming smile when he invited her in was an encouraging sign. She sat down opposite him and waited.

"Do you have any idea how talented you are?" he asked, almost casually.

Trudy was at a loss for words. She shook her head.

"You are amazing," Packard continued. "I watched a rehearsal the other afternoon. Your patience with the Beckett sisters was remarkable."

"Well," said Trudy when she found her voice, "I was trained as a piano teacher. Maybe that explains it."

"Could be," Packard mused. "And speaking of the piano, why don't you use our staff musician, Gordon Fleming? I can understand your accompanying yourself in rehearsal, but on the air . . ." He let his words trail off.

"Oh, Gordon's a terrific pianist," said Trudy. "It's just that I feel better, more in control, when I'm playing."

Packard smiled, as if her words confirmed something he already knew. "I've been going over our schedule for the fall," he said, "and I think we should make you a regular on *The Evening Serenade*."

Once again Trudy was speechless.

Packard waited.

"Well?" he asked.

Trudy managed to control her excitement enough to give him an answer. "I don't know what to say," she said. "I'm flattered, of course. But it would be a lot of work . . . not that I mind the work . . . but . . . the Beckett sisters have other jobs."

"Which brings me to another point. The sisters are good, but they aren't trained musicians. And I don't much care for the name 'The Chansonettes.' Sounds a little pretentious to me."

"What are you suggesting?" Trudy asked, becoming concerned at the turn the conversation had taken.

"I'm suggesting auditions; I'm suggesting a new group and a new name. There might even be a show of your own in the future."

"Can I think it over?" Trudy asked, hoping the question didn't sound presumptuous.

"Of course," he assured her. "But I'd like your answer within a week or so. We're going to need to get things organized, work out a time slot, co-ordinate publicity—details like that."

As it happened, the Beckett sisters were understanding. They weren't about to give up their jobs for the uncertainties of a radio career. But they did encourage Trudy to try her luck.

"You've got the talent," said Paulette. "Someday you'll be famous, and we can say we knew you when you were just starting out."

"What about your teaching, here in Wheatley?" Genevieve asked. "Surely you won't have time for both?"

"My classes will be winding down over the summer. That'll be the perfect time to make the break."

"Your students will be disappointed."

"I know, and I'll miss them. But maybe the radio show won't work out. Who knows? I could be a flop and find myself back here in the New Year."

Trudy might have had doubts, but the Beckett sisters had none. They were convinced Trudy had a bright future in radio.

Auditions for Trudy's new group were held in the CKOK studios, and more than a dozen singers applied. Trudy led them through their numbers while Packard listened from the control room. In the end, two young women, Edith Alexander and Isobel Large, both from Windsor, stood out from the others. Edith was an accomplished soloist in her church choir, and Isobel, like Trudy, taught piano. Trudy was disappointed they couldn't find a third candidate who measured up, because most of her arrangements involved four voices. But in the end it was decided they'd make do as a trio rather than compromise on talent in the search for a fourth member.

Trudy hadn't decided on a name for the new group. She discussed it with Packard. French names were out; she could tell that from his reaction to "The Chansonettes." A choice had to be made soon, because plans for the new schedule were in their final stages.

One morning on the way to work, an idea popped into Packard's head. The group was now a trio. Why not call it the Triolettes? "Trudy and the Triolettes." It had a nice ring to it.

After rehearsal that afternoon he suggested the name to the group. Both Edith and Isobel were enthusiastic. Trudy wasn't sure, but she couldn't come up with anything better, so in the end that was it. They became The Triolettes.

One of the biggest challenges Willis faced that spring was the conversion of the 25 Hz of power provided by the province to the 60 Hz required by the new 5000-watt transmitter. It was beyond his expertise. Fortunately, William Carter had a suggestion.

"It's not a unique problem," he said. "I've talked to engineers at a couple of American stations. KFI in Los Angeles faced the same situation. They're going to send me the details about how they overcame it."

The solution would prove costly. The converters that were required were huge, and to accommodate them the building had to be enlarged. Willis found himself putting in long hours with Carter as the two men coordinated renovations and the installation of the new equipment. When completed, the site would be impressive. With the massive converters in place it would look more like a power plant than a radio station.

Fortunately for them, approval from the Canadian Radio Broadcasting Commission, while still just a formality, had not come through in time for the launch at the beginning of May. This gave Willis and Bill Carter more time to implement the many technical modifications that were required. While the delay was a disappointment to the rest of the staff and to *The London Free Press*, which had published a special edition to mark the change, Trudy welcomed it. She was always the perfectionist, and it gave her more time to fine-tune the arrangements. The launch date was re-set: CKLW would move to 840 kilocycles in the first week in November.

Willis and Trudy had survived her parents' scrutiny, and by July it was time for a trip to Leamington so that his folks could meet their son's new girlfriend. Trudy was apprehensive, but Willis assured her she would face none of the obstacles he'd endured in Wheatley.

"They're going to love you," he said with complete confidence.

His words proved to be prophetic. Anna Maria had secretly fretted over her son's single status. He was twenty-eight, an age when most of his friends were married. But her concerns dissolved the moment she met Trudy.

"She's delightful," Anna Maria confided to her son once the introductions had been made. They were in the kitchen preparing fresh lemonade. "And I notice she's calling you Willis. That's a good sign. I never really thought of you as Wheeler."

In the living room Clyde was proudly showing Trudy their new radio. "I almost feel I know you," he said. "Your voice comes across so clearly on *The Evening Serenade*."

"Willis deserves some of the credit. He's a whiz on the technical side."

At dinner that evening Trudy envied the ease with which the Littles had accepted her. Why, she wondered to herself, did her own parents, particularly her father, have to be so difficult?

On the drive back to Windsor, Trudy told Willis how much she'd enjoyed the visit. "I was worried that your dad might ask me about baseball," she said, smiling. "I played softball as a kid, but I don't know anything about the professional game."

"I don't follow it either," said Willis. "Maybe I should. Then I'd have more in common with Dad."

"I wouldn't worry about it. He seems awfully proud of you. They both do."

"I guess you're right," Willis shrugged. He still harboured a faint regret that he hadn't played baseball as a boy. Clyde would have loved it if he had.

Trudy, for her part, was thinking about her father. She'd deliberately withheld word of her decision to stop teaching piano. Instead she let her parents assume that this responsibility would keep her in Wheatley. Secretly, she'd already begun looking for an apartment in Windsor.

Chapter 35

Windsor and Leamington, 1933

Trudy's folks moved to Ethel in September, and J.B. was already complaining to Blanche about the shortcomings of his new parsonage.

"It's too damned small," he muttered.

Blanche knew what would follow. Her husband had found fault with every new home they encountered, and then he set about renovating. He was handy that way, and within a year or so the house would usually be more to his liking. Another concern was his beloved bees. He would have to move the hives from Wheatley during the winter, when the bees were hibernating, and he was fretting about their ability to withstand the change.

In September Trudy found a one-bedroom apartment on Erie Street East in Windsor. Willis was delighted. It meant they'd be seeing a good deal more of each other, especially given that Trudy would be spending a lot of her time at the station, rehearsing. She hadn't told her parents, leaving them with the impression that she was still boarding with the Beckett sisters in Wheatley. She hoped this deception would buy her enough time to get properly settled in her new apartment.

One evening after she and Willis had eaten the simple supper she had made, they were relaxing on a second-hand sofa she had bought for her living room.

"What do you think your father will do when he finds out?" Willis asked.

"Oh, he'll have one of his temper tantrums, for sure, but Mom will listen patiently and eventually calm him down."

"She is such a wisp of a woman, it's hard to imagine her standing up to your dad."

"Don't let her size fool you. She is one tough cookie and can defend her castle when it comes down to it," said Trudy, then started to giggle.

"What's so funny?"

"I just had a thought. I have my own castle now. This apartment is my castle, and I'm strong enough to defend it against any siege J.B. might mount."

"Would you like to enlist my services as gatekeeper?"

"Sir Willis Little Lancelot," said Trudy, laughing. "It does have a nice ring to it." She leaned over and kissed him lightly on the cheek. "My knight in shining armour."

Trudy didn't get a chance to be the one to break the news to her parents about her move to Windsor. Blanche learned about it in a letter from Agnes Philmore, the mother of one of Trudy's former pupils. She was convinced that her eight-year-old son, Brian, was a prodigy, and lamented Trudy's decision to abandon teaching. Where was one to turn, Agnes wondered, since there were no other piano teachers in Wheatley?

Blanche held off discussing this development with J.B. until after the Sunday services. Over the years she'd learned that this was the best time to approach him whenever there was difficult news to be faced. They'd finished eating and he was reading a church circular.

"I've had a letter from Mrs. Philmore in Wheatley," she began, and something in the tone of her voice alerted J.B. to the fact that he had better pay attention.

"And?" he said, his brow furrowed.

"It appears that our Trudy has given up teaching piano."

"What?" J.B. bellowed. He rose from his chair, and the circular fluttered to the floor. He began pacing. "Just quit? After all the money we spent on her education?"

"I'm afraid so," said Blanche. "And apparently she's moved to Windsor."

"And what, pray tell, is she planning to do in Windsor?"

"Mrs. Philmore didn't say, but I think it probably has something to do with radio."

"Oh, fine," said J.B. venomously, "here we are in the midst of a depression with thousands out of work and she chucks a perfectly good job to . . ." He searched for the right words. ". . . to sing those silly pop songs. It makes no sense."

Blanche tried to placate her husband. "Maybe it's just temporary. Trudy's always been impulsive."

J.B. slammed his palm down on the dining room table with such force that the cutlery rattled. "She couldn't wait to get out of the house and go off to Toronto to study piano. Couldn't take her teacher training closer to home. No, that would never do. It had to be the best. Nothing but the best for our Trudy." He continued, his voice dripping with sarcasm. "The Royal Conservatory of Music in Toronto. 'Royal'!—as if she was some sort of princess."

"Well, Edith went to Toronto," Blanche reminded her husband. "I think Trudy was simply following in her footsteps."

"Yes," huffed J.B., "but Edith did something with her education. She became an opera singer. She didn't waste her talent singing silly pops songs on the radio."

Blanche decided to remain quiet. Eventually, she knew, her husband would run out of steam.

But J.B. wasn't through. "I'll bet that Willis Little is behind this." He stopped pacing and looked at Blanche. "Willis Little! His father is nothing but a poolroom hustler, for God's sake. Trudy should have known better, she should have had nothing to do with his sort."

"Now, Josiah," said Blanche, her voice firm, "we don't know that Willis is involved."

"Well, I hope he isn't," J.B. continued, "for his sake and for hers."

The telephone call was surprising. Willis usually phoned his folks on Sunday night, making sure that the long distance charges would be on his bill, not theirs. But this call was different. It was from his mother to his office at the station, and it was on a Friday afternoon.

"Is there any chance you could come home this weekend?" she asked.

Willis detected the tension in her voice. "What's the matter?"

"Oh," she said, "it's not something I want to discuss on the phone."

"Is it Dad?" Willis asked, fear gripping him.

"No, no," said Anna Maria, sensing his worry. "Your father's fine. But something's come up and we'd like a chance to discuss it with you in person. Bring Trudy, if you like."

"Trudy has rehearsals all weekend, but I can make it." Willis knew it was fruitless to press his mother for more information. "I'll drive down tomorrow."

When Willis arrived, shortly before noon, both Anna Maria and Clyde were seated outside on the front porch. This was unusual, since his father normally worked on Saturdays.

He mounted the steps. "Mom? . . . Dad?"

Anna Maria rose and greeted him with a prolonged embrace. Something was clearly wrong.

"Dad, what are you doing home?" he asked his father.

Clyde father shrugged and was about to reply when Anna Maria interjected, "Why don't you sit down? I'll get you some iced tea and we can talk." She didn't wait for a reply, but hurried inside.

Willis sat down on a cane rocker beside his father. "What is it?"

Clyde turned to look at his son. "It's the business, I'm afraid."

Willis waited for him to continue.

"The bank has foreclosed on the mortgage," said Clyde. "I've had to file for bankruptcy."

The news was a shock to Willis. His father had mentioned as recently as Christmas that business at the pool hall had dropped off, and these days foreclosures and bankruptcies were everyday occurrences, but Willis had hoped that the money he'd been sending home every month would keep his father solvent.

"I had no idea things were that bad," he said. "Isn't there something we can do? Maybe I could get a loan in Windsor."

Clyde shook his head. "I'm afraid it's too late for that."

Anna Maria returned and handed Willis his iced tea. "It wasn't just the slowdown in business," she said. "Your father has always been a soft touch."

Clyde busied himself filling his pipe.

Willis looked at his mother, waiting for her to continue.

"So many of the men borrowed from him. And there were others he extended credit to for tobacco, or cigarettes. And when the hard times hit, they weren't able to pay him back." Willis noted the faintest hint of reproach in her voice, tinged with resignation. "When the bank manager looked at the books . . . well, that was the final straw."

"What will you do?" Willis asked his father.

"Jack Ridley, who owns the bowling alley, has asked me to look after things part-time in the evenings," he said, lighting his pipe. "That should help keep the wolf from the door."

"And we thought," said Anna Maria, tentatively, "that since you're so rarely home these days, we might rent out your room."

The full impact of the news was beginning to sink in. "Sure," said Willis, nodding in agreement. Until now he'd been blind to the situation, living in Windsor, working in a business that seemed immune to the ravages of the Depression. "I can always bunk down on the sofa when I visit," he said. He felt the need to inject a little optimism into the conversation. "You know me. I can sleep just about anywhere."

"That's true," said Anna Maria with a rueful smile. She patted Clyde's hand affectionately. "You're just like your father," she said.

Chapter 36

Windsor and Detroit, 1933

On the afternoon of October 5 Malcolm Campbell, the president of CKLW, called a general meeting of the staff. He stood before them as they filed into Studio A. When everyone had settled in, he began to read from a prepared statement.

"This morning I received a communiqué from The Canadian Radio Broadcasting Commission in Ottawa." He paused, letting the suspense build. Would there be another delay? His expression gave nothing away. Then he looked up from the papers in his hand and smiled. "I am pleased to be able to say that we have finally been given formal approval to go ahead with our expansion plans."

Cheers, applause, backslapping, and a general babble rose from his audience as the staff shared their feelings of relief that things were on track. When the noise subsided, the president continued. "On November 6, promptly at midnight, CKLW will move to 840 kilocycles, with five times its previous signal strength. I want to congratulate all of you for your hard work, and particularly for the patience you've displayed during this extended delay. To celebrate, I've booked the ballroom at the Dearborn Inn in Detroit for our annual Christmas party this year. You're all invited, of course, and please bring your families and friends along. They all deserve our thanks for standing by you during these last few difficult months."

Willis found Trudy as the crowd filed out of the studio. "Do we have a date?" he asked.

Trudy feigned surprise. "A date?"

"You know what I mean," said Willis.

Trudy continued to tease. "Know what?"

"There is going to be this party," said Willis, deciding to play along.

"Not really a party, but a dinner dance. I just wondered if you'd be free that evening?"

"And when is this . . . dinner dance . . . going to be held?"

"I'm not sure exactly. Next month sometime, probably close to Christmas."

"Well, let me know when you have a date and I'll check my calendar. It gets filled up pretty quickly during the holidays."

"I'll do that," said Willis with a wink. "Maybe we should invite your parents? Mac did say families were welcome."

"Over my dead body," said Trudy.

The increased signal strength of CKLW extended the reach of the station well into Michigan, beyond Detroit. In Canada it meant better reception for the listeners in London. On the business side, ads were placed in all the newspapers in Detroit, Windsor, and London announcing CKLW's shift from 540 to 840 on the radio dial. Additional salesmen were hired, and businesses, stores, and restaurants in Detroit were canvassed seriously for the first time in the search for new advertisers.

The Commission's approval also meant that CKLW was part of the fledgling Canadian radio network. In 1933 the CRBC provided a mere two hours of Canadian content a week. Tuesday evenings, from nine until ten, listeners were able to hear the orchestra of Geoffrey Waddington in a program broadcast from Toronto. On Thursday evenings there was another orchestral offering, *One Hour with You*, this time from Montreal, with Giuseppe Agostini conducting.

CKLW already had an agreement in place with CBS, and authorities had little choice but to allow this affiliation to continue. It gave CKLW the best of both worlds. The station was now part of the new Canadian network, but could pick and choose among CBS programs as well. Among these was one of the first soap operas, *The Romance of Helen Trent*, an afternoon show designed to appeal to housewives. American vaudeville comedians were making their way into radio, among them Fred Allen, whose weekly half-hour comedy show had become a big hit. CBS also broadcast baseball games, including the World Series. This became a source of particular pride for Clyde Little, who liked to boast to his friends that it was his son's work that made it possible for them to enjoy the games in the comfort of their homes.

The evening hours attracted the largest audiences, and the schedule consisted of a mixture of news, musical programs, comedy shows, mysteries, and dramas. CKLW blended locally produced programs with those from CBS and, of course, the CRBC. Political pressure was building on the Canadian Radio Broadcasting Commission to provide more homegrown programs, and it was understood that they would be added to the mix as they became available. Everyone at the station hoped that eventually some of the CKLW shows would be picked up by the Canadian network. After all, there were listeners right across the country, not just in Toronto and Montreal.

With Trudy living in Windsor, Willis found himself dividing his nights between her apartment and his room. At Trudy's insistence, he had kept his place on Donnelly Street for propriety's sake. In the back of her mind was the spectre of her father unexpectedly descending on them. She didn't want to give him any more ammunition in what was now a running battle over her decision to give up teaching in order to pursue a career in radio. Until now he had been concentrating on her career, but Trudy was sure that in time his focus would shift to the suitability of Willis as a boyfriend. She hadn't had a face-to-face confrontation with her father since she'd moved. The town of Ethel was a safe distance from Windsor, but J.B. had written and phoned, haranguing her for what he termed her rash decision. He wouldn't drive down, he said; he insisted that she come to see him. So far she'd refused, but she knew that a showdown was inevitable.

If it hadn't been for their respective parents, Trudy and Willis would have been living an ideal existence. Trudy's trio was working out well, and Willis, lost in the world of vacuum tubes and antenna specifications, loved the challenges of his work. But his father's business failure still haunted him, and he knew that Trudy was troubled by her estrangement from her family.

One evening in her apartment they discussed these concerns. There had been some good news from Leamington. Anna Maria had managed to rent out Willis's old room, so there would be a little extra money coming in.

"That's great!" Trudy enthused. She envied the concern Willis had for his parents. "Didn't your folks ever object to anything you did?" she asked.

Willis thought about this for a moment. "Not really," he said. "There were some times as a teenager when I got into trouble, but it was all minor stuff."

"You're so lucky," said Trudy, shaking her head.

"What is it with your father?" Willis asked. "Why is he such an SOB?"

"I told you some of this already, but I think it really began with Doris, my oldest sister. When she was eighteen she got pregnant and ran off to get married. Her husband was a jerk, so when their second child arrived she walked out on him and came running back to J.B. and Blanche. She brought the new baby with her and left the older boy in the care of her husband's parents. I think it was the shame this brought on the family that led to our moving here from New Brunswick. I was too young to understand what was going on."

"I guess something like that would shake you up, particularly if you were a minister. What about your other sisters?"

"They all cleared out as soon as they finished high school. Doris remarried, and now she lives in Essex. Alice went directly into nursing; she's married too, and she works in Detroit. It was my sister Edith who blazed the trail for me. She was very talented, and went off to the Royal Conservatory of Music in Toronto when she was eighteen. I followed her there. Claire's the one who's closest in age to me; she's a couple of years older. She went to Alma College in St. Thomas and became a librarian here in Windsor."

"Edith's the opera singer?" Willis prompted.

"Yep, she's studying abroad, and she's the apple of Dad's eye. He is such a snob! He brags about Edith, but thinks I'm wasting my talent singing pop songs."

"And your brothers?"

"Wendell and George," said Trudy with a sigh. "Every failure to meet his exacting standards was followed by a trip to the woodshed. I can still hear them wailing. They took off as soon as they could. Wendell works in a bank, and George went into business for himself. Wisely, he moved away to Cincinnati."

"What a difference between our two families," Willis mused. "Me, an only child, and you, one of seven."

"We would have been eight," said Trudy with a touch of sadness. "Mom lost a son, stillborn, when I was just two."

"If he'd lived, you'd have a younger brother," said Willis.

"I often wonder," Trudy mused. "Maybe . . . who knows? . . . if he'd survived it might have changed how J.B. treated me."

"How so?"

"Well, he would have been a distraction." Trudy smiled at the thought. "I wouldn't have been the baby in the family."

"Am I going to meet any of these sisters or brothers of yours?"

"You will, in time. For now I'd rather keep you a deep dark secret. Once the introductions start, word will get back to my folks. They're still digesting the fact that I've given up teaching. I want them to get over that before I drop the other shoe."

"The other shoe?" Willis laughed. "I've been called a lot of things, but never a shoe."

"Think of yourself as a comfortable old shoe," said Trudy with a smile.

"I guess I can live with that," said Willis good-naturedly.

Chapter 37

Detroit and Windsor, 1933

Trudy decided it was time to risk introducing Willis to her sister, Alice Kingerley. Alice and Trudy were close, in spite of the difference in their ages, and Trudy knew that Alice could be counted on to avoid any mention of Willis in her letters or phone calls home to J.B. and Blanche in Ethel.

In late November Trudy and Willis were invited to dinner at Alice's home in a suburb of Detroit. As they were driving across the Ambassador Bridge it began to rain. The car radio was tuned to CKLW, and by chance the music that was playing was Ethel Waters' version of "Stormy Weather."

"I hope that's not a bad omen," said Willis.

"Don't fret," said Trudy. "You'll love my sister."

"She's the nurse."

"Yep, at the Women's Hospital in Detroit. Her work involves delivering babies, but the doctors there also deal with birth complications."

"What about her husband?"

"Al's a salesman for a copper wire company. I don't think the Depression has hurt his business."

When they arrived at the Kingerleys', Willis gave a low whistle. The home was an impressive new two-storey brick house, set back beyond a wide expanse of lawn. In the driveway was a gleaming new Buick sedan.

"Boy," said Willis as they hurried through the rain, "this Al must be a one very successful salesman."

The front door opened as they climbed the steps and a woman appeared. "Trudy!" she called out, with obvious affection. "And this must be your new beau. Come in, get out of the rain."

They entered the hall, and a tall, handsome man appeared. Trudy introduced Willis to Alice, who gave him a welcoming hug, and then to Al. The two men shook hands.

"Your timing is great," said Al with a grin. He had a deep, pleasant voice. "I was just about to mix drinks. I know Trudy will have a rum and coke, but what's your pleasure?"

"Rum and coke would be fine for me, too," said Willis.

They were ushered into the living room. It was tastefully decorated with modern furniture and wall-to-wall carpeting. A radio played softly, and Willis noted with satisfaction that it was tuned to 840.

As the evening progressed Willis relaxed. It was clear that this couple wasn't judging him. Alice, in particular, was warm and friendly. In looks, it was clear she was Trudy's sister—not quite as pretty, but a Champion nonetheless.

"Do you hear any news from the folks?" Trudy asked. "How are Mom and Dad doing?"

"Not so good," said Alice with a shake of her head. "Frankly, I'm worried. Dad seems in a bit of a funk. I haven't talked to him, but Mom called the other night to say he misses Wheatley."

"I'm sure he'll adjust," said Trudy.

"I hope so." Alice decided to change the subject. "How are you making out with your new singing group?"

"They're great," said Trudy. "It's wonderful working with trained musicians. It makes my job a lot easier. But I do miss the Beckett sisters. We began just as four friends who liked to sing. With the new girls there isn't that kind of easy informality—not yet, at least. It's all . . . professional, I guess you could say."

"And your own program? That must be a lot of work."

"It is, but I love it."

Al quizzed Willis about technical matters related to broadcasting.

"It was a challenge in the beginning" Willis said, "getting the station launched, then the switch to a higher frequency, but now it's settled into a routine. I just make sure everything is running smoothly. I order new equipment and put out any fires that crop up."

"Speaking of fires," said Al as they were sipping their coffee, "want to see a match burn twice?"

"Sure," said Willis.

Al produced a package of cigarettes, offered one to Willis, and took one for himself. Then he brought out a book of matches, lit his own cigarette, and reached across the table to light Willis's. He blew out the match. "That was once," he said with a grin, then touched the still hot, but extinguished match to Willis's wrist.

"Ouch!" said Willis, withdrawing his hand.

"And that was twice," said Al with satisfaction.

"Al, that's not funny," said Alice, her voice sharp. "You could have burned Willis."

"Not the way I did it," he explained. "I gave the match time to cool just enough so it wouldn't burn, but it was still hot enough to make the point."

Willis examined his wrist. "He's right," he admitted. "No damage done."

Alice laughed nervously. "Please forgive my husband. I'm afraid his sense of humour can be embarrassing at times."

It was almost ten o'clock when Willis and Trudy said goodbye and made their way to the car. The rain had let up, and as they drove off Trudy asked Willis for his impressions.

"I liked them," he said. "Especially Alice. She made me feel right at home from the beginning. There was none of the hostility I felt when I met your folks."

"Well, they liked you, too," said Trudy. "Alice told me so when we were clearing the dishes in the kitchen."

"So I passed the test?"

"With flying colours."

Trudy's sister Claire, who was also married, was living in Windsor. Although Claire was younger than Alice and nearer Trudy's age, she and Trudy weren't as close. Trudy was hesitant about introducing Willis. She was worried that Claire might mention him to J.B. or Blanche. She'd asked Alice to keep news of her boyfriend quiet, but wasn't comfortable asking Claire to do so.

Willis was curious. "What's Claire like?" he asked.

"She was always quiet, the shy one in the family."

"And her husband?"

"Ed? He's the complete opposite to Al Kingerley. He's a little rough around the edges. He works for Ford, but he used to be a cop. He was in the RCMP out west."

"Oh-oh!" said Willis in mock horror. "I'll have to watch myself around him."

Trudy laughed. "Just behave yourself and you'll be fine."

"That's no fun," Willis teased.

"I'll be going home for Christmas. That's when I'll break the news that you are still very much in the picture. Then we can arrange something with Claire and Ed."

The CKLW dinner dance at the Dearborn Inn was held on December 20. Willis and Trudy gave a lift to Isobel and Edith, the other members of Trudy's trio. It was a cool, crisp evening, and as they drove west they could see that most of the suburban homes had been decorated for Christmas.

The Dearborn Inn was also festooned for the holidays. The balcony railings, the elegant main entrance, and the arched windows of the stately three-storey Georgian building were draped with strings of coloured lights.

"It's gorgeous!" Trudy exclaimed.

"Go right in, sir," said a uniformed porter. "I'll park the car for you." He took the ignition keys from Willis and handed him a ticket stub.

Inside, they checked their coats and made their way through a spacious lobby. Trudy was wearing a formal gown of blue satin while Willis had rented a classic black tuxedo for the occasion.

"May I see your invitation?" asked a pretty young woman at the door to the ballroom. Willis handed her the engraved card. "Ah, yes," she said, "table eight, over there on the right."

The room was spacious. A parquet dance floor was surrounded on three sides by rows of tables. At the back of the room, on a slightly elevated stage, a pianist was playing softly. Trudy recognized the tune: it was "Did You Ever See a Dream Walking?" They sat down and took in the room. The lights in the crystal chandeliers had been replaced with coloured bulbs, a Christmas tree was on the stage beside the piano, and garlands of mistletoe adorned the walls. The lighting was appropriately subdued.

"This is beautiful," said Isobel in a whisper.

"Gorgeous," added Edith.

They were among the first guests to arrive, but soon the ballroom began filling with people. The event was a joint venture, with guests from both CKLW and WSPD, its sister station in Detroit. They were joined at the table by Elton Plant, a singer who had given up a promising musical career for a position in advertising, and Gordon Fleming, the station's pianist. It was clear that the seating arrangements had been carefully thought out, with common interests as the determining factor. Willis was out of his depth when it came to music, so he was relieved when they were joined by the station's chief engineer, Bill Carter, and his wife.

Willis glanced down at the menu and began to read it. You would not guess, he mused, that the country was in the midst of a depression.

ORANGE BLOSSOM COCKTAIL

Chilled Half Grapefruit, Maraschino Iced Celery Queen Olives Roast Young Tom Turkey Chestnut Dressing Cranberry Sauce Mashed Potato New Green Beans

HAUTE SAUTERNE

Pineapple and Cottage Cheese Salad Mayonnaise Frozen Pudding Assorted Fancy Cakes Demi Tasse

CHAMPAGNE

Cigars and Cigarettes

The food was excellent, the wine was chilled to perfection, and the service was discreet. After the plates had been cleared away, a band moved into position on the stage, replacing the pianist.

"It's Wayne King!" said Trudy excitedly.

"You're right!" said Isobel.

"I wonder if Dorothy Janis is here," said Edith.

"Who's Dorothy Janis?" asked Willis.

"She was a silent movie star," Elton explained. "She and Wayne King were married last year."

The band began to play and Willis and Trudy rose to dance. Willis was thankful the song, "It Was Only a Paper Moon," was slow-paced. Although Trudy had coached him patiently, he was never comfortable dancing to up-tempo music. At least not when he was sober.

"This is nice," Trudy whispered as they glided across the floor.

Holding her closer, Willis murmured, "Better than nice."

Fortunately for Willis, Elton Plant took over when the pace of the music picked up. He had come alone, and was more than happy to have Trudy as a partner.

As the evening wore on, Willis, fortified by several rye and ginger ales, took over from Elton and managed a few of the more intricate steps. He found that Trudy inspired confidence in him on the dance floor—although there were times when he wondered who was actually leading.

The band began to play "Stardust," the song that signalled an end to the evening. Willis and Trudy rose to dance. They negotiated their way to the centre of the floor and, locked in an embrace, swayed gently to the music.

"I'd like to ask you something," Willis whispered.

"Yes?"

"Will you marry me?"

"Willis, you're drunk."

"Tipsy, maybe, but not drunk."

"Can I think it over?"

"Sure, take all the time you need."

"I don't think I'll need much time."

Photo Gallery

Anderson Home - Wilcox, Pensylvannia

Anna Maria Anderson and Thomas Clyde Little -1900

Clyde, Anna Maria and Willis - 1920

Clyde Little - Leamington, Ontario, 1925

Ralph Curtis & Wheeler Little - Queen's University - 1929

Willis (Wheeler) Little - Windsor, Ontario - 1934

Willis Clayton Little - 1935

Trudy Champion Little - 1935

Willis Little - Ottawa - 1942

Willis and Andrew Little - 1943

Chapter 38

Leamington and Windsor, 1933–1934

Christmas at the Littles' was subdued. Clyde was still brooding over the pool hall bankruptcy. He was working evenings at the local bowling alley, but it wasn't the same as running his own business. Having his son home for the holiday improved his mood, though, and because their boarder was away for the holidays there'd be no sleeping on the sofa for Willis. He reclaimed his old room.

As usual, the family exchanged gifts on Christmas Eve. Willis gave Clyde a baseball autographed by Charlie Gehringer and Hank Greenberg, two Detroit Tiger stars. He knew that his dad, who had played second base in his youth, thought Gehringer was the best second-baseman in the major leagues. He'd had help from Trudy in finding something for his mother. She had picked out an antique cameo broach. Anna Maria smiled knowingly when she opened the present.

"You have such good taste," she said wryly.

Willis received his annual supply of hand-knitted wool socks of the kind he favoured, and somehow Clyde had managed to find him a carton of Old Kentucky cigarettes.

"How'd you manage to get these?" Willis asked. "I usually import them from the States."

"Connections," said Clyde cryptically.

Willis broke the news of his engagement over dinner on Christmas Day.

His mother wiped tears of happiness from her eyes.

"You're one very lucky man," said Clyde, a little surprised by Anna Maria's unusual show of emotion.

"And Trudy is a lucky young woman," Anna Maria quickly added.

"I think we're both lucky," said Willis.

"Has she told her folks yet?" Anna Maria asked.

"She's going to," said Willis. "She's driven up to Ethel to spend Christmas with them."

"That should be interesting," said Clyde. "From what you said, they didn't exactly welcome you with open arms in Wheatley."

"True," Willis admitted. "I'm sure there'll be some angry words, but we're prepared to go ahead with or without their approval."

"Have you set a date?" Anna Maria asked.

"Not yet. Sometime in the spring."

In Ethel, Trudy waited until after the turkey dinner on Christmas Day to tell her folks about her engagement.

J.B., as she expected, exploded with rage at the news. "You have some nerve," he said, "ruining our holiday with this cockamamie idea!"

Trudy looked to her mother for support, but Blanche remained silent.

"Willis Little will never be welcomed into this home," J.B. shouted, "and neither will you if you marry him!"

"Fine," said Trudy, her own anger mounting. "If that's the way you feel, I won't stay another minute!"

She left the table, fled to the sanctuary of her room, and packed her things. When she came back downstairs and headed for the door, Blanche intercepted her.

"He's just blowing off steam," she said. "Give him some time and he'll come around."

"I don't understand," Trudy sobbed. "You'd think I was marrying the town drunk, the way Dad carries on. Willis is such a fine man, well educated, and with a good job and a bright future."

"I know," said Blanche, "but your father is depressed. Things aren't working out for him here."

"Well, that's too bad," said Trudy, still fuming, "but it doesn't excuse his behaviour."

Back in her apartment in Windsor, Trudy called Willis. "We need to talk," she said. "I hate to ask, to take you away from your folks, but can you come to Windsor?"

"What is it?"

"Not over the phone, Willis."

When he heard the tone of her voice he chose not to argue with Trudy. "I'll be there this evening."

Clyde and Anna Maria were understanding.

"I'm pretty sure it's her parents," Willis explained. "She didn't say as much but that's all it can be."

"You leave this very minute," said Anna Maria. "Trudy wouldn't ask if it wasn't serious. And call when you have the time."

As he drove to Windsor Willis mulled over the situation. What was really at the root of the Champions' objections to their engagement? It had to be more than just his father's occupation. This wasn't class-ridden England, after all, and it was what he himself did for a living, not what Clyde did, that counted.

In the early days of radio, the Jehovah's Witnesses and the Catholic Church had waged a war of words on the air. You could still tune in fire-and-brimstone sermons from south of the border. Maybe that was it. Maybe J.B. was simply old-fashioned, and viewed radio as a threat to the established order.

And, of course, the fact that Willis was a Mason didn't help.

At the apartment Trudy threw herself into his arms, sobbing. He didn't know what to say, so he didn't say anything. He stroked her hair softly. It was several minutes before she settled down.

She sat on the sofa, shaking her head. "It was awful," she said, "It was like a scene from one of those old-fashioned melodramas."

"How so?"

"You know, the part where the daughter is banished." She deepened her voice. "'Never darken my door again.'"

Willis smiled in spite of himself. "That bad?"

"Yes, that bad. And what really teed me off was that Mom didn't stand up to him. She let him rant and rave as if it was par for the course."

Willis reached out and took Trudy's hands in his. He looked into her eyes. "Does it really matter?" he asked. "If he's such a bastard do you really care if he approves of me?"

"I suppose not," said Trudy. "It's all so unfair. But you're right. We don't need his approval. In fact I don't really care if I never see him again."

Willis rose and went to the sideboard. "Drink?"

"God, yes, a stiff one."

Willis poured rum into one glass and rye into another. In the kitchen he added ice. Then he topped up Trudy's drink with Coke and his own with ginger ale. When he returned with the drinks he sat down beside Trudy.

"Let's set a date," he said.

"A date?" Trudy looked confused.

"A wedding date."

"Yes!" she said enthusiastically. "I'll drink to that."

Trudy and Willis went to Leamington to welcome in 1934. They had turned down invitations to parties in Windsor, because Trudy wanted to be with a family to start the New Year. J.B. and Blanche might not have been willing to accept them, but Anna Maria and Clyde were delighted at the prospect. They sat around the radio in the living room listening to Guy Lombardo's orchestra. On the stroke of midnight, Willis popped the champagne and everyone exchanged hugs and kisses.

Anna Maria raised her glass. "To Trudy and Willis. May the New Year bring nothing but happiness."

"To Trudy and Willis," Clyde echoed.

As she sipped her champagne, Trudy wondered why her parents had to be so difficult. Watching Anna Maria and Clyde she was filled with envy. They were so welcoming, so open, they brought tears to her eyes.

For propriety's sake, Willis bedded down on the sofa and Trudy slept upstairs in his bedroom.

Back at the station, word of their engagement spread quickly. They had decided on a June wedding, but hadn't picked an actual date. Trudy proudly showed off her diamond engagement ring and Willis accepted congratulations from just about everyone on the staff.

But their news was soon eclipsed by other concerns. *The London Free Press* was unhappy with the partnership and was threatening to pull out. At issue was the lack of London news and programming.

Willis had foreseen this problem. It was hard enough to serve two cities, Windsor and Detroit, without trying to please a third. An effort had been made to include London stories in newscasts, but it was a small university town, and there simply weren't very many newsworthy events. The studio facilities in London weren't on a par with those in Windsor, so when London artists were invited to perform, they often had to make the long trip to Windsor. Another problem was the CBS affiliation. The owners of the London newspaper complained that CKLW sounded more American than Canadian.

The partnership had been formalized the previous May, but it had been for a one-year term only, with both parties free to reassess the deal when the trial period ended. In May of 1934 *The London Free Press* withdrew and linked its services with CFPL, a station based in London. This meant less money for CKLW, but the loss was offset by increased advertising revue from Detroit.

In the same year there was an increase in the programming provided by the Canadian Radio Broadcasting Commission. This did address some of the *Free Press*'s concerns about American domination of the station, but of course it was too late.

Trudy's engagement to Willis had a profound effect on J.B. His depression deepened. The main symptom was his reluctance to make his usual modifications to the parsonage. "It's too damned small," he had complained, but he hadn't done anything about it. There had been talk of remodelling the kitchen, but so far it was just talk. He continued to fret about the future of his beloved bees. Their hives had been trucked in from Wheatley, and he was worried that they might not survive in the spring. He continued to preach each Sunday, but Blanche noted a lack of passion in his sermons. Fortunately, no one in the new congregation seemed aware of that particular change, but Blanche was aware that his sour mood was turning off some members of the Ethel congregation who had previously involved themselves in church affairs.

"You don't seem your old self," she commented one Sunday afternoon, risking a reprimand.

"Just leave me alone," he muttered, and retreated to the rocking chair in the living room.

His melancholy condition prompted Blanche to seek advice, secretly, from Alice. "He just isn't himself," she complained, "and when I suggest a visit to the doctor he simply refuses, insisting there's nothing wrong."

"How's his appetite?" Alice probed.

"Oh, he eats everything I prepare," Blanche said, "but without much gusto. And he's lost weight."

"Is he sleeping well?"

"No, now that you mention it, he isn't. He seems restless. I often hear him pacing around during the night."

Alice wondered if these changes in her father were the product of the move to a new and unfamiliar parish, or a by-product of his standoff with Trudy. In some ways his symptoms reminded her of the "baby blues" she had seen in new mothers after they'd given birth.

Chapter 39

Windsor and Detroit, 1934

At CKLW, the New Year was greeted with heightened optimism. The people of Detroit were tuning in to the new station in increasing numbers. The CBS affiliation certainly helped. Soap operas were proving a hit with housewives, and schoolchildren were following late afternoon radio serials such as *Jack Armstrong, the All-American Boy*; *Cowboy Tom*; and *Buck Rogers*. The evening programs that showcased such stars at Bing Crosby and Kate Smith were also popular.

The Triolettes were featured every night, so Trudy was busier than ever. Willis, though, found his role diminished. In the first few days after the increase in signal strength there had been technical problems, but by now most of those difficulties had been overcome. The move into the new studios in the Guarantee Life Building was complete.

With more time on his hands, Willis busied himself researching broadcasting innovations. A constant stream of new equipment was becoming available—more sensitive microphones, more refined turntables, and improved components for the transmitters. It was his job to evaluate these products and, if they showed promise, to recommend their purchase. Not all of his suggestions were approved, but a good many were, and he took immense pride in the quality of the station's technical output.

A date was set for the wedding. It would happen on June 15 whether or not J.B. approved. If he continued to be opposed to it, they would marry in a civil ceremony in Detroit. Trudy hadn't been in contact with her parents since fleeing at Christmas, but Alice had been pleading her case with Blanche.

Trudy's single-bedroom apartment on Erie Street was small, so the search began for more spacious accommodation. In March, Willis found a new two-bedroom apartment on Ouellette Avenue. "What do you think?" he asked Trudy after she'd looked it over.

"I think it's perfect," she said. "We can use the second bedroom as a study when we don't have guests. There's even room for a piano."

"The apartment is available now," said Willis. "I thought I might give up my place on Donnelly and move in right away."

"That's a great idea," said Trudy. "Some of the wall colours are a little bland. Maybe you could paint the place before I move in. The smell of fresh paint always makes me sick. I guess it's something I associate with Dad—he was forever repainting rooms when I was a kid."

"No word on that front?" Willis asked.

"None. Alice seems concerned. She says J.B. isn't his usual self these days. Seems moody and withdrawn."

"What does Claire think about it?"

"Oh, they don't have much to do with J.B. and Blanche. Dad wasn't that thrilled with Claire's choice of Ed as a husband."

"Seems dissent runs in the family."

"Yes, but Dad didn't put up the same amount of resistance when Ed and Claire were married last year. Ed Pritchard works for Ford, and I guess Dad thinks that that has more status than CKLW."

"When do I get to meet Claire and Ed?" he asked.

"I've tried to arrange something, but I think Claire is worried that if we have dinner together it might get back to J.B., and he'd take it out on her."

"What does her husband think of all this?"

"I think Ed has washed his hands of the Champion family. The less he has to do with us the better."

"But you'll invite them to the wedding?"

"Oh, sure. If we have a church wedding, that is—they can't duck out of that. I'm not so sure it would be the same if we were to have a civil ceremony."

In the end, the threat of a civil ceremony brought J.B. around. He simply couldn't tolerate the idea of a daughter of his being married outside the church. He wouldn't conduct the service himself, he said, but he grudgingly agreed to attend, provided a minister sanctified the marriage.

When Alice relayed this information to Trudy, she laughed bitterly.

"No one asked him to officiate. He's the last preacher I'd want for our wedding."

Alice and Al Kingerley attended church regularly in Detroit. They liked their minister, Reverend Herbert Rhodes, and suggested that he be asked to conduct the service. They also offered to host a reception after the ceremony. The offer was accepted, and Alice arranged for Willis and Trudy to meet with the minister in the first week in June.

"He seemed awfully young," said Trudy as they left his office.

"Nothing like your father?"

"Lord, no. And I'm thankful for that."

"Are you sure he'll show up?"

"Reverend Rhodes?" said Trudy, teasing.

"No, you know who I mean."

"Oh, Dad will be there all right. Kicking and screaming, no doubt, but Blanche will make sure he's there."

"I wonder . . .," mused Willis.

"What?"

"You know the part of the ceremony where the minister asks if anyone has just cause . . . ?"

"And you're worried that's when Dad might raise a fuss?"

"You did say kicking and screaming."

"No, he'll keep his mouth shut. He wouldn't want to embarrass himself."

On the day before the wedding Willis drove Trudy to Alice's in Detroit. Anna Maria and Clyde had already arrived from Leamington and would stay overnight with their son in the apartment on Ouellette Avenue. J.B. and Blanche had taken a room in the Fort Shelby Hotel in Detroit.

The ceremony was scheduled for three o'clock, and the weather cooperated. It was a warm late-spring day with just a slight breeze off the river.

The Methodist church the Kingerleys attended was an imposing building. Its Gothic Revival exterior was clad in grey limestone trimmed with rows of dark travertine. A majestic steeple rose high above the sloped slate roof. An array of stained glass windows gave colour to the vast interior. A balcony looked down on the rows of pews. When they had visited the church to meet the minister, Willis was surprised by its size. It was at least twice as large as J.B.'s previous church in Wheatley, and he hoped it would impress the old man.

Although Reverend Rhodes had offered the services of the church organist for the service, Trudy wanted Edith Alexander, from her trio, to perform. The minister had agreed, and Edith's rendition of Bach's "Jesu, Joy of Man's Desiring" welcomed the guests as they arrived.

Trudy and Willis wanted a simple ceremony. They invited only a small group of friends from work and, of course, their families. In all, twenty-four people attended.

Alice was the maid of honour, and Willis asked Ralph Curtis, an old friend from university days, to be best man. Trudy refused to wear white, choosing instead a bias-cut wedding gown of cream satin and a matching veil. Willis wore a new dark blue wool suit for the occasion, with a crisp white shirt and his Queen's University tie, done in a Windsor knot.

Heads turned when the music shifted from Bach to Mendelssohn. Trudy made her way slowly up the aisle on her father's arm. Willis was shocked at J.B.'s appearance. He had lost a good deal of weight and looked frail. He was no longer the intimidating presence Willis had encountered in Wheatley.

The music faded and Reverend Rhodes asked, "Who giveth this woman to be married to this man?"

J.B. replied, his voice barely audible. "I do," he said, and turned from Trudy to join Blanche and the other members of the Champion family.

The rest of the ceremony was simple and straightforward. Willis noticed that the question "Does anyone have just cause . . .?" was never asked.

When vows had been exchanged and the wedding ring had been slipped onto Trudy's finger, Reverend Rhodes concluded, "Forasmuch as you, Willis Clayton Little, and you, Anna Gertrude Champion, have consented together in wedlock and have plighted your faith and truth to each other in the sight of this company, now, therefore, in the name of the Father, and of the Son, and of the Holy Ghost, I pronounce you husband and wife. Amen. Whom, therefore, God hath joined together let no man put asunder."

Willis lifted Trudy's veil and placed a chaste kiss on her lips. There was a scattering of applause. The couple turned, and Handel's *Water Music* filled the church as they made their way down the aisle.

Outside, Willis and Trudy were showered with rice, and there were hugs, kisses, and handshakes. Eventually Trudy broke away from their friends and managed an awkward introduction.

"Mom, Dad, I'd like you to meet Willis's folks, Anna Maria and Clyde."

J.B., grim-faced, extended a hand and Clyde shook it. It reminded Willis of the way boxers touched gloves before a fight.

Anna Maria bent and kissed Blanche lightly on the cheek. "Don't they make a lovely couple!" she remarked, glancing at Trudy.

"Indeed," said Blanche noncommittally.

Fortunately the awkward silence was broken when Trudy's oldest sister, Doris, came forward and introduced herself. "And this," she said proudly, "is my husband Jack Elseley."

Trudy beckoned Willis to join them.

In Doris Willis recognized the Champion look. Pretty, but tough. Her plump, older husband smiled good-naturedly.

"So, I finally get to meet some other members of the Champion clan," said Willis dryly as he greeted the Elseleys.

"And last, but not least," said Trudy, as Claire and her husband joined the group, "Claire and Ed Pritchard."

"Any others?" Willis asked mischievously. Claire was also pretty, but without the spirit he sensed in Doris. Ed Pritchard was short and wiry. He had a thin moustache, and when they shook hands Willis noted a tattoo on his arm.

"I'm afraid my two brothers couldn't make it," Trudy said. "But they did send their best wishes."

"Our other daughter, Edith, is abroad, studying opera," J.B. intoned.

Alice and Al Kingerley joined them. "This is turning into a family reunion," laughed Alice. "But instead of standing out here talking, why don't we head back to our place for the reception."

Much to Willis's relief, J.B. and Blanche did not attend the reception. They had to get back to Ethel, they explained, because J.B. had services to conduct the next day. But Anna Maria and Clyde stayed on, charmed by the attention they received and a little awed to meet, in person, some of the announcers and performers they listened to regularly.

Fortunately, prohibition had ended the previous year, so champagne was readily available and the Kingerelys' bar was fully stocked. By early evening the party was in full swing. The radio was turned to CKLW, and music, the main staple of its Saturday evening programming, filled the room.

Clyde and Anna Maria excused themselves at about nine. Two of

Willis's high-school friends were returning to Leamington and offered the Littles a lift. This was a relief to Willis, who by this time was feeling no pain.

He stood with Trudy in the front door and waved goodbye as his folks drove off.

"They seemed to enjoy themselves," said Trudy with satisfaction.

"Oh, they did," said Willis as he lit a cigarette.

"I was so relieved that Mom and Dad weren't here," said Trudy. "I know that that sounds awful, but they would have put a damper on the party."

"Your father didn't look too well."

"I know. Frankly, I was surprised." Trudy dropped her eyes. "I even found myself feeling a little sorry for him, if you can imagine."

"It was good to finally meet your other sisters," said Willis, to change the subject.

"What did you think?"

"Claire seemed shy, but not Doris. She's feisty."

"And Ed and Jack?" Trudy prompted.

"Ed seemed a bit standoffish, but Jack was friendly enough."

Trudy and Willis had decided against an immediate honeymoon; they looked ahead to a summer stay at a Lake Erie cottage instead. After most of the guests had left, they accepted Alice's invitation to stay overnight in Detroit rather than make the return trip to Windsor.

On Sunday morning they slept late, and by the time they woke up Alice had the coffee ready in the kitchen. Willis had a hangover and passed up the coffee, opting instead for a glass of water, which he gulped down with two Aspirins. Trudy hadn't had much to drink the previous evening, so she enjoyed a pancake breakfast.

The conversation turned to J.B.'s health. "I'm worried," Alice admitted. "Before they left I made Mother promise she'd get him to see a doctor."

"He certainly wasn't himself," agreed Trudy. "I wasn't sure what to expect when they accepted our invitation. I guess I thought there might be some fireworks, an outburst of some kind, but he was as meek as a lamb. It wasn't like him at all."

"I suppose we should be thankful," said Alice, "but it is troubling."

"How old is J.B.?" Willis asked.

"He'll be sixty this year," said Alice.

"Getting on," said Willis thoughtfully.

Chapter 40

Windsor, Ethel, and Port Stanley, 1934

The concerns about J.B.'s health proved well founded. Although the Champions left immediately after the wedding so that J.B. could conduct his regular Sunday services, he did not do so. When he awoke the next morning, he simply refused to get out of bed. Blanche placed a hurried call to a retired minister, who agreed to fill in.

At noon she went to see how her husband was feeling. "Just leave me alone and get the hell out of my bedroom," he shouted, and buried his face in his pillow.

Blanche called the family doctor and he agreed to come by in the afternoon. Dr. Pope was a short, energetic man in his mid-fifties. He listened to Blanche's account, then mounted the stairs to J.B.'s bedroom. Blanche decided to remain downstairs.

The doctor rapped gently on the door, and when there was no reply he let himself into the room. The curtains were still drawn, and it took a moment for his eyes to adjust to the gloom. He approached the bed.

"Reverend Champion?" he asked softly.

"Go away," J.B. muttered.

The doctor pulled a chair to the bedside and turned on a table lamp.

"Reverend Champion," he said, his voice firmer.

J.B. had turned to face the wall. "Why can't you just leave me alone?"

"We need to talk. I'm here to help," said the doctor, and put his hand gently on J.B.'s shoulder.

"No one can help."

"Come on, now, let me have a look at you." He reached down to his bag and withdrew a thermometer.

J.B. reluctantly turned over on his back. The doctor inserted the thermometer in J.B.'s mouth, then removed a stethoscope from his bag, slipped it beneath J.B.'s pyjama top, and listened to his heart and breathing. Next he took J.B.'s blood pressure. He returned the equipment to his bag, reached across to remove the thermometer, and studied it. "Temperature's normal," he said, almost to himself. "And your heart and lungs seem fine."

"I'm fine," J.B. said stubbornly. "Just tired, that's all."

"Well," said the doctor, "I suppose tired is one way to describe it. I understand you drove to Windsor yesterday for your daughter's wedding?"

"Drove back from Windsor after the wedding," J.B. corrected him. "We drove down on Friday."

"Weddings can be stressful," the doctor suggested.

"Stressful?" J.B. rasped. "Stressful? That's an understatement."

Doctor Pope, sensing a sore point, moved the conversation to safer ground. "Why don't you rest in bed, take a few days off. Then, when you're feeling up to it, I'd like to see you at my office. We need to do some tests."

"Sure," said J.B., then turned his face to the wall.

Downstairs Blanche was waiting anxiously, fearing the worst. "What is it?" she asked the doctor when he appeared.

"I can't say for sure. He seems healthy enough. Has anything like this happened in the past?"

"Never," said Blanche emphatically. "It is so unlike him to miss a service. I've seen him preach with a fever when he should have been home in bed."

"I'd like to run some tests—blood, urine—before I can be sure, but I'd say it's a simple matter of exhaustion."

"Exhaustion?" It wasn't a word Blanche associated with her husband. "So you think he'll come out of it?"

"I'll know better when we get the test results. In the meantime, just let him be. Make sure he gets plenty of water and whatever food he can manage."

"You'll come by again?"

"Tomorrow," the doctor promised.

Word of her father's condition was relayed to Trudy in a phone call from Alice.

"What does the doctor think?" she asked her sister.

"He's not sure. He's calling it exhaustion for now, but he wants to do some tests."

"I always thought of J.B. as indestructible."

"Me too," said Alice. "I'm going to drive up there next weekend to talk to the doctor myself."

"Do you think it has anything to do with the wedding?" Trudy asked hesitantly. "I know he was reluctant to come . . ."

"Oh, I doubt it," Alice said. "It's probably a combination of things. He doesn't like his new parish, he isn't happy with the parsonage, and to make matters worse his beloved bees didn't survive the move from Wheatley."

"Oh, I didn't know," said Trudy, surprised.

"Well, you had the wedding to worry about," said Alice. "We didn't want to bother you about that."

"You'll call when you know more?"

"For sure," said Alice. "The doctor said he doesn't think there is anything physically wrong. But I'll call you on the weekend."

When Trudy told Willis he shrugged. "I thought my father would be the one to have serious health problems, with his diabetes."

"Me too," said Trudy. "I guess we have to face the fact that our folks aren't getting any younger."

"Exhaustion?" Willis mused. "What kind of diagnosis is that?"

"It's a mystery."

Trudy and Willis took a week off in August to enjoy their delayed honeymoon in a small summer cottage in Port Stanley, on Lake Erie. It provided the perfect escape from work and family concerns. Neither Willis nor Trudy was a great swimmer; they preferred to sun themselves on the sandy beach and merely dip in the lake occasionally to cool off. A rowboat came with the cottage, and Trudy prepared a picnic lunch a couple of times so that they could row to a more secluded beach for lunch. They spent the afternoon making love on a blanket.

"It's different," said Trudy.

"Different?"

"Now that we're married."

"Better?"

"Much better."

Alice and Al Kingerley visited them, sleeping overnight on the screened porch. The focus of Port Stanley's economy was changing from commercial fishing to tourism. Fishing boats still went out each morning, but there were fewer of them, and the catch was rarely worth the effort. On the second day of their visit Al came back from the town pier with some lake perch, and Alice pan-fried them for lunch.

The perch was a first for Willis. "These are delicious," he said. "I wonder if we can find them in Leamington."

"I don't know," said Alice. "We had them as a regular treat when we lived in Wheatley. But they have to be cooked freshly caught to be at their best."

"I'll have to ask my mother about them next time we're there."

"How are your folks?" asked Alice. "They seemed to enjoy the wedding."

"They're in fair shape," said Willis. "I think Dad is finally reconciled to losing his business, but his eyesight is deteriorating and that's a worry."

"Diabetes can do that," Alice acknowledged. She wondered if Willis realized that the day would likely come when Clyde would lose his eyesight completely. Since there was nothing that could be done about it, she decided she'd keep that information to herself.

"But, you know," added Trudy, "I've never heard him complain. I'd hate to have to put up with all those injections."

"Dad is pretty stoic," said Willis.

"And speaking of dads," said Trudy, "has there been any change in J.B.'s condition?"

"Not really. He gets up for meals now, but he picks at his food and he's always complaining. He still sleeps a good deal, and hasn't mentioned work. Last week he even missed a service. And I don't think he reads his Bible these days."

"Sad," said Trudy, and shook her head.

Eventually tests were run, and the doctors concluded there was nothing physically wrong with J.B. The church brought in a replacement and granted J.B. extended sick leave. There was no improvement by the fall, so Doris and Alice made the trip to Ethel to meet with Blanche at the doctor's office. It was Alice, with her nursing background, who led the discussion.

"Any idea what the problem is?" she asked.

"No, and frankly we're at a loss," said Dr. Pope. "We've seen this sort of depression in some of the veterans returning from the war, but rarely in a civilian."

"What can we do?" asked Alice.

"Well, I understand that the church might grant him a leave of absence, or even early retirement. It might be best if he were in more familiar surroundings. Is it possible he could return to Wheatley? That might be a start. He never seemed happy here in Ethel."

Alice looked at the others. "I suppose so," she said. "We can look for a place in Wheatley."

That's what they decided to do. Essex is relatively close to Wheatley, so it was Doris who said she would begin the search for a suitable home. The church granted J.B. a leave of absence with the understanding that if his health improved he might be given a new parish.

Trudy had not gone to Ethel, fearing her presence might aggravate the situation. When Alice called with the news that J.B. and Blanche would be moving back to Wheatley, she asked, "Do you think that will help?"

"It can't hurt," said her sister. "At least it's something. At least in Wheatley he'll be among old friends. And, of course, he'll be closer to his family."

At this news, Trudy felt a sudden, unexplained wave of nausea.

Chapter 41

Windsor, 1934

When Willis and Trudy returned from Port Stanley, Garfield Packard asked to see Trudy in his office. When she appeared he motioned her to a chair.

"I think the time has come," he said with a smile.

"Time for what?" Trudy wasn't sure where this was leading.

"Why, time for you to have your own program, of course."

Trudy was too surprised to say anything. The idea had been mentioned as a possibility the previous year, when Packard had asked her to replace the Beckett sisters and rename her group, but at that time Trudy had thought it was just talk, something Packard said to justify the changes he was asking her to make.

"Well?" he asked.

"I'm flattered," Trudy managed to say.

"How would you feel about a weekly half-hour show, on Saturday evenings at six-thirty?"

Again Trudy was speechless. Whenever she'd fantasized about her own program, she'd thought in terms something shorter, perhaps fifteen minutes in length, something for late night listening, but now Packard was suggesting a half-hour weekly program in a time slot with the potential for a large audience. Not in her wildest dreams had she imagined this.

"Do you think we're good enough?" she asked. "Saturday night is pretty popular."

"I think you're better than good enough," said Packard. "I waited this long because we were still working out the details of our relationship with the Canadian Radio Broadcasting Commission. Now, with that settled, I think it's time we put some Windsor talent on the network. So far, Toronto and Montreal are providing most of the music. We need to change that."

Again Trudy was silent. The idea that her trio would be heard across Canada was both exciting and more than little daunting.

Packard lit a cigarette and waited.

Trudy found her voice. "When would you like us to start?"

"How about the first week in October?" Packard asked. "That will give you a little more than a month to work out the details."

"That should be enough time. Who'll be the announcer?"

"I'm not sure. How would you feel about introducing your own music?"

"Gosh," said Trudy, "I never thought about it. I just assumed . . ."

"Well, we can assign an announcer if you like," said Packard. "But think it over. Until now you've been a singer, a performer. If you introduce your own songs your listeners will develop a more personal relationship with your program. It will be more intimate; they'll feel they're getting to know you."

Trudy had to admit there was logic to the suggestion. "Let me work on the music first," she said. "When I've got that figured out I can practise announcing during some of the rehearsals. We can decide then."

Packard nodded in agreement. "Then it's settled," he said. "And, of course, with a regular weekly program we'll have to draw up a contract. No more appearance fees."

Trudy had been so preoccupied with thoughts of the musical requirements she faced that she hadn't thought about the financial arrangements.

"Of course," she said, and waited.

"That's not my bailiwick," Packard said, "but I will put in my two cents worth with the money folks. I'm pretty sure you'll be pleased."

Willis was going over some specifications for new vacuum tubes when Trudy rushed into his office. Surprised, he looked up from his papers.

"Look at you," he said with a grin, "like the cat that swallowed the canary."

"I've just come from a meeting with Packard," she said breathlessly.

"And . . .?" Willis prompted.

"He's offered me my own program."

"About time," Willis commented.

"A half-hour show, at six-thirty on Saturday evening."

"Terrific," said Willis, continuing to grin.

"Terrific?" said Trudy. "It's more than terrific, it's amazing! I can't believe it."

"Believe it," said Willis, still smiling.

"Willis Little," said Trudy, "you knew!"

"Let's say that the little bird you just swallowed told me a little something before he met his fate."

"You knew and you didn't tell me?"

"I knew something was in the wind, but none of the details."

"And you didn't think to pass it on?"

"That would have spoiled the surprise."

"How long?"

"How long what?"

"How long have you known?"

"Calm down, Trudy. I only heard this morning." Willis got up from his desk and moved to close the door. He took Trudy in his arms. "I think this news calls for a celebration." He kissed her, and would have continued the kiss had she not pushed away.

"Not here," said Trudy. "Someone might come in."

"Nothing wrong with a man kissing his wife."

"True," said Trudy, and moved back into his arms with a sigh.

The days went by in a blur, with Trudy working long hours with Isobel and Edith on the new program. The first order of business was to come up with a name, so that the publicity people could prepare some promotional material. Trudy had a theme song in mind, a ballad called "I Love You Truly," but it wasn't an appropriate name for the new show. There was some talk of calling it "Trudy and the Triolettes," but Garfield Packard thought that that name was too strongly associated with the group's performances on *Evening Serenade*, and he wanted a fresh start.

One night in their apartment Willis was whistling idly to himself as he installed some bookshelves in the living room. Trudy came in to inspect the work and stopped at the door.

"What's that song?" she asked.

"I'm not sure," said Willis. "Must be something I heard at the station."

"Whistle it again," Trudy prompted.

Willis did his best. Whatever the song was, Willis's rendition left a lot to be desired, but Trudy picked out the melody and began humming it. That was when the words came to Willis.

"Something about a breeze," he said.

"That's it!" said Trudy. "Willis, you are a genius."

Willis shook his head. "I am?"

"The song," said Trudy. "I recognize it. It's 'Breezin' Along with the Breeze.' Josephine Baker recorded it back in the twenties."

"And?" Willis was puzzled.

"It would be a perfect theme song," said Trudy excitedly. "And the perfect name for our show."

Willis looked pleased.

Trudy shook her head, "The lyrics are on the tip of my tongue, but for the life of me I can't recall them." She looked at Willis and cocked her head. "Want to take a little trip?"

"A trip? Where?"

"To the station, that's where. I bet they have the record there."

"But Trudy, it's almost ten o'clock. Can't this wait until tomorrow?"

"It can, but I'll toss and turn all night trying to remember the words."

"We can't have that," said Willis.

They quickly found the record at the station and were soon back in the car. Willis shook his head and laughed. "When you get a bee in your bonnet . . ."

"But Willis, don't you see? It's what we've been looking for all week. A suitable song, a suitable name for our program."

"I guess so," said Willis with a grin.

When they returned to the apartment Trudy went straight to the gramophone. She took the disc from its sleeve and examined the label. "It was written in 1926." She placed the record on the turntable and gently lowered the needle onto the groove. The mellow sound of Josephine Baker's voice filled the room.

I'm just breezin' along with the breeze . . .
Trailing the rails, I'm roamin' the seas.
Like the birdies that sing in the trees,
Pleasin' to live, livin' to please.

The sky is the only roof I have over my head, And when I'm weary, Mother Nature makes me a bed.

I'm just goin' along as I please,
Breezin' along with the breeze.

Trudy tilted her head and listened to the music. The singer was backed by a jazz band. The quality of the sound left something to be desired, but the lyrics were clear. Willis had been sitting on the sofa. When the music stopped Trudy lifted the needle and played the record a second time. She extended her hand to Willis and he rose and took her in his arms. As they danced she whispered into his ear, "My hero. A man who can barely carry a tune comes riding to my rescue. Willis Little, I love you."

At the station the next day she played the record again for Garfield Packard.

"What do you think?"

"I think you've found it. 'Breezin' Along.' It's a brilliant choice. And it sums up the mood of the program. Light, breezy, and, above all, entertaining."

Edith and Isobel were enthusiastic too. Soon Studio A was filled with the sound of the Triolettes rehearsing the song. While Baker's rendition was upbeat and brassy, Trudy's arrangement was a little slower, with subtle backing from her piano.

Garfield Packard watched from the control room. By chance Willis dropped in to see how things were going.

"She has no idea how talented she is," said the manager. "With that voice and her attention to detail, she could go a very long way."

"Trudy's older sister is an opera singer," said Willis. "Her father is a bit of a snob, thinks she's wasting her musical gift singing pop music."

"Well, her father may have to change that opinion. I think his pop-singing daughter is going to be famous someday soon. She'll certainly eclipse her sister when it comes to public recognition. We're fortunate to have her with us now, but when she becomes better known I doubt we'll be able to keep her."

"I plan to keep her right here," said Willis. "And I've always loved her voice—but then I'm a little prejudiced."

After Trudy had mapped out the songs she wanted to feature in the first program, it was time to write the continuity, the introductions to the individual numbers.

She felt a moment of panic at the thought. She had always been a little nervous about singing, but had learned to live with that. Music was where she felt at home, where she was most confident. Announcing was another matter entirely.

"I've never had any training for on-air announcing," she protested.

"None of our people have," said Packard. "Some of the announcers at the network level trained as actors, but at the local level it's mostly men and women who have appealing voices and speak clearly. Give it a try."

Trudy reluctantly sat down in front of the microphone. When the red light flashed she began to read.

"Good evening," she said, "and welcome to the inaugural broadcast of *Breezin' Along*, a half-hour of music featuring Trudy Little and the Triolettes."

At this point she stopped and looked into the control room window. "Sorry," she said, "I'd written this for someone else. Can we try it again?"

Packard's voice came over the studio intercom. "Sure," he said, "but take it a little more slowly and remember to breathe between sentences. Think of it as a conversation with Willis or a friend."

Trudy started again. "Good evening," she said, "and welcome to *Breezin' Along*, the inaugural broadcast of a half-hour of music from radio station CKLW in Windsor. I'm Trudy, and I'll be here along with Edith and Isobel every Saturday evening at this time with a selection of tunes. There'll be something old, something new, something borrowed, and occasionally something blue. No, it isn't a wedding, but there will be romance in the air."

She stopped and looked once more to the control room.

Packard's voice came over the intercom. "Bravo!" A moment later he entered the studio. "That was terrific," he said, "but I have an idea. Since you always stand when you sing, I'd like you to try it again, standing. I think you'll breathe more easily that way."

They went over it again and Packard was right. Trudy did sound better standing, and gradually her fears began to dissipate. To no one's surprise she spent the rest of the day going over and over the script. It was clear that she was going to be as meticulous about announcing as she'd been about singing. Eventually Packard called a halt to the process.

"Take a break, Trudy. Rome wasn't built in a day."

"You think it's going to work?" Trudy asked.

"I'm sure. More than sure. But you can pick up where you left off tomorrow. I'll make sure the studio is available."

Willis had come into the studio a couple of times during the afternoon to observe. As they were driving home Trudy asked him what he thought.

"You were great," he said, "and I think Packard is right. You sound as if you are in the same room as your listeners, talking to them. Much better than the more formal approach of the staff announcers."

"You're sure?" Trudy asked.

"Absolutely," said Willis with conviction.

Trudy leaned across and gave him a kiss on the cheek. He turned to smile, but she chided him: "Eyes front, soldier. We don't want to end up in a ditch."

Willis laughed. "Don't worry, I won't ditch you."

Chapter 42

Essex and Windsor, 1934–35

Since J.B.'s replacement in Wheatley was already settled in the parsonage there, it was decided that he and Blanche should relocate to a modest house in Essex, where Doris and Jack lived. That way Doris would be able to help out if they needed assistance.

Blanche drove them down from Ethel in the family's 1931 Dodge sedan. The trip took more than five hours, and J.B. sat beside his wife in silence for the entire journey. When they arrived, Doris and Alice were waiting and helped unpack the car. They showed J.B. to his room at the back of the house, and without a word he lay down on the bed, fully clothed, and closed his eyes.

Over tea in the kitchen Blanche brought them up to date on their father's condition. "I think he was disappointed we weren't going back to Wheatley," she said, "but it's hard to be sure, because he doesn't say much."

"I don't think Wheatley would have been a wise choice," said Alice. "It wouldn't have been the same. You wouldn't be in your old home, and I don't think it would be fair to the new minister there."

"Oh, I'm sure you're right," Blanche admitted. "And it's good to know we have family close by. I always felt a little isolated in Ethel."

"Well, Jack and I are here," said Doris, "and Alice, Claire, and Trudy are only an hour's drive away."

"I know a doctor in Windsor," Alice added. "He's had some success with patients like Dad."

"Would he be willing to come here?" asked Blanche.

"I can ask," said Alice, "but he usually sees people in his office." She decided she wouldn't mention the fact that the doctor was a psychiatrist.

On her way back home to Detroit Alice stopped in Windsor to have dinner with Willis and Trudy.

"How's the old man?" asked Trudy.

"Not much to report," said Alice. "He's still depressed. Didn't say a word to Blanche during the drive down. And he retreated to his room as soon as they arrived."

"Sad," Trudy mused.

"True," said Alice, "but I think that getting them out of Ethel was important. That's a first step. But let's not dwell on J.B. I want to hear all about your new program."

"Well, the big news is that we've finally come up with a name for the show."

"You have?"

"Maybe we should wait till we're on the air . . .," said Trudy with a playful smile.

Alice crossed her arms defiantly. "I am not leaving here until I know."

"Okay, okay. We're going to call it 'Breezin' Along,' after the song."

"'Breezin' Along'!" said Alice excitedly. "That's a great name. And I know the song. It's perfect."

"The credit actually goes to Willis," said Trudy.

Willis shifted uncomfortably in his chair.

"How so?" Alice asked.

"We were racking our brains and couldn't come up with anything. One evening, here in the apartment, I heard Willis whistling a tune. It was 'Breezin' Along,' and I knew right away it was what we were looking for."

"Well done, Willis," said Alice with a laugh.

The big day was the first Saturday in October. Trudy arrived at the studio early and spent the morning going over the technical side of things with Willis. It wasn't really necessary, because they'd carried out a full rehearsal the day before, and everything had worked properly, but Willis humoured Trudy, realizing that it would keep her busy, give her something to do. Isobel and Edith came in after lunch and they rehearsed the full program twice more. By then it was late afternoon, and Willis suggested they take a walk to kill the hour before the show. Trudy agreed. Indian summer had arrived, so they didn't need coats as they strolled along Victoria Street.

"Do you really think we're ready?" Trudy asked.

"Yep," said Willis patiently. He'd lost count of how many times he had answered that question.

"You're sure?"

"I'm sure." Willis linked her arm in his. "Positive."

When the On Air light flashed in Studio A at 6:30 Trudy began to play the piano. After a few bars the technician in the control room cued Trudy and she spoke into the microphone.

"Hi there, radio listeners, welcome to *Breezin' Along*." She played softly in the background as she continued to speak. "I'm Trudy, and along with Isobel and Edith we are the Triolettes, and we'll be bringing our music into your homes every Saturday evening at this time." At this point the three women began to sing:

> *We're just breezin' along in the breeze . . .*
> *Trailing the rails, roamin' the seas.*
> *Like the birdies that sing in the trees,*
> *Pleasin' to live, livin' to please.*

Trudy took the next refrain alone, accompanying herself and singing in a clear contralto that betrayed none of the nervousness Willis knew was there. He stood with Packard in the control room. They were mesmerized.

> *The sky is the only roof I have over my head,*
> *And when I'm weary, Mother Nature makes me a bed.*
>
> *I'm just goin' along as I please,*
> *Breezin' along with the breeze.*

Edith and Isobel joined in again, repeating the opening refrain. When the music finished, Trudy spoke again.

"We hope you liked that little number. We are coming to you from radio station CKLW in Windsor. Our studio is on the tenth floor of the Guarantee Trust Building, overlooking the Detroit River. The view is spectacular—and that reminds me to extend a special hello to our American listeners in Detroit."

"God, she's good," Packard murmured in the control room. "I knew she'd be better than any announcer at introducing her own music."

Trudy started playing again and then spoke once more. "If you look out your windows tonight you'll see something familiar in the sky. It's the inspiration for our next selection." Trudy began to sing, a solo this time. "Blue moon, I saw you standing alone . . ."

Isobel and Edith joined her on the second stanza.

The rest of the program featured songs that were popular that year, Cole Porter's "Don't Fence Me In," "I Get a Kick Out of You," "Stompin' at the Savoy," "The Very Thought of You," and "The Gypsy in My Soul." The pattern varied. On some of the songs the trio was featured throughout. Others were a blend of trio and solo voices. For "Stompin' at the Savoy" there was no vocal, just Trudy's piano. Finally, as time grew short, Trudy introduced a farewell number.

"We're going to close as we began," she said, "with another song inspired by the night sky."

This time Trudy sang the entire song alone, playing her piano so softly that it caressed the words:

It must have been moonglow,
Way up in the blue.
It must have been moonglow
That led me straight to you.

When she finished, she glanced at the studio clock and saw that her timing was perfect: she had forty-five seconds left.

"We hope you've enjoyed our music," she said, "and that you'll invite us back into your homes again next week—same time, same station: CKLW, 860 on your radio dial. Until then, we'll be breezin' along out of here." She began to play, and all three voices harmonized:

We're just breezin' along in the breeze . . .
Trailing the rails, roamin' the seas . . .

When the last note had faded, Trudy slumped over the piano with an exhausted sigh.

Isobel and Edith looked questioningly at the control room window. Willis stood behind the glass, beaming, holding up two thumbs. The sound technician's hands clapped a silent tribute.

The door to the studio opened and Garfield Packard bounced in. "Terrific!" he bellowed. "Sensational!"

Trudy got up from the piano. "You really think so?"

"You were magnificent!" he said. "All of you!"

Trudy hugged Isobel, then Edith. "I can't thank you . . ."

"Don't be ridiculous," said Edith. "We should be thanking you."

"This calls for a celebration," said Packard. "The Lasalle Hotel at eight o'clock. Champagne for everyone."

When the others had left the studio, Trudy gathered up her sheet music. Willis entered quietly.

"That was something!" he said.

Trudy turned. "You liked it?"

"Liked it? I loved every minute of it."

Soon she was folded into his arms. "I'm exhausted," she whispered.

"I shouldn't wonder," he murmured. "All the work you put in."

"Can't we just go home?" Trudy asked. "I feel like I could sleep for a week."

"We'd better put in an appearance at the hotel. But we don't have to stay very long if you're not up to it."

Of course by the time they made their appearance, Trudy's fatigue had dissipated. She was, in fact, the life of the party. "I've never been so nervous," she laughed. "Just before we signed on I was sure I was going to be sick. The thought ran through my mind that it might be the first time on radio anyone heard a performer upchucking!"

"That would have been one helluva a way to launch a program," Packard said dryly.

By midnight the champagne had taken hold. The last of the adrenalin was gone. It was time to say good night.

As Trudy and Willis rose to leave, Packard raised his glass." A final toast to Trudy, our rising star."

Glasses were clinked and champagne sipped.

"See you Monday," Trudy called as they made their way to the door.

Chapter 43

Windsor and Leamington, 1934–35

With *Breezin' Along*, Trudy began to receive her first fan mail. In the past no one had heard her speaking voice on the radio, just her music. Now, by announcing her own show, she had established an indefinable link with her listening audience, just as Packard had predicted.

The letters were personal, as if they'd been written to a friend, someone the writer knew. One radio reviewer, writing in the *Border Cities Star*, noted that along with her "obvious musical talent" Trudy demonstrated "a clever dramatic ability." There were also telephone calls to the station praising the program. And there were requests for the trio to perform in concerts and weddings. It was all heady stuff for Trudy, Isobel, and Edith.

"I was hoping folks would like the show," Trudy said to Willis, "but all this attention is a little overwhelming."

"You deserve it, honey. No one works harder."

For Willis, this was a period of relative calm. There were no innovations in radio that fall; all he had to do was keep the existing equipment running smoothly. He was relieved that there were no major challenges for him to tackle, because Trudy was putting in long hours, and he wanted to be there for her, quietly in the background, as her career blossomed.

Word from Essex was encouraging. While J.B. remained quiet, he seemed to be coming out of his depression. He expressed the hope that in the spring he might begin raising bees again—a positive sign. He saw a local doctor, but when Alice suggested a trip into Windsor for an appointment with "a specialist," he refused.

"Nothing wrong with me," he insisted. "Just tired." When he learned that the specialist was, in fact, a psychiatrist, the old J.B. surfaced. He made a scene, stamping his foot like an angry child and shouting that Alice had no business "meddling" in his affairs. Psychiatrists were for "crazy people," he said, and he certainly wasn't crazy.

Blanche was secretly relieved at this eruption. It was the first time since they moved to Essex that her husband had shown true emotion. She hoped it was an indication that he was on the road to recovery. Trudy was both disappointed and relieved at this state of affairs. It would have been nice if her father had been a little more accommodating, but she'd never been entirely comfortable around J.B., and, in truth, the less she had to do with him the better. She did miss her mother, though.

The word from Leamington was discouraging. Clyde's eyesight was deteriorating, and it wasn't clear how much longer he could continue managing the bowling alley. Trudy and Willis spent the Christmas holiday with his folks, and they were both shocked at Clyde's condition. He could no longer even read the newspaper. At first he had tried to keep this from Anna Maria, but she hadn't been fooled for long. He now reluctantly allowed her to read to him, particularly the sports pages. His beloved Detroit Tigers had won the American League pennant that season with the best winning percentage in the team's history, but had lost the World Series to the St. Louis Cardinals, four games to three. He was looking forward to the coming season, convinced it would be the Tigers' year.

Because both Trudy and Willis were doing well, they insisted on buying the tree that year. After an hour shivering in the cold, Trudy finally found one to her liking at the local market, and they tied it to the roof of the car. Back home, Willis wrestled it through the front door and into the living room.

"My, my," Anna Maria remarked. "That's some tree!"

"Where do you want it?" Willis asked, slightly out of breath.

"Oh, over there, in the corner," said Anna Maria. "Let me clear some space first."

That evening, two days before Christmas, they decorated the tree with ornaments that Anna Maria had accumulated over the years. She had even convinced her father to part with some of the Swedish trinkets they had used in Wilcox when Anna Maria was a child. It was a nostalgic time for Anna Maria, with each ornament evoking its own particular memory. Clyde sat in the rocking chair, the decorations in a box on his lap. Trudy and Willis watched, sipping rum-laced eggnog as Clyde handed the ornaments to his wife and she carefully hung them from the branches. The radio, tuned to CKLW, filled the room with Christmas carols. When Anna Maria was finally finished, she turned to Willis.

"I'm not tall enough to reach," she said, handing him a star.

He took the star from his mother and, extending his arm, deftly dropped it into place at the top of the tree. "There," he said. "What do you think?"

Anna Maria stood back, examining her work. "I think it's beautiful. And now I'd like some of that eggnog."

The family exchanged gifts on Christmas Eve. The first thing they opened was a present from Anna Maria's father in Wilcox. Andrew Anderson had sent them a hand-carved wooden nutcracker.

Tears came to Anna Maria's eyes. "We brought this from Sweden when I was a girl. We weren't permitted much on the voyage. I was told I could select one memento from home. I chose this." She handed it to Clyde.

He ran over his hands over the carving. "The wood must be hard, to crack nuts."

"Oh, it works," Anna Maria assured him. "I can remember Mother using it." She paused to wipe the tears away. "And it's carved in the shape of a dog's head. In the good times we had a dog—a St. Bernard, in fact."

"We'll test it out tomorrow," Clyde said.

Trudy and Willis had made Anna Maria promise to keep things simple. Although they themselves could afford expensive gifts, they knew that Anna Maria and Clyde couldn't. To save embarrassment all around they'd agreed to confine themselves to one gift apiece, plus whatever they could fit into Christmas stockings. Willis gave his father a hammered brass humidor for his pipe tobacco. For Anna Maria he had selected a pair of fleece-lined slippers. Trudy gave her mother-in-law a cut glass table ornament, and for Clyde there was a pair of slippers that matched those Willis had given his mother.

When Willis took the present from his mother from beneath the tree he held it up and shook it. "I wonder what this could be?" he said mischievously.

"Don't be such a tease," Anna Maria admonished.

Willis tore open the wrapping paper, exposing three pairs of hand-knitted woollen socks. "Now that's a surprise," he laughed, and turned to Trudy. "Mom has been knitting me socks all my life. I've yet to buy a pair." His second gift, from his father, was a Ronson cigarette lighter. "Oh, Dad," he said, shaking his head, "it's just what I needed. No more matches for me." He promptly lit a cigarette with his new acquisition.

When it was Trudy's turn, she picked up a tastefully wrapped package. "It's so pretty I hate to open it," she said. The accompanying card, when she read it, brought tears to her eyes. In Anna Maria's hand were written the words "For Trudy, who brought light into our lives." It was signed "Anna Maria and Clyde."

She wiped her eyes and reluctantly opened the package. It was a fine lace tablecloth. Trudy protested, "I thought we promised to keep things simple."

"Well, it's from both of us, so in a sense it is simple."

"It's beautiful," said Trudy, and held it up admiringly. "Just beautiful." She rose and gave Anna Maria a hug, then bent down and kissed Clyde on the cheek. "Thank you," she whispered.

Trudy and Willis welcomed the beginning of 1935 at the Kingerleys' in Detroit. They sat around the radio in the living room listening as Guy Lombardo and His Royal Canadians ushered in the New Year with their version of "Auld Lang Syne."

"To Trudy and Willis," said Alice, raising her glass. "May the New Year bring bigger and better things."

"To Trudy and Willis," Al echoed.

The next morning at breakfast Alice told them that Claire's husband, Ed Pritchard, had been laid off by Ford. "I would have told you last night, but I didn't want to spoil the evening."

"What will they do?" Trudy asked.

"Fortunately Claire still has her job at the library," said Alice, "and I hear that Ed has taken work, of a sort."

"What kind of work?"

"He's at a service station in Windsor, pumping gas."

Trudy shivered at the thought. "We're so lucky, you know," she said. "Both Al and Willis have jobs, and so do you and I."

"We are indeed fortunate," Alice replied.

In the first week of the New Year Garfield Packard asked to see Trudy in his office. Always the worrier, she wondered what the station manager wanted. When she pushed through the door she could tell from the smile on his face that whatever it was, she could relax.

"Sit down, sit down," said Packard.

"What is it?"

"Good news. Great news, in fact."

"And what is this great news?"

"The network is picking up *Breezin' Along*. You are going to be heard all over Eastern Canada."

The news did send a thrill through Trudy, but she wasn't completely surprised. She knew that the show was proving popular in the Windsor-Detroit area, and she knew it had been Packard's hope that Windsor would find a voice on the Canadian Radio Broadcasting Commission.

"When?" she asked.

"Next week."

"Wow!" Trudy hadn't expected this so soon. "Do they want any changes in the format?"

"Nope, they love it just the way it is. You should have heard the network programmer. He went on and on about how unique you sounded."

"Unique? I guess that's a compliment."

"Trudy, they love you in Toronto, and soon they're going to love you all the way to Charlottetown."

When Trudy told Willis he gave a whoop of joy. "You're putting us on the map," he said excitedly. "A network foot in the door for CKLW."

"I'm not sure I like the idea of being a foot in the door," said Trudy in mock dismay.

"Oh, it's just a manner of speaking," said Willis, a little chagrined. "I'm so damn proud of you, you know that."

"I know," said Trudy, "I was just teasing."

Trudy was all pins and needles the day of her program's network debut. But it was familiar territory to her by now, and she took comfort in the knowledge that no changes in routine were required. Part of her wished that Packard had withheld the news until after the first network broadcast, but she realized that wasn't a possibility. Too many people were involved for the station management to be able to keep it a secret. All week she'd been receiving congratulations. People at the station were proud of the work they were doing, and were delighted with the added recognition a slot on the network would provide.

The show itself went off without a hitch, and by the end of the month the excitement had waned. The only difference Trudy noticed was that the fan mail was now coming from Toronto, Montreal, Saint John, Halifax, and Charlottetown as well as Windsor and Detroit.

In February Packard asked to see her again. This time she approached his office with more confidence. And it proved to be justified.

"I've had a request from one of the newspapers," he said. "They want to do a feature story on you."

"On me?"

"Well, on you and the program. They'd like pictures and an interview. I said I'd get back to them with a time and place. Any suggestions?"

"Gosh!" said Trudy. She thought a moment, then replied, "Why not have them come in on a Saturday, when we're rehearsing? There won't be many people working, so it won't inconvenience anyone. And there will be plenty of chances for pictures. But give me an advance warning, because I'll have to let Edith and Isobel know."

"I will. And that's a great idea. It probably means the story would run on Sunday, and that's the best time for publicity."

Packard set up the interview for the following Saturday and told Trudy it would be done by Daryl Weston, the *Border Cities Star*'s music critic.

"I hope he isn't a classical music snob," she said, with the memory of her father's criticism still smarting.

"No, he's a fan. Said he likes the program," Packard reassured her.

When the critic arrived he immediately put Trudy at ease. He shook her hand and smiled. "They didn't warn me you had the looks to go along with that magical voice."

"I do my best," said Trudy, laughing at the flattery.

Weston was of medium height, probably in his forties, a little on the plump side but with fine, delicate features and a mop of unruly hair. He'd studied music, he said, but hadn't the talent to make a living professionally. With two teenaged sons, he found that journalism provided a more stable income. They talked for the better part of an hour, then Trudy introduced him to Isobel and Edith. A photographer arrived and took shots of the trio with his Speed Graphic and an individual picture of Trudy. When the photographer left for another assignment Trudy asked the critic if he'd like to watch their rehearsal.

"That would be great," he said.

"I think we can find an extra chair for you in the control room."

The trio began rehearsing their songs, a little tentatively at first, aware that they were being judged by someone with a musical background. But within a few minutes they'd forgotten about the critic and were into the music. An hour later, when they took a break, they learned that Weston had left a few minutes earlier.

"What did he think?" Trudy asked the sound technician.

"Oh, he liked what he heard, all right. He was even tapping his foot to the music."

"That's a good sign," Trudy agreed.

The program that evening was, to Trudy's ear, one of their best to date. Packard seemed pleased as well.

"Any idea when the story will appear?" she asked.

"Weston said that unless all hell breaks loose it should be in tomorrow's paper."

Trudy tossed and turned all night, going over the details of the interview. She hoped she'd impressed Weston, and that he'd been listening to the program that evening.

When the newspaper finally arrived Trudy turned to the entertainment page, her heart pounding and her hands shaking. There was her picture and the headline "*Breezin' Along*." She read on:

> At an early age, Trudy came to Western Ontario with her parents. Asked when she commenced her musical studies she replied, "as soon as I had all the bars pulled out of my crib. I guess the folks thought I was cut out to be a harpist."
>
> But like many other youngsters, piano lessons were inflicted on her at the time when she got more fun out of twirling 'round on the piano stool. Trudy stuck to her guns, however, and in 1930 she marched out of the front doors of the Toronto Conservatory of Music with an ATCM degree. Before this event there had been a long list of church and concert engagements in which she sang and played her own accompaniments, and from this work she commenced organizing small voice groups and teaching piano lessons. She made her radio debut with her own trio of girls in 1933 and today it is one of the best and most professionally finished trios on the networks.
>
> She writes all her own arrangements and each member of the trio reads music expertly. Hard work is what she thrives on and those associated with her know all about the long hours of rehearsal. Trudy comes from a musical family. All of the members are accomplished musicians or singers. At the present time her sister Edith is a headliner in opera in Germany. For the past eight years she was leading dramatic soprano at Breslau.

> Trudy's only hobby, when she has time to indulge in it, is painting and sketching and she has turned out some really remarkable artwork that portrays bright, cheerful subjects.

Trudy let out a long sigh of relief.

"What's the verdict?" Willis asked.

"I am so relieved," she said, and handed the newspaper to Willis. "Please read it . . . out loud. It will be music to my ears."

Chapter 44

Windsor and Detroit, 1935–1936

In the spring of 1935 J.B. Champion officially retired. He'd been granted a leave of absence the previous fall, but when his health failed to improve the church insisted he step down permanently.

"Thirty-nine years of service," he grumbled, "and I'm pushed out like this."

Blanche had hoped the move to Essex and the break from the responsibilities of regular work would allow her husband to bounce back, but, while his mood had improved slightly, it was clear he was in no condition to return to the pulpit. "Now you'll be able to devote more time to your bees," she said, trying to console him.

"I suppose," he grudgingly admitted.

J.B. hadn't seen Trudy since the wedding, and Blanche knew better than to raise the subject. The news that Ed Pritchard had been laid off by Ford didn't seem to surprise him. "The man has no education, what do you expect?"

In April Alice called Trudy with news that came as a surprise. "I want you to be the first to know," she said excitedly. "I'm expecting."

"Oh, Alice, that's wonderful. When are you due?" "December. Might even be a Christmas baby."

"Now that would be some present," Trudy laughed. She knew Alice and Al had been trying to start a family.

"You'll be next," Alice teased.

"Bite your tongue," Trudy chided. "I'm not ready to be a mother."

As the months passed, *Breezin' Along* proved as popular on the Canadian Radio Broadcasting Commission as it had been locally. The network newspaper, *Radio News*, featured a picture of the trio on the front page of its edition of August 5.

"I'm so relieved," Trudy said to Willis. "Isobel and Edith finally get the recognition they deserve." She'd been disappointed for them when the earlier publicity in the *Border Cities Star* had featured a picture of her alone. The *News* also profiled Trudy, noting that she had "an inexhaustible fund of ready wit and a refreshing attitude to life in general." The article described her as "a tonic for radio listeners."

Willis was content to work in the shadow of his wife. His official title at CKLW was Chief Control Engineer, and his main concern was quality control. It was his task to make sure the equipment was functioning properly. From time to time there would be minor on-air malfunctions, but these were rare. Willis ran routine tests every week, and usually discovered any problems before they became liabilities. After a microphone malfunctioned during a newscast he installed backup mikes in each studio. These were usually older models that had been replaced, but they still worked well enough to be used in an emergency.

Since CKLW served both Detroit and Windsor, it had more listeners than any other station in the Canadian Radio Broadcasting Commission network. And in the United States there were only three cities with stations that had larger audiences—New York, Chicago, and Los Angeles. So there was money available for equipment, and Willis was proud to say that CKLW could hold its own with the best broadcast facilities on the continent.

Willis could make this claim because, while he had yet to visit Los Angeles, he had been to New York and Chicago to inspect stations in those cities. In fact it was in New York that he finally got the chance to meet Trudy's sister Edith, who was visiting from Germany. Trudy accompanied him to New York, and they lunched at the Russian Bear Restaurant on Second Avenue.

It was immediately clear to Willis that Edith was nothing like Trudy. There was a family resemblance, to be sure, but the similarities were insignificant. While Trudy was lively and impetuous, Edith was formal and imperious.

He'd been warned. "She's a diva," Trudy had explained.

"What's a diva?"

"In the world of classical music the most talented singers are often called that. These are the performers who are at the top of their profession. They're treated like royalty, particularly in Europe."

Willis found he had little in common with Edith, so he contented himself with enjoying the meal, an excellent filet mignon with mixed vegetables, while the two sisters chattered on about the various members of the Champion family.

He did manage to ask Edith about Hitler, who in 1935 was beginning to emerge as a force in Germany, but his sister-in-law maintained that she wasn't interested in politics and didn't have an opinion.

When the meal was finished Willis left the sisters. "I'm afraid I have to go," he said, excusing himself. "I have an appointment with the folks at CBS. I'll catch up with you later at the hotel."

Since CKLW was a CBS affiliate, the New York technicians were more than happy to show Willis around. He was impressed. They had a half-dozen large, beautifully appointed studios, all featuring the latest equipment. New York was the heart of the network; it was where the majority of the programs were produced. This fact was reflected everywhere, from the vast entrance lobby to the offices of the various staff members. Willis suddenly understood how his father had felt during his try-out with the Detroit Tigers. That had been major-league baseball, and it was a different world from the barnstorming world in which Clyde excelled. Willis was looking at major-league radio, and it dwarfed the CKLW station.

When he returned to the Chelsea Hotel, where they were staying, Edith and Trudy were seated in the lobby, talking.

Edith smiled and beckoned Willis to sit. "Come, join us." There had been a subtle shift in the behaviour of his sister-in-law. She no longer seemed quite as distant or aloof.

"Edith is performing this evening," Trudy said, "and she's invited us to be her guests."

"Great," said Willis, wondering what he was in for. He'd heard broadcasts of the Metropolitan Opera from New York, and they had never appealed to him. "Where is she singing?"

"Carnegie Hall," said Trudy, with obvious pride.

Willis was impressed. He didn't know anything about opera, but he had heard of Carnegie Hall. It had been the location for many of the major music productions featured on CBS.

"Isn't this exciting?" whispered Trudy as they were shown to their seats that evening in the spacious auditorium.

Willis nodded, taking in the dimensions of the hall, his engineer's mind engaged by the acoustic details.

The concert featured three singers, Edith, the tenor Charles Hackett, and the baritone Lawrence Tibbett. Each would sing two solos, and they would join forces for three more selections. Edith had chosen Gounod's "O Divine Redeemer," and "Come Unto Me," from Handel's *Messiah.*

Trudy was enthralled with the performance, but Willis found it less compelling. He wondered at a world that lavished such esteem on opera, when to his mind Trudy's singing was far more entertaining. However, he kept these thoughts to himself after the performance, when they were permitted to visit Edith in her dressing room.

"You were magnificent," Trudy enthused. "The whole evening was a delight."

Edith smiled. "I thought it went well," she admitted. "I was particularly impressed with Lawrence. He has an amazing range for a baritone."

"I've never heard better sound," Willis managed. "Now I know why Carnegie Hall is so famous."

"You should come to Europe," Edith said. "There are concert halls like this one in every city."

"Some day," said Trudy, laughing and nudging Willis playfully.

"Some day," he agreed.

"But right now," said Edith, "I'm starving. I've made dinner reservations for us at a Bavarian restaurant that was highly recommended by a friend back in Germany."

The taxi let them off at a busy intersection in North White Plains. The restaurant was Maxl's Rathskeller. Willis wondered at the name. To him it sounded like "Rat's Cellar," but he was too hungry to be put off by the thought.

Trudy glanced at the menu and realized it was all in German. She decided she would ask Edith to order for them, and Willis agreed. When the waiter appeared he was soon engaged in a lengthy conversation with Edith. When he had taken their orders he turned to Trudy. "Your sister's German is remarkable," he said, speaking English with a slight accent. "I'd never have guessed she was an American."

"Canadian," Edith corrected him.

"Ah," said the waiter, as if that explained everything.

Edith had ordered something different for each of them. Trudy had a smoked beef specialty, *Selchfleisch*, with sauerkraut and dumplings. For Willis she chose the traditional Wiener schnitzel, with *Mohnnedeln* poppy noodles. Her own meal consisted of a beef ragout, *Beuschel*, and *Speckknodel* dumplings made with bacon bits.

The misgivings Willis had had about the name of the restaurant vanished as he began to eat. The food was, in a word, delicious. It was accompanied by Bavarian beer. At Edith's suggestion they all ordered the same dessert. Maxl's was famous, she explained, for its apple strudel. Soon after it arrived they understood why. The light pastry was layered with apple, cinnamon, and raisins. It was the perfect finale to the heavy main course. Schnapps and coffee rounded out the meal.

It might have been the euphoria of the performance or the combination of beer and schnapps, but back in the hotel that night, Trudy didn't care. When Willis reached for her in bed she took his hand and put it between her legs. It was the first time she'd done this, the first time she'd led in this dance of darkness. With her other hand she guided him into her. He moved slowly at first, trying to delay the energy building up within him. He wanted to please her and that always took time.

"Don't wait, Willis," she breathed into his ear. "I'm ready."

Her words were electric. A spark coursed through him, and he stroked harder, deeper, and felt her rise to meet his thrusts.

"Christ!" he whispered hoarsely, and the charge exploded within him, driving them both rocking into the heat of the August night. Beneath him, Trudy moaned, her legs wrapped tightly around his hips.

That is how a new life began. That night, as when two sticks are rubbed together, a flame flickered and caught. Not so different from the way millions and millions of other lives begin, but different nonetheless, with fire's capacity for infinite variety. And its uncertainty. Would it sputter and die out? Catch and provide warmth? Or rage out of control, consume everything? Such is the fascination of the flame.

In the morning, Trudy awoke and shivered. It had cooled off during the night. She snuggled closer to Willis. Then she remembered—he hadn't used a condom. The warmth of his sleeping body comforted her and calmed her fears. In time she slept again.

Back in Windsor, Trudy used the trip to New York as the theme for her program. She featured the trio in "Manhattan," and "Give My Regards to Broadway." Then she played a piano version of *Rhapsody in Blue*, the George Gershwin composition inspired by the sound of a locomotive leaving Grand Central Station. She brought the trio back for "Stompin' at the Savoy," and finished with a solo rendition of "Autumn in New York."

"We should send you off more often," Packard joked after the show.

"Paris in the spring would be nice," said Trudy with a laugh.

Waves of nausea woke Trudy on Labour Day, and she barely made it to the bathroom in time. "Must be the flu," she said to Willis, but in the back of her mind she wondered if it wasn't something else. Alice had experienced morning sickness that spring and Trudy thought that what she was experiencing seemed remarkably similar to what her sister had described. She took her own temperature and found that it was normal. That would have been a comfort on past occasions, but not this time. She decided she wouldn't mention her fears to Willis. He'd just received word from his mother in Leamington that his father's eyesight had deteriorated so rapidly over the summer that he'd been forced to give up his evening work managing the bowling alley. Trudy didn't want to add to Willis's worries. Because it was a holiday, both she and Willis had the day off. Lying in bed she remembered what day it was. She found no humour in the possibility that she might become aware she was pregnant for the first time on Labour Day.

By the second week in September the test results had confirmed that Trudy was expecting. It was not good news. She was not ready to be a mother—in the future, sure, but not right then. She had her career ahead of her and didn't want anything interfering with her chances of greater success.

Willis, on the other hand, was quietly pleased at the prospect of beginning a family, even a little proud of it. He kept these feelings to himself, though, because his wife didn't share them, and he didn't want to upset her. There was no talk of abortion, but Trudy did try a variety of old-fashioned remedies for her condition. Castor oil, hot baths, even vigorous exercise failed to produce the desired result, and by the second month she reluctantly began to accept the fact that, like it or not, she was going to have a child.

At first her sister was the only person she confided in. Alice's morning sickness had come to an end in the third month of her pregnancy, so that, at least, was consoling information for Trudy.

"It's not so bad," Alice insisted. "My baby is due in December, yours in April. They'll be cousins."

Trudy tried to make the best of it. She buried herself in work. Her plan was to take the summer off after the baby arrived and resume her career in the fall.

Chapter 45

Leamington and Windsor, 1935

By October Trudy could no longer hide her pregnancy. Soon the word had spread, and she smiled bravely through a round of congratulations from the CKLW staff. She assured Garfield Packard that she intended to continue with *Breezin' Along* once she'd had her baby.

Trudy had not told her parents, but Willis felt it was time to let Anna Maria and Clyde in on the news. He hoped it would prove a tonic for his father, who was still trying to come to terms with the loss of his eyesight and his work. They arrived in Leamington on a warm fall afternoon, and both Clyde and Anna Maria were waiting for them on the porch as they pulled up in the car.

The moment Anna Maria saw Trudy getting out of the car and recognized her daughter-in-law's condition, she put her hands to her mouth. "Land o' Goshen," she exclaimed, and hurried down the steps to embrace Trudy. "Congratulations!" she exclaimed. "I'm so thrilled."

"What's going on?" asked Clyde from his rocker on the porch.

"We have a surprise for you, Dad," said Willis as they climbed the steps. "You're going to be a grandfather."

Clyde's now sightless eyes filled with tears. Willis wasn't sure if it was happiness or regret. He realized his father would never see his grandchild, but he'd watched his father gradually loose his sight over the years and he knew instinctively what to do.

"Give me your hand," he said, and motioned Trudy to move closer. He took his father's hand and placed it on Trudy's stomach.

"Feel that?" he asked.

When Clyde smiled, Willis said, "That's the next generation, due for delivery sometime in April."

Trudy bent down and kissed Clyde on the forehead. "That's wonderful," he whispered.

The weekend was uneventful. It was hard to judge the mood. It was clear that Trudy's pregnancy was welcome news to Anna Maria and Clyde, but Trudy had to hide her own regrets about it. And of course there was general concern about Clyde's condition. After the initial flurry of excitement, Clyde lapsed into uncharacteristic silence, puffing on his pipe, seemingly lost in thought. The only time he showed any emotion was during the afternoon broadcast of the fifth game of the World Series from Wrigley Field in Chicago, in which the Cubs defeated his beloved Detroit Tigers 3–1.

"Damn," he said in disgust, and turned off the radio.

"Don't give up, Dad," said Willis. "The Tigers are still alive. They'll win tomorrow when they're back home at Navan Field."

"I hope so," Clyde mumbled.

That night as they went to bed, Trudy remarked on the change in her father-in-law. "It's sad," she said. "He's so quiet. Not his usual self at all."

"I'm worried," Willis admitted. "I think he's still trying to come to terms with his loss. I just hope that in time he'll accept things. And we are going to have to give them more financial help."

"We don't have much luck with fathers," said Trudy, shaking her head. "From what Alice tells me J.B. is still in a funk, and now your dad is hit with this. It doesn't seem fair."

"No," Willis agreed, "it doesn't."

Back in Windsor the next day, Willis was relieved when the Tigers defeated the Cubs to win their first World Series. He called his father to share the news. It was clear from Clyde's voice that the victory had roused him from his lassitude. Willis hoped that the change would last.

At the station, there was a new development that would have profound ramifications for Willis. The CKLW owners, unhappy with their CBS affiliation, decided to throw in their lot with the fledgling Mutual Broadcasting System. Mutual had been formed in 1934, joining stations in Newark, Chicago, and Cincinnati. The addition of CKLW gave them a presence in Detroit. Mutual had developed a number of innovative programs, notably *The Lone Ranger* and *The Adventures of Superman*. They were also aggressively into sports programming and carried the World Series, baseball's All-Star Game, and Notre Dame football.

Packard assured Trudy that *Breezin' Along* would be welcomed by

Mutual and that while the station's American content would change, it was still part of the Canadian Radio Broadcasting Commission's network and valued local programming.

But the Mutual network did not have the financial resources of CBS, and it was soon apparent to Willis that the days of buying top-of-the line equipment for the station were numbered. Mutual began to insist that he use suppliers of their choosing, which led to some decisions that he felt compelled to question. Although they were pinching pennies, there were times when he was ordered to buy expensive parts from certain American companies when he could have obtained equivalent equipment for less money in Canada.

Eventually he brought this to the attention of Garfield Packard. It reminded him of the conversation about expenses he'd had with Major Cory in the station's early days. Back then he'd been told that money was not an issue, because *The London Free Press* had just signed on as a partner. They had pulled out, but the slack had been taken up by CBS.

"I just don't get it," he said to Packard. "We're supposed to be cutting costs, but when I propose a purchase that would save us money I'm told to forget it, to buy from a more expensive source."

"I'll look into it," Packard promised.

Then, in November, he asked to see Willis in his office.

"Close the door," he said.

Willis complied and looked at the station manager. Something was clearly wrong. "What is it?"

Packard shook his head sadly. "I'm afraid I have some bad news." He looked down at the papers on his desk. "We're going to have to let you go."

Willis was stunned. He reached for a chair and sat down across the desk from Packard.

"I don't have much choice," the manager said. "Word has come down that we no longer need a control engineer."

"But that doesn't make sense. Every station worth its salt has one. The work I do is essential."

"I agree, but that task is going to be shifted to Bill Carter."

"He's capable, all right," Willis said, "but he already has his hands full."

"I'm sorry Willis, but I have no choice in the matter."

Suddenly it dawned on Willis. "I know what this is all about," he said angrily. "It's those American suppliers. The ones I've been ordered to use. Someone's getting a cut on those transactions. That's it, isn't it?"

Packard's face reddened. "I'm going to forget you said that," he huffed. "And if you want a recommendation from me in the future you'll keep those thoughts to yourself."

Willis rose from the chair, "You know what you can do with those recommendations," he said as he headed for the door. "You're the last person I'd ask for anything." He stormed out of the office.

That evening he broke the news to Trudy.

"I don't believe it," she said. "There must be some mistake."

"Believe it," said Willis angrily.

"Start from the beginning," she said, trying to make sense of the news. "Tell me exactly what happened."

Willis recounted the conversation. When he told Trudy about his suspicion that someone was taking a cut from purchases, she looked up.

"I hope you didn't mention that."

"Damn right I did."

"Oh, Willis," she said, "that was a mistake. What difference does it make? If you'd kept your mouth shut about that there might be some hope. I mean . . . I could have lobbied for you. But with an accusation like that you've burned your bridges." She shook her head. "What are we going to do?"

"I'll find something," said Willis. "Don't worry."

"Don't worry? Don't worry? Of course I'm going to worry. In case you haven't noticed, I'm pregnant. And jobs aren't exactly growing on trees these days."

With that Trudy fled to their bedroom and closed the door.

Willis lit a cigarette and wondered if his wife was right. Maybe if he'd kept his mouth shut he'd still have a job. He was pretty sure his complaints about financial inconsistencies were behind his dismissal. But he'd been raised to tell the truth, to call things as he saw them. True, he was angry and lost his temper when Packard broke the news, but that didn't change the facts. And he wasn't about to apologize. He went into the kitchen and took a bottle of O'Keefe's Old Vienna from the icebox. He sat down and drank from the bottle. It was only then that he noticed that the radio was on, tuned to CKLW, playing music softly in the background. He reached over and turned it off. From the bedroom he could hear his wife's muffled sobs.

Chapter 46

Windsor and Leamington, 1935–36

Trudy pleaded her husband's case to Packard, but to no avail. He said he had no choice, and pointed out that there had been other staff cutbacks when the station switched its allegiance from CBS to Mutual. Trudy did not mention Willis's financial suspicions. Members of the staff commiserated with her, but she could tell they were reluctant to do more than offer their sympathy. Jobs were important, the Depression was still gripping the country, and no one wanted to rock the boat.

Alice relayed news of Trudy's pregnancy to Blanche and J.B., but they had made no effort to get in touch with their daughter, and now, with Willis out of work, Trudy wasn't about to initiate contact. She could just imagine her father berating her for marrying Willis. For her part, she found it hard to forgive her husband for the accusations he had made to Packard about CKLW finances. She was convinced he would still have a job if he had kept quiet about his suspicions.

Trudy and Willis decided they'd keep the facts from Anna Maria and Clyde. They would be spending Christmas in Leamington, and that would be soon enough to bring his parents up to date. Willis even hoped that by then he might have found other work.

However, he still hadn't found anything by the time they drove down from Windsor in December. They found Clyde in a marginally better mood than he was at the time of their previous visit, although he was still quiet and withdrawn. They withheld the actual news about Willis's situation until the day after Christmas.

"Oh, my," said Anna Maria, "what a shame."

"I'll find something," said Willis, with more confidence than he felt.

"I'm sure you will," replied Anna Maria, but Trudy detected an uncharacteristic note of concern in her mother-in-law's tone.

"Well," said Clyde, "you're certainly not alone. It seems to me that every second person I know is out of work."

On the morning of December 31 the phone rang and Anna Maria answered. She beckoned to Trudy with a smile. "It's for you."

It was Al Kingerley. Alice had gone into labour and he was calling from the hospital. "The doctors say everything is fine. They expect the baby later today, maybe this evening."

"You'll let us know the moment you hear?" said Trudy.

"You'll be the first on the list," Al promised.

The rest of the day Trudy paced and waited. All of her concern was focused on Alice. Her own problems were forgotten for the moment. She picked at her food and refused to leave the house when Willis suggested a walk. "I want to be here when the call comes," she said firmly.

It came at about eleven o'clock on New Year's Eve. "It's a girl," Al announced excitedly. "Eight pounds."

"And Alice?" Trudy asked.

"Exhausted, but fine," said Al with a chuckle.

"Have you decided on a name?"

"We're leaning toward Joan."

"'Joan Kingerley,'" said Trudy. "It does have a nice ring to it."

Later that night Trudy and Willis welcomed the New Year with Anna Maria and Clyde. The news from Detroit had certainly been welcome. As they sat in the living room listening to the countdown from Times Square they raised their glasses of eggnog.

"May 1936 be a better year," said Anna Maria.

"I'll drink to that," said Willis as the clock struck midnight.

In the new year Willis continued looking for work. He found freelance work repairing radios and public address systems, but these odd jobs didn't produce much in the way of a steady income. Scientific Sound, a company in Kingsville, employed him as a consultant, but he was only called in when there were specific tasks to perform. This, like his freelance work, fell far short of full-time employment. His physical condition limited his options: the ankle sprain he had suffered at Queen's had led to a chronic back condition and a slight limp, which was more pronounced when he was tired.

Trudy was still working at CKLW, and her salary was sufficient for their necessities, but for the first time since his graduation from Queen's Willis was unable to help his parents out financially. Anna Maria rented out a second bedroom, and when Clyde's old baseball teammates became aware of how tough things were for the Littles they would drop by from time to time with food—eggs, or bacon, or the occasional chicken or leg of lamb. Anna Maria had put up preserves the previous fall and winter, and these and the vegetables from the well-stocked root cellar kept meals on the table. Anna Maria had struck up a friendly relationship with the engineers who shuttled the trains back and forth along the siding to the Heinz plant, and she provided them with cold drinking water from her backyard pump in the summer. They, in turn, made sure the Littles had a steady supply of old railway ties to burn in their wood stove. Fortunately it was a mild winter, but there were many days when the house was chilly, and the meals were often served cold.

Trudy immersed herself in work. Fortunately her morning sickness had abated. It was awkward at first, working with the man who had let Willis go, but after her initial unsuccessful intervention on her husband's behalf his name was not mentioned. She also took on extra assignments singing weekends with local dance bands, and spent a good deal of her spare time with Alice and her new baby. By March she was in the eighth month of her pregnancy and she had to stop working.

One Saturday morning in April she was sitting at the breakfast table while Willis looked through the newspaper, checking for employment possibilities.

"Have you given any thought to working outside your specialty?" she asked.

Willis looked up. "What do you mean?"

"I mean if there isn't any radio work out there, maybe you should try something else. Ed Pritchard is pumping gas."

"Well, what do you expect? Ed has no education."

"And because you do, you think that sort of work is beneath you?" Trudy asked angrily.

"No, that's not what I meant," said Willis, his own anger building. "But the money I'd make pumping gas wouldn't even pay the rent."

"And your freelance radio work does?"

"That's not fair," Willis countered. "It takes time to build up contacts."

"In case you haven't noticed, Willis, we don't have a lot of time. In another month we'll have another mouth to feed." Trudy rose from the table and stormed out of the kitchen.

Willis sighed and lit a cigarette. There had been more bickering lately, and he had put it down to Trudy's pregnancy. She'd been moody and depressed. But perhaps she did have a point. He went back to the newspaper, and this time scanned all of the employment opportunities. Nothing looked promising. He'd been shocked the other day to see an old high-school friend selling apples on a street corner. He'd turned away before he was recognized. Would the day come when he'd be reduced to that? He hoped not.

It was still dark on the morning of April 18 when Trudy nudged Willis awake. "I think it's time," she said. "My contractions are getting stronger."

Willis rolled out of bed, switched on the light, and checked his watch. It was shortly after 4:00 a.m. He looked at Trudy. "Are you sure?"

"I'm sure."

"You call Alice," said Willis, "while I warm up the car."

When he returned Trudy was waiting in the living room with the bag she'd packed previously to take to the hospital. Willis helped her down the stairs and into the car.

Once he had Trudy comfortably settled in the passenger seat Willis put the car into gear and drove off, slowly and carefully, pushing the excitement from his mind. There was almost no traffic at this time of night, and Willis hoped they would make it across the bridge to the Detroit Women's Hospital with plenty of time to spare.

The sleepy customs officer on duty asked them if they had anything to declare. Willis looked at Trudy and smiled.

"No, nothing," he said, and they were waved on through.

"Not yet, at least," said Trudy when they were out of earshot. She started to laugh, but the pain of a contraction cut her short.

Alice was waiting for them at the hospital. She opened the car door and helped Trudy out.

Willis leaned across the front seat. "You go on in. I'll park the car and be with you in a minute." He pulled the door closed.

Trudy was already in bed by the time he reached her room. Alice smiled at Willis. "Any trouble getting here?"

"No, there was hardly any traffic, so we made good time."

As he spoke, an intern came through the door. "Hello Alice," he said. He looked down at his clipboard, then at Trudy.

"This must be your sister." He reached for her wrist. "How are we doing?" he asked.

As she started to reply, Trudy was gripped by a contraction.

"Well, I guess that answers that," he said, and turned to Willis.

"I'm afraid I'll have to ask you to leave for a minute if you don't mind. I'd like to examine her."

Willis headed out the door and reached for the package of cigarettes in his shirt pocket. He hadn't smoked in the apartment during Trudy's pregnancy, because the smell made her nauseous. When he reached the waiting room he lit up and took a satisfying stream of smoke into his lungs.

A few minutes later Alice joined him."The intern says everything's under control. Trudy's obstetrician, Dr. Kennedy, has been called and should be here in plenty of time for the delivery. It could be a long night, though."

Trudy, left alone in her room, felt fear rising in her throat. She had been worried throughout the pregnancy, but tonight the fear was different. It seemed to come from deep inside, and it had started with the first contractions. It was as if she was having two sets of contractions, one physical, the other mental. When the mental contractions hit, all the doubts and difficulties of her life were magnified, and overwhelmed her.

The plan had been for her to take six months off after the baby was born, but with Willis out of work everything had changed. They'd managed to save some money, but it wasn't going to last very long. Willis's parents couldn't help, and there was no way she'd ask her folks for anything. The thought of J.B. waving his finger at her, preaching eternal damnation, filled her with fear. She shuddered, and the shudder brought on a contraction. The pain was sharper now, more focused. She tried to remember the breathing exercises she'd learned, but the pain was making it difficult.

Willis made his way back to Trudy's room to check on her progress. There was very little he could do. His visits soon became part of a pattern: when he was in the room with Trudy, they spoke only when she needed something—a drink, a cool cloth for her forehead. She had withdrawn into herself, and Willis, who had never been good at small talk, watched her silently from a chair. Between contractions he would pace the hospital corridor or go to the waiting room for a smoke. Alice had gone home to get some sleep, promising to return in the morning. She was sure the baby wouldn't be born before then. The periodic visits by the nurse on duty were the only other interruptions in the pattern: forty-five minutes with Trudy, the walk to the waiting room, the cigarette, and then the walk back.

The night seemed anything but routine for Trudy, although a pattern was developing for her, too. With each contraction she withdrew more and more deeply into herself. She was only vaguely aware that Willis was there, in the chair, providing her with occasional sips of water. She felt herself drowning in despair. She was being drawn to a place she didn't want to go, a place she had been before but had pushed from her mind. The memory of that place was like the memory of a forgotten dream . . . the feelings were there, but the details weren't. The contractions were drawing Trudy closer and closer to some important truth, and she was terrified.

By the time Alice returned to the hospital the contractions were just minutes apart. Trudy was wheeled into the delivery room, where the obstetrician was waiting. The nurse on duty had warned Dr. Kennedy that while the labour was progressing normally, the patient was behaving strangely.

What the doctor didn't know was that the pain of childbirth was pushing Trudy into the past, forcing her back to something that had been buried long ago and forgotten. The elusive details of what had seemed a nightmare were returning to her, and she was becoming aware that it wasn't a dream at all, but a memory. She began to scream and thrash about on the delivery table. The nurses were barely able to restrain her. The doctor nodded to the anesthesiologist and the mask was clamped over Trudy's nose and mouth. In the moments before she lost consciousness the past became the present—the hands holding her down . . . the mask silencing her . . . and an unbearable pain between her legs.

Alice knew that something was wrong. She had never seen anyone behave quite this way in the delivery room. There had been something in Trudy's eyes . . . a fear unrelated to the pain of birth. Alice relaxed a little as Trudy slipped into unconsciousness, and the doctor reached for the forceps. Usually this troubled Alice; she always worried about the risk to the baby. Today, though, she was more worried about her sister. Something wasn't right. When the baby emerged, bloody but intact, and screaming loudly, the doctor looked up at her.

"Your sister has a boy, Alice," he said. "Congratulations."

A few minutes later Alice went out to the waiting room. Willis crushed out his cigarette and got to his feet. Before he had a chance to ask she told him, "You have a son, Willis. A fine, healthy, eight-pound boy."

Willis felt his knees give slightly, and his eyes filled with tears. He embraced Alice. "And Trudy, how's Trudy?" he asked.

"She's fine. It was a rough delivery, but she'll be all right in a little while." Alice hoped she was right. Trudy had seemed confused when she regained consciousness, but that sometimes happened.

"Would you like to see your son?" she asked Willis.

"You bet," he said, and Alice led him down the corridor to the nursery.

Alice was worried. Her sister wasn't herself. Three days after her son's birth Trudy was still lethargic and listless. She barely touched her meals, and seemed all too ready to relinquish bottle-feeding to the nurses on duty. Willis, too, was puzzled, and his excitement at becoming a father for the first time was tempered by concern for his wife. She seemed off in some remote world of her own.

On the fourth day Willis and Alice were sitting together, smoking and drinking coffee in the hospital cafeteria. He'd spent the morning with Trudy. Alice was on duty that afternoon and evening, so he was planning to go home and catch up on some sleep.

"What does her doctor say?" he asked.

"Nothing new," Alice replied. "He thinks it's just the baby blues, and it'll pass with time."

"Baby blues?" Willis had never heard the term. "How long does it usually last?"

"It can vary . . . a few days, sometimes a few weeks." She didn't share her suspicion that this might be something more complicated. She didn't want to worry Willis needlessly. He had enough on his mind—out of work, his savings drying up, rent due on the apartment. Alice looked at her watch and rose from the table. "I'm on duty in a few minutes. I'll look in on Trudy first. Go on home, I'll call if there's any change."

Willis retrieved his fedora from the tabletop and made his way out of the hospital. Alice checked the chart at the nursing station. There was nothing new. She walked down the hospital corridor to Trudy's room, opened the door, and froze.

Trudy was framed in an open window, crouched on the sill, poised to jump. She was unaware of her sister's presence until Alice carefully closed the door and moved into the room.

Trudy turned.

"Stay right there, Alice," she said, her voice that of a frightened child. "Don't come any closer."

Alice stopped. "What are you doing?" The room was on the sixth floor.

She had to get her sister off the window ledge. "Come on, Tru," she said, calling her by a nickname she had used when they were children. "Let's talk. Like old times." She moved carefully to the foot of the bed, sat down, and patted a place beside her. "Come sit here with me. We haven't talked in a long time."

There was a moment of silence. Alice held her breath. Trudy looked out the window, then back to Alice. She shuddered, then a sob shook her body and the tension broke. "No," she said, her voice no longer that of a child, "we haven't." With tears in her eyes she got down from the windowsill and moved to the bed. Alice wrapped her arms around her sister. They sat together, locked in an embrace, Trudy crying and Alice shaking with relief.

Chapter 48

Detroit, Windsor, and Leamington, 1936

It was ten days before the doctors agreed to release Trudy from hospital, and then only on the condition that Alice would supervise her sister's care from her own home in Detroit. They felt that the risk of suicide still existed, although there had been no further attempts. During her time in the hospital Trudy had been withdrawn and uncommunicative. She ate sparingly, and only when Alice or Willis were present. She seemed indifferent to her newborn son. No one knew what to make of the situation. Specialists were called in, but were at a loss to explain why the birth had resulted in such a fundamental change in Trudy's personality. A psychiatrist said it was clear that Trudy had suffered a nervous breakdown, and recommended therapy. Whatever it was, the event had reduced the lively, fun-loving, wisecracking young women into a person without any of the old spark. She had lost weight, and only slept when she was heavily sedated.

The boy had been born with a slight crinkle in his left ear. The doctor thought it might have been caused by something in the womb, possibly the umbilical cord. It wasn't clear whether it was permanent or something the child would outgrow. It was barely noticeable, but Trudy was obsessed by it and worried that it might be a permanent defect.

Willis, who didn't share his wife's concern, tried to reassure her. "It's nothing," he said, "just a tiny crimp."

This preoccupation with one tiny defect was an indication of how fragile Trudy was. Willis sought to deflect it, and tried just about everything else he could think of to bring his wife out of what the doctors were calling her "state." Replies to his questions, when they were answered, were monosyllabic. Flowers went unappreciated. He brought in a radio, hoping music might help, but she asked him to turn it off.

When she was finally permitted to leave the hospital, Willis helped her to his car, and Alice followed, carrying the baby. Trudy and Willis had decided on a name before the birth. If it was a boy, he would be called Andrew, after Anna Maria's father, and Clyde, after Willis's dad. So Andrew Clyde Little made the trip from the hospital to Alice's home in suburban Detroit, howling all the way at the top of his lungs.

When they had settled Trudy in the guest bedroom and provided the new arrival with a bottle, he finally stopped crying. Having fed his son in the hospital, Willis had become quite accomplished at the task. He would soon learn the fine art of diaper-changing from Alice.

Gradually Trudy's spirits began to improve. She took her meals with the family and engaged in conversations. By the end of the first week in Detroit, Alice suggested that Trudy might make a quicker recovery if she was back in familiar surroundings, in the apartment in Windsor. She offered to make the trip with them to help make sure her sister was comfortable.

It was raining, and on the approach to the Ambassador Bridge they were delayed by a flat tire. When Willis got back behind the wheel after changing to the spare he was soaked. For some reason Trudy found this hilarious. It was the first time they'd heard her laugh since the birth, and it was such a welcome sound that Willis instantly forgot his discomfort. He exchanged a meaningful glance with Alice and they drove across the bridge and into Windsor in silence.

Alice's suggestion that Trudy's condition might improve once she was back in the Ouellette Avenue apartment proved wise. It was fortunate that Willis was still unemployed, because he was able to provide full-time care for his wife. His cooking skills left a lot to be desired, but he was able to manage a few simple meals. Eventually Trudy began to take over, first making breakfast and then, within a week, assuming all the cooking responsibilities. Willis wasn't sure whether it was a reflection on his own inadequacies in this department or a sign that she was getting better. She showed no inclination to leave the apartment, and Willis didn't force the issue, so he did the food shopping while his son was sleeping.

She still seemed indifferent to Andrew, but watched with curiosity as Willis fed him and changed his diapers. In those first few weeks Willis assumed the role of the mother, which proved an important factor in his son's early development. His constant presence was a comfort to his young son.

Slowly, almost reluctantly, Trudy began to show an interest in the child. She would hold Andrew for short periods of time, and one day Willis noticed she was humming softly. It wasn't long before she took over some of the feeding. Willis would heat the formula on the stove and hand it to Trudy. Willis felt a surge of love for both of them as he watched her cradling her son in her lap and holding the bottle to his hungry mouth. And with love came optimism. Perhaps in time things would return to normal, and the teasing, cajoling Trudy he knew and loved would be back. Perhaps.

The doctors at the Women's Hospital in Detroit had suggested that Trudy see a psychiatrist if her condition didn't improve. But with neither Willis nor Trudy working there was no money for therapy. The hospital bills had wiped out their savings, and it was a struggle each month just to pay the rent. Willis continued to receive a few freelance assignments, usually involving public address systems, but these weren't the full-time employment he was seeking. And when he left the apartment for these jobs he worried about leaving Trudy alone with the baby. Trudy herself was in no condition to return to work.

In the first week in May Willis's mother visited from Leamington to help out with the new arrival. Anna Maria was shocked at Trudy's appearance, and when she shared her concern with Willis he simply shook his head.

"I don't what to do," he said. "The doctors in Detroit were mystified." He didn't mention therapy to Anna Maria, because he knew that his folks had money problems of their own. "They say she has the 'baby blues,' and that the condition usually clears up with time."

Anna Maria said she'd heard of the "baby blues," but had never encountered anyone suffering from them. She recalled the delight she had experienced after the birth of Willis. "At least the baby seems healthy," said Anna Maria as the sound of Andrew's crying filled the apartment. "He certainly has a fine set of lungs."

"I knew babies cried," said Willis, "but I think this little guy is setting some kind of record."

As Anna Maria packed to leave at the end of a week she apologized that she couldn't stay longer. "I have to get back to your dad," she explained. "A nurse is coming in to give him his insulin injections, but I worry just the same. Maybe if you came to Leamington and stayed with us I could lend a hand."

"It's possible," Willis admitted. "But where would you put us? You have two boarders as it is."

"We'd manage somehow," Anna Maria assured him.

When the weather warmed up Willis finally managed to talk Trudy into venturing outside. Alice and Al had given them a baby carriage, and on sunny days they bundled up their young son for a walk around the block. Trudy was reluctant to spend much time away from the apartment, and seemed to grow more anxious the longer they were out, but each day Willis extended the length of the excursions, and by July Trudy seemed more relaxed.

It was then that he suggested a drive to Leamington to see his parents.

"Oh, I don't know," said Trudy, fear reducing her voice to a whisper.

"I think it would be good for you," said Willis. "And Dad has been asking about his grandson."

"Let me think about it," Trudy said.

Andrew was not making life easy for the couple. He cried a good deal of the time, unless he was being held or fed. The medical philosophy of the day wasn't helping. Doctors suggested that newborns had to be trained, and that they should be fed on a regular schedule. If they cried at a time when a feeding wasn't scheduled, their parents were supposed to let them cry. Trudy was adamant about following this regimen, and Willis was willing to go along, up to a point. However, when the racket became too much for him he would take the baby in his arms and walk him around the apartment, humming softly. This worked most of the time.

One day in June, after they'd let Andrew cry for more than an hour, Trudy went to the crib. "Oh, my god," she shrieked, and Willis came running.

There were bloodstains on the blanket, and Andrew's belly button was extended and bleeding. Willis took over. He lifted Andrew from the crib. "Find some extra towels," he said. "We've got to get him to a doctor."

In the car, Trudy sobbed as she held her son. Willis decided the best bet would be to return to the hospital in Detroit. Surely they'd know what to do. The bleeding seemed to ease off. At customs, the agent sensed the crisis and waved them through. Willis drove the car to the emergency care entrance and rushed his young son inside, while Trudy followed. An intern greeted them, took a quick look at Andrew, and asked them to take a seat in the waiting room.

By this time Trudy was hysterical, but the intern assured them that he'd seen this condition before and it wasn't serious. He promised that a doctor would be with them soon. Fortunately, the physician who had delivered Andrew was on duty, and within minutes he was with them. They were ushered into a small outpatient surgery.

"What we have here is an umbilical hernia," Dr. Kennedy explained. "It isn't common, but it does happen, especially when newborns cry a good deal. The exertion pushes out the knot that was tied when the umbilical cord was cut." He put a reassuring hand on Trudy's shoulder. "Why don't you wait outside? I'll stitch this little fellow up and he'll be fine. You'll be able to take him home in an hour or so." Willis led Trudy back to the waiting room.

"What are we going to do?" she sobbed. "Maybe it was wrong to let him cry so long."

"We were only following the doctor's orders," Willis assured her, but he too was wondering about the wisdom of the regular feeding schedule they'd been trying to maintain.

A nurse returned with Andrew in less than an hour and placed him in Trudy's arms. Dr. Kennedy appeared moments later. "He's right as rain," he said soothingly. "I put in a couple of stitches and applied a dressing. Just change it every day and come back in a week, when we'll remove the stitches."

"What about his crying?" Willis asked. "Could it happen again?"

"I doubt it, but I can prescribe a mild sedative until things heal. And a week of extra attention won't spoil him. Once the stitches are out you can put him back on a regular feeding schedule."

On the drive back to Windsor Trudy was silent. Andy, sedated, nestled quietly in his mother's arms.

The combination of medication and the extra coddling worked their magic. The incessant crying stopped, and Trudy dutifully changed the dressing each day. By week's end, with the stitches removed, their son had, to all appearances, a normal bellybutton.

Willis had other worries. They were behind in their rent, and although the landlord had been understanding, it was clear that his tolerance was wearing thin. Willis had decided that they'd eventually have no choice but to accept Anna Maria's offer and move to Leamington. In the meantime he had to break through Trudy's travel fears. He felt that a visit to his folks would be the first step, and managed to overrule his wife's objections. As they drove to Leamington on a Saturday morning in July, the movement of the car lulled Andrew to sleep. Trudy was fretful, constantly picking at her nails and checking the car bed to make sure her son was okay.

At the house on Talbot Street Anna Maria and Clyde were waiting on the front porch. Willis took Andrew from the car bed and carried him up the walk. He'd looked forward to this moment from the day he'd learned Trudy was pregnant.

"Hold out your arms, Dad. It's time you met your new grandson."

Clyde did as he was asked, and Willis deposited his son, still sleeping and swaddled in a blanket, into his father's arms. Clyde's face creased into a smile, and he gently ran his fingers over the baby's fine features. He turned his now sightless eyes to his son. "He's a beauty," he said simply.

"He is," Willis agreed.

While Anna Maria and Trudy were watching this scene unfold Anna Maria wrapped one arm around her daughter-in-law and held her close. "Things are going to look up," she whispered into Trudy's ear. "You'll see."

In the weeks that followed, Willis and Trudy made several trips to Leamington, and gradually Trudy began to relax at the prospect of moving there. Her mood was improving, but slowly. She had promised the people at CKLW that she would return to work in September. In August, with Willis looking after Andrew, she went back into the studio to rehearse. She was still gaunt and hollow-eyed, but in an effort to minimize the effects of her weight loss she'd had her hair cut dramatically short in a style that was popular. Isobel and Edith felt that the result did not flatter her, but they kept this opinion to themselves. Trudy did her best to appear bright and cheerful at this reunion, and the trio managed a couple of hours of rehearsal.

"Not bad," Trudy commented. "A trifle rusty, but nothing we can't fix with a little effort." She bade the women goodbye and returned to the apartment. "I'm exhausted," she said to Willis, and went into the bedroom. When he looked in minutes later, she was on the bed, still dressed, fast asleep. Willis gently removed her shoes and covered her with a light blanket.

Trudy's return to the studio meant that she could begin to draw her regular performance fees. This provided sorely needed cash, and they were able to pay some of the back rent they owed their landlord. Willis wondered whether the move to Leamington, something he had thought inevitable, would be necessary after all. His own earning capacity was limited, though, because someone had to look after Andrew when Trudy was away, and they couldn't afford a babysitter. Even the mention of the subject set Trudy off.

"He's too young," she said. "No way I'd trust a teenager."

"We might find someone older," Willis countered.

"Right," said Trudy, "and where would we find the money to pay her?"

Willis began to wonder if Trudy was a little too protective of their son. And he didn't like her tendency to give the child the mild barbiturate the doctor had prescribed whenever he was restless and had trouble getting to sleep. When he was alone with Andrew he broke all the rules. He picked him up when he cried and fed him when he seemed hungry. He was captivated by this small bundle of life he'd fathered. He loved to take him for long walks in the baby carriage, and was never in such a hurry that he wouldn't stop to let anyone who seemed interested take a peek at the baby. He loved the compliments these occasions produced.

The day for Trudy's first live show was soon upon them. Trudy had always suffered from stage fright, but over the years she had come to accept it. Once the performance began it had always disappeared. But this time it was different. To the untrained ear, the program was smooth and professional, but Isobel and Edith sensed a difference. When they finally signed off, Trudy slumped to the keyboard. "Thank God that's finished," she said. "Do you think anyone noticed?"

"Noticed?" Isobel asked.

"I was petrified," Trudy said. "My hands were shaking so much that I wasn't sure what keys I was playing."

Edith tried to reassure her. "You were fine," she said. "You've been away for a while. It's natural for you to be nervous. You'll be okay once we're back in the swing of things."

But Trudy wasn't fine. Her stage fright persisted, and it didn't abate as it once had during the programs that followed. She was at a loss to explain the change. There was more pressure on her now that Willis wasn't working, but deep down she knew that that wasn't the reason. Something had shifted within her when she'd given birth. Her confidence was gone—lost forever, she feared. She soldiered on through September, but finally gave up after just six shows. She couldn't take the strain, and was afraid that if she didn't stop she would slip back into depression.

And a new symptom had surfaced. She'd begun experiencing panic attacks. They could occur without warning at any time during the day or evening. When they did, she found herself paralyzed with fear but unable to say what it was that was frightening her.

In the fall, one of Anna Maria's boarders moved away, so there was a spare bedroom in Leamington. Willis suggested that they move in with his parents.

"I don't know . . .," Trudy said.

"We'd be saving the rent money, and with Mom to help out with Andy I could devote more time to job-hunting."

"I hate to impose on them."

"Nonsense. They'd love to see more of you and their new grandson."

"If you're sure . . ."

"I'm sure."

In October they moved to Leamington. Most of the furniture from the Windsor apartment was put into storage and they took only their personal possessions, Andy's crib, his baby carriage and a few toys.

By now it was clear to Willis that his early optimism about finding work had been unrealistic. In the past whenever he'd been dealt a personal setback he'd been able to bounce back quickly. At Queen's when he'd flunked his second year he hadn't been expelled, but had been allowed to repeat the year. And he had learned from the experience. After his internship with General Electric had run its course, a job had been waiting at Northern Electric. And when he was let go by Northern Electric in 1932 his friend at Queen's had come to the rescue and found him work teaching. Unemployment was a new experience for Willis. There simply weren't any jobs open that required his skills. And when he moved outside his field of expertise and applied for other work he was turned down, rejected. Then Trudy had stopped working and her health had become a major concern. One afternoon they had a particularly bitter exchange of words.

"If you'd kept your mouth shut you'd still be working at the station," she reminded him.

"But what they were doing was wrong," Willis protested. They'd been over this too many times to count.

"And you saw yourself as the judge and jury."

"I just did what I thought was right. I didn't know it would cost me my job."

"Well, you should have given us some consideration. You aren't alone in the world anymore, in case you missed that fact."

"I've told you I'm sorry."

"Sorry doesn't put food on the table."

Willis felt his anger flare. "Your decision to pack it in didn't help."

With this, Trudy burst into tears, ran up the stairs, and slammed the bedroom door. The noise woke Andrew and he began to cry.

Willis strode out of the house and climbed on his bicycle. He rode as fast as he could, his legs pumping, his heart pounding. When he reached the public beach he let his bike fall to the sand and sat down, panting. When he'd regained his breath he walked out on the pier and stood facing the open water of Lake Erie. For the first time in his life he felt defeated, useless. Was this how Trudy had seen things in the hospital when she threatened to jump from the window? It would be so easy to dive from the pier and just swim until he was too exhausted to return. But images of Trudy and Andrew left to fend for themselves flooded his mind, and he shook the thought from his head. Something would turn up. It had to. He walked back along the pier, found his bike, and returned to his folks' home on Talbot Street. The trees were turning in the slanting sun, but the colours were lost on Willis.

Radio in Canada had been largely unregulated in the late 1920s. Just about anyone with money could set up a station and begin broadcasting. Many Canadians who lived close enough to the United States were tuning in to American stations. In 1928 the federal government was subjected to repeated on-air attacks from two stations, one run by the Jehovah's Witnesses and the other by the Roman Catholic Church. Prime Minister Mackenzie King appointed a Royal Commission to look into the possibility of regulating radio broadcasting in Canada. It was chaired by Sir John Aird, the president of the Canadian Bank of Commerce. King had expressed his belief in the potential of network radio when he addressed the country on the occasion of Canada's Diamond Jubilee in 1927. Fifty-seven stations across the country were linked in an ad hoc network that carried the Prime Minister's speech.

The Commission held hearings across the country. Although it was troubled by the religious attacks, it was more concerned about the powerful influence American stations were having on Canadian culture. In 1929 it recommended the formation of the Canadian Radio Broadcasting Commission. It proposed a network of seven powerful stations to be set up across the country. The cost would be 2.5 million dollars. Unfortunately, the market crash that year and the depression that followed delayed implementation of the plan.

King decided to postpone a decision concerning the radio network until after the 1930 federal election. However, his Liberal government was defeated by the Conservatives, R.B. Bennett became prime minister, and the project seemed to be dead. A group of dedicated Canadians lobbied hard, pressing the Conservative government to act. It was not clear whether the regulation of radio should be a federal concern or a provincial one until, in 1932, the Privy Council in London ruled that broadcasting in Canada was a federal matter. This cleared the way for the formation of the Canadian Radio Broadcasting Commission.

CKLW was part of this early network, and it had given Trudy the chance to be heard across Canada. But when the CRBC was used by the Conservatives to broadcast partisan political programming during the 1935 general election campaign, it drew the ire of the opposition Liberal party. When the Liberals won the election and Mackenzie King became prime minister again, he scrapped the CRBC and introduced legislation in Parliament to establish a new body, the Canadian Broadcasting Corporation, which was born in November 1936. Its mandate was to make radio available to Canadians everywhere, no matter how remote their location. It was also expected to be non-partisan, and while it was funded by the federal government, as a crown corporation it would be free from political interference.

The moment he heard about the creation of the CBC, Willis applied for work with the new Corporation. He had all the required credentials, and was immediately offered a temporary position, to begin in January 1937. There was a catch, though. The work was in Toronto; it would require a move to a new city.

It was decided that Trudy should remain in Leamington for the time being. When the time came for Willis to begin working for the CBC he would go to Toronto alone and rent a room there. If he was able to secure a permanent CBC job, Trudy and Andrew would join him.

VI Toronto

Chapter 50

Toronto, 1937

Willis took the train from Windsor to Toronto. He knew that Trudy liked to drive, so he had decided to leave the car behind in Leamington. He hoped that having a car at her disposal would help lift her spirits. In Toronto, his first priority was to find a room. He'd been to the city twice before, when he was a student at Queen's. He had made the trip as a fan of his university's football team, the Golden Gaels, who had a heated rivalry with the Toronto Varsity Blues. But both times he had come to town the night before the game and returned to Kingston the next day, so he had had little time to explore the city.

When he arrived at Union Station he found a newsstand and bought a copy of *The Toronto Star*. He turned to the Rooms-to-Let section and began scanning the ads. However, he soon realized he would need a map, because he had no real sense of the city. He checked his bag in one of the station lockers and walked to a nearby gas station, where he bought a map and got some change from the attendant for the calls he planned to make.

He crossed the street to the Royal York Hotel and sat down in the lobby, where he read through the ads again and circled a number of promising choices. Then he checked the map for locations. The CBC studios were located at 805 Davenport Road, but rooms in this area were expensive. He did find a promising possibility to the north, on Lawrence Avenue. The ad was for a room only, with no board, which suited Willis. He deposited his nickel in the pay phone in the hotel lobby and called the number. He was told that the room would be five dollars a week. The landlady assured him that her house was close to a streetcar stop, and that he could easily commute to the CBC studios in less than half an hour.

Willis went back to Union Station, retrieved his bag from the locker, and hailed a cab. It was Sunday and traffic was light. He asked the driver take him to the Davenport Road location.

When they got there the cabbie asked, "Want to get out here?"

"No," said Willis, "I just wanted to have a look. It's where I'll be working."

"The radio station?" asked the driver.

"Yep."

"Eveready used to manufacture batteries in that building," said the cabbie, "but they moved."

Willis gave the driver the address of the rooming house, and they drove north on Bathurst Street. When they arrived, Willis paid the driver and got out, hefting his two suitcases. Before him was a modest two-storey brick dwelling, set back from the street. There was a light covering of snow on the lawn. On his right he could make out a low-cut hedge that bordered the property. He mounted the steps to the front porch and rang the bell.

A woman opened the door. "You must be Willis," she said, and beckoned him in. "I'm Mrs. O'Reilly." She glanced at her watch. "You made good time."

"I did," said Willis. He put down his bags and shook the woman's hand. Middle-aged and stout, she had a turned-up nose and a ruddy complexion.

"Would you like a cup of tea?" she offered. "Or a coffee?"

"Coffee would be great," said Willis.

"Go and take a seat in the parlour," she instructed as she headed off for the kitchen.

Willis sat down in an easy chair. At one end of the room there was a tiled fireplace with neat stacks of wood and kindling beside it, ready for use. Light filtered in through lace curtains on a bay window that faced Lawrence Avenue. Wallpaper of a subtle floral design had been chosen to blend with the patterned upholstery of the easy chair and the matching sofa. A watercolour landscape painting hung above the fireplace, and magazines were arranged on a coffee table. It was, he decided, a comfortable room. Noting an ashtray, he lit a cigarette.

Mrs. O'Reilly returned to the room carrying a tray, which she placed on the coffee table. "Do you take cream and sugar?"

"Yes," he said, "both."

Mrs. O'Reilly motioned to the tray as she took a seat on the sofa opposite him. "Why don't you add your own? You know how you like it."

Willis poured some cream, spooned in sugar, and stirred. He noticed that Mrs. O'Reilly was also drinking coffee, but hers was black.

"Now," she said, "tell me a little about yourself."

Willis explained why he was in Toronto and the nature of the work he expected to do. He mentioned that his family had stayed in Leamington, and that he would likely be away some weekends visiting them.

Mrs. O'Reilly told Willis that she rented out three bedrooms on the second floor, of which two were taken, one by a schoolteacher and another by a salesman. She showed Willis the room that was available. It was at the front of the house, facing Lawrence Avenue, where traffic could be busy, so it was the least desirable room. But it was large, and Willis, never bothered by noise when it was time to sleep, decided he'd take it. He paid a month's rent in advance and unpacked his bag.

On his first day of work at the CBC a good deal of his time was taken up with filling out forms. He noted that his salary would be $2,500, the same as he had received at CKLW. He spent the rest of the day touring the local station, CBL. His guide was Jack Walker, a man about his own age. Like Willis, Walker had studied electrical engineering, but in the United States. In addition to showing him around, he briefed Willis on the politics of the newly established Corporation.

"There's the old guard, the people who worked for the Canadian Radio Broadcasting Commission, and the new folks, the ones who were brought in when the Commission gave way to the CBC. You'll find that some of the veterans are resentful, particularly the ones who'd been in positions of authority. They don't like the idea of taking orders from newcomers. These things will probably work themselves out eventually—at least I hope so."

"Are you in the old guard or the new?" Willis asked, with a mischievous smile.

Walker laughed. "Oh, I'm like you, a rookie."

Willis learned that the main challenge they had encountered when they converted this battery manufacturing plant into a radio station was the sound of the streetcars from the Toronto Transit Commission's nearby Hillcrest Yards. The conversion had been successful, but as a result, the studios had been so effectively soundproofed that additional microphones had to be installed elsewhere in the building in order to provide the realistic background ambience that was often needed for radio dramas.

It was mid-afternoon by the time Willis was introduced to the network's Chief Design and Construction Engineer, Harold Smith.

"Welcome aboard," he said as he shook Willis's hand. Willis judged him to be in his forties. The man radiated a combination of confidence and enthusiasm.

"Thank you, sir," said Willis, trying to infuse his own words with the same degree of energy. "It's great to be a member of the team."

Smith looked down at some papers on his desk. "I'd like you to spend a week here at Davenport getting the feel of things," he said. "Then we'd like you to have a look at some property near Hornby, to make sure it's a suitable site for our powerful new transmitter."

"Hornby?" Willis asked.

"Yes, it's out in the country, to the west of Toronto. Do you have a car?"

"I do, but not with me. I left it with my wife, back in Leamington."

"Well, I'll introduce you to some of the engineers working on the transmitter plans. You can probably arrange transportation with them."

It was soon clear to Willis that in joining the CBC he was moving into a different broadcasting environment. CKLW had been a major station, but the Corporation was a vast network of stations. It would provide him with a fresh and exciting set of challenges.

The CBC had inherited the facilities of its predecessor, the Canadian Radio Broadcasting Commission. It was expanding to outlets and repeater stations across Canada. Toronto was the centre for English-language programming, while Montreal served the French population, mainly in Quebec and New Brunswick, but in other communities as well, wherever there were pockets of French-speaking Canadians.

When Willis arrived there were eight CBC stations, and they provided six hours a day of network programming. There were 132 full-time employees, ten of whom were producers and fourteen were announcers. The Corporation signal was able to reach radios in forty-nine percent of Canadian homes, although the vast majority of listeners were still tuning in to the more popular American stations. There were plans to expand the network to the point that it would eventually reach the entire population.

There was occasional friction among members of the technical staff, but it was mild compared to the infighting among the executives and the programmers. In addition, because it was a crown corporation and funded by taxpayers, there were inevitable squabbles with the federal government over funding. A 1937 plan called for the construction of 50,000-watt superstations in Toronto and Montreal, with two additional powerful stations in the Maritimes and the Prairies. It was hoped that when a new transmitter in Vancouver was completed, the CBC's reach would expand to eighty-nine percent of the population. The government had agreed to fund the stations in Toronto and Montreal, but it was delaying funding for those in the Maritimes and the Prairies.

Jack Walker, who had shown Willis around the Davenport Road facilities, introduced him to Russell Hill, the man responsible for surveying the Hornby site. Hill, an architect, had taken soil samples to make sure the ground at Hornby was solid enough to support the 645-foot-high transmission tower planned for the station. He offered to drive Willis to Hornby.

"Have you been out to the site before?" Hill asked.

"No, this will be my first time. What about you?"

"I've had my people look it over, and we've done some preliminary tests. It looks good, with lots of clay in the surface soil. We're about ready to sign off on it if you think the location is right from a technical point of view."

When they reached the site they got out of the car and walked through an open field. The land was flat and overgrown with weeds and brush.

"What's our elevation here?" Willis asked.

"We're well above sea level," said Hill, "and about two hundred feet higher than Lake Ontario. No worries about the signal interfering with ships."

"Elevation isn't really as critical as you might think," said Willis. "It's more important for the surrounding countryside to be more or less flat."

"I've wondered about that," said Hill. "I guess that's why the Niagara Escarpment wasn't chosen."

"Too rocky," Willis explained. "The soil is sandy, and it's a heavily wooded area. Not the best location." He took a portable receiver from the case he was carrying and set it on the ground.

"What's that?" Hill asked.

"It's a modified radio. Picks up stations but registers other kinds of electrical interference as well."

"Such as?"

"Well, factories, airports, even highways can cause problems. And the earth itself has a magnetic field. It can scramble signals."

Willis turned on the receiver, and when it had had a chance to warm up he began turning the dial. There was a sudden burst of music, and Willis reduced the volume.

"That must be Hamilton," said Hill when the song finished and an announcer introduced the next number. "Out here we're about halfway between Toronto and Hamilton."

Willis took out a pad and pencil. He turned the dial across the full frequency of the band, stopping to write down the stations as he encountered them and record their signal strength. But he was just as interested in the static he encountered between the stations; in fact he listened to it with more care than he did to the actual radio signals.

While he was carrying out these tests Hill wandered off, stopping from time to time to pick up and examine rocks. By the time he returned, Willis, satisfied that he'd done what he could, had turned off the receiver and was packing it back into his bag.

"What do you think?" the architect asked.

"I'm not a hundred percent sure," Willis replied. "I'd like to come back tonight if you're up for it. Radio signals are stronger after dark. We've had our engineers out here before, and I'm really just confirming some of their findings, but before I give them my blessing I'd like to make sure I've covered everything."

"Sure," said Hill as he looked at his watch. "What time would you like me to pick you up?"

"Would eleven or eleven-thirty be too late?"

"Hell, no. I'm curious. You going to run the same tests?"

"Yep," said Willis, "and I think you'll be surprised at how different the results will be."

The two men returned to the car and drove back into the city.

Willis spent the afternoon discussing his notes with Harold Smith. The chief engineer seemed pleased with the fact that everything Willis had recorded confirmed the conclusions they had already reached. The Hornby site seemed ideal. But he admitted that no one had yet checked the nighttime readings.

"I don't expect we'll turn up any serious problems," said Willis, "but it's better to be safe than sorry."

"Agreed," said Smith.

When Willis and Hill returned to the field that evening they carried flashlights, but there was a full moon so they found they didn't need them. They stopped at the place where they had done the earlier testing and

Willis took out his receiver. He turned the dial, and found many more stations than he had encountered previously. At one point they heard the clear voice of an announcer from Cincinnati.

"That's amazing," said Hill.

"They must have one helluva powerful transmitter down there," Willis commented, "but then you never can tell. Radio signals are weird. One night at Queen's we picked up an Australian station."

After a few more minutes of testing Willis shut off the receiver. "Good," he said. "No nasty little surprises."

Back in the car Hill suggested they stop at his place for a drink. "I think we've earned it."

"I'll drink to that," said Willis with laugh.

The next morning he reported his positive findings to Smith.

"That's great," his new boss said, rubbing his hands together. "Now let me show you the plans we have for our new transmitter." As Smith talked, Willis realized that he was a kindred spirit, as caught up in the potential for radio as he was.

One of the main hurdles they faced involved converting the alternating current provided by the hydro company into the more powerful direct current required by the transmitter. Willis had practical experience in converting AC to DC power, and he also understood the problems involved when it came to boosting signal strength.

"You certainly know your stuff," Smith commented after they'd discussed the situation.

"We were fortunate in Windsor," said Willis, deflecting the praise. "A station in California had already solved most of the issues."

"I have a feeling we should probably check out the more powerful stations in the United States before we get too far along in our design."

"Not just the U.S.," said Willis. "The Europeans are ahead of the Americans in a good many areas."

Smith nodded in agreement. "That's true," he said.

Every second weekend Willis took the bus from Toronto to Leamington so that he could spend time with his family. He found that Trudy was still depressed, but detected what he took to be some positive signs. He wondered if this was just wishful thinking on his part.

When he arrived at the Davenport studios on the first Monday in March, Smith, the Chief Engineer, greeted him with a smile.

"Come into my office," he said. "I have some good news for you."

Willis followed him along a corridor, his mind racing with possibilities.

"Have a seat." His boss indicated a chair. He opened an envelope and extracted a sheet of paper. "I just received word from Ottawa. Your application for permanent employment has been approved." He handed the letter across the desk to Willis. "Congratulations."

It was the news he'd been waiting for, but he was so overcome with emotion that he was momentarily speechless. He looked down at the letter, which was signed by Ernest Bushnell. He noticed with satisfaction that his salary would be more than he had been making at CKLW.

Now, at last, he could bring Trudy and Andy to Toronto.

"May I use your phone?" Willis asked when he had regained his composure. "I'd like to share this news with my wife."

"Of course." Smith pushed the phone across the desk and rose to leave the room.

"It will be a long-distance call," Willis said apologetically.

"I think that under the circumstances the CBC can afford it." Smith smiled and closed the door gently behind him. Willis was reminded of the time he'd asked to use George Ketiladze's phone at Queen's to call home with news of his CKLW appointment.

Chapter 51

Toronto and New York City, 1937

Trudy arranged to leave Andy with her sister Alice in Detroit so that she and Willis could look for a house in Toronto. As they drove back across the Ambassador Bridge into Windsor, Willis turned to Trudy.

"How are you feeling?" he asked.

"Excited, I guess, and a little frightened too."

"At least you know Toronto," he said. "It's where you studied music."

"Yes, but I was on my own then. Things are different now."

Willis noticed a hint of sadness in her voice. He chose to ignore it. "In my limited time there I've come like the city. It's easy to get around. And I've been reading the papers. There are lots of houses to rent."

"I hope so."

They headed east along Route 2 and Trudy turned the radio on. It was tuned to CKLW.

"Can we change the station?" Willis asked, still bitter about how his time at CKLW ended.

"Sure." Trudy turned the dial until she found some music on a Detroit station. Shep Fields was singing the current hit "That Old Feeling." Trudy absent-mindedly hummed along.

Since Trudy was familiar with the city from her time at the Royal Conservatory of Music, they soon narrowed the search to the northwest part of the city. That would cut down travel time when Willis was needed at the Hornby site, but it would also be an easy commute to the Davenport offices.

Trudy checked the area while Willis was at work. A realtor showed her a number of houses, but nothing seemed quite right. She had given herself a week to find a place and by Friday she was beginning to despair, but then the agent called her at Willis's boarding house.

"Something has just come on the market," the agent said.

Trudy grabbed the keys to the car and drove to an address on Cranbrooke Avenue. It was just a few minutes from Lawrence and a block west of Yonge Street, the city's main north-south artery. It was a bungalow, set back from the street on a gently sloping lawn, and it had the look of an English cottage. Trudy felt a pulse of excitement. Inside there was a spacious living room, two bedrooms, and a modern kitchen. Trudy felt a faint stirring as she toured the rooms. There was a fenced-in back yard and—the clincher—a swing set and a sandbox. It was perfect.

She called Willis and he arranged to get away from work so that he could see for himself. As he sat in the cab he told himself that if Trudy had made a choice, he would go along with it. During his quick inspection he was pleased to note that the basement offered the possibility for a workshop. The rent was within their budget, and before the hour was out he had signed the lease and given the agent a deposit. Trudy drove him back to his office and then returned to the boarding house to pack for their return journey to Leamington.

Willis called a moving company in Windsor and arranged to have them pick up the furniture they'd had in the Ouellette Avenue apartment that was now in storage. They were to move into the Cranbrooke Avenue house on the first of May. Trudy and Andy would remain with the Littles until then, but she wanted to be in Toronto when the movers arrived, to supervise the arrangement of the furniture. She planned to drive from Leamington the day before the move.

"Are you sure you don't want me to come down?" asked Willis. "It's a long haul, and you'll have Andy."

"I'll be fine," Trudy assured him. "It's only when we're driving that I can be sure he'll sleep. There's something about the moving car that seems to soothe him."

Willis was relieved. He took Trudy's willingness to make the trip without him as a positive sign.

At work, Willis and his colleagues were finalizing details for the transmitter. Forty thousand dollars had been set aside for the construction of the building, which was to be located near the base of the antenna. A similar complex was planned for Verchères, outside Montreal. It would broadcast French-language programming to Quebecers.

No one at the CBC had actually worked on a 50,000-watt transmitter, so Willis was chosen to travel to New York City, where he had contacts with people in the Mutual Broadcasting System from his days at CKLW. He was to take a first-hand look at the facilities of WOR, a Mutual station that had upgraded to 50,000 watts two years earlier. He would make the trip in the last week in May.

The move from Leamington went smoothly. Trudy and Andy drove to Toronto on Friday, the thirtieth of April, and stayed with Willis on Lawrence Avenue. She was at their new house on Cranbrooke the next afternoon when the van arrived from Windsor. Since it was a Saturday, Willis had the day off and was available to supervise the arrangement of the furniture. He was unable to take a more physical role because of the chronic back problem he'd had since his time at Queen's. When the movers left, he and Trudy slumped down on the sofa in the living room, exhausted by all the decisions they'd had to make. Andy was asleep and the house was silent. Willis reached over and turned on the radio. He tuned it to CBL, and classical music filled the room.

"Well," he said, "here we are."

"Finally," said Trudy, and rested her head on his shoulder.

In the days that followed, Trudy's mood varied. There were times when she was upbeat and optimistic, and others when she was subdued. One night she confided to Willis that her anxiety attacks had increased in frequency. She'd first had these paralyzing episodes in which she was filled with fear for no apparent reason when she was a teenager, but they had subsided when she left home to study music in Toronto. The attacks returned without warning after Andy's birth, and were now a constant worry. She'd had a particularly difficult few weeks after she received news from Germany that her sister, Edith, was expecting a baby.

"I'm worried about her," Trudy said. "I wish she'd come home to have the child. Things are so unsettled in Germany."

"It's normal to be anxious about Edith," Willis reasoned. "Anyone would feel the same way. I bet your other sisters are worried too."

"Oh, they are, "Trudy admitted. "But their anxiety isn't the same as mine. It doesn't make them helpless. I'm such a mess."

"I think we should find you a doctor here in Toronto," said Willis. "There has to be someone who can help."

"I'll call Alice," said Trudy. "She may know someone."

Alice did indeed know a doctor in Toronto, a psychiatrist who had interned in Detroit but had moved back to Canada. Trudy called his office and made an appointment.

Because she hadn't found a babysitter yet, Willis took time off to look after Andy on the afternoon of the appointment.

When Trudy returned to the house Willis looked at her expectantly.

"At least I have a name for my problem," she said, a note of resignation in her voice. She hung up her jacket and sat down opposite him. "It seems I'm suffering from postpartum depression."

"And what, pray tell, is postpartum depression?"

"It's a fancy word for the baby blues."

"What about the anxiety attacks?"

"Same thing. Seems they sometimes go hand in hand with the depression."

"And what did the doctor suggest?"

"For now he's given me a prescription." She took a slip of paper from her purse. "Something called Sodium Amytal. I'm to take it whenever I feel the anxiety coming on."

"What about a long-term solution?"

"The doctor says postpartum depression varies from patient to patient. It can last for just a month or for as long a year. He thinks my case was complicated by all the uncertainty about work—you being laid off and me with my stage fright."

"Are you going to see the doctor again?"

"Yes, he wants me back once a week so that he can monitor my condition over time."

"Did you like him?" Willis asked.

Trudy thought a moment before answering. "He seemed okay. Hard to judge from one appointment. He asked a lot of questions. I did most of the talking."

"What sort of questions?"

"Oh, everything from childhood illnesses to family history. He asked about our marriage . . . were we happy . . . what you do for a living. It was pretty wide-ranging. And he did a physical checkup—blood pressure, pulse, the usual stuff."

"If you're going to see him every week we'd better find someone to babysit Andy. I can't take that much time off work."

"I know, I know," said Trudy with a sigh. "It's just that I'm terrified at the prospect of someone else looking after Andy. I know that doesn't make much sense, but it's how I feel."

"I'll ask at work. Maybe someone has a daughter who could come in."

"No," Trudy said empathically. "I'm not going to leave Andy with a teenager."

"An older woman then?"

"Yes, someone with experience."

Eventually a neighbour across the street offered to help out. Mrs. Frey had three children of her own, but they were all in school, so she could easily spare a couple of hours a week if the appointments were in the morning. Trudy called the doctor and set up a schedule. She would see the psychiatrist on Tuesdays at ten.

"You'll get better," Willis assured his wife. "Other women must have gone through it."

"I wouldn't wish it on anyone else," said Trudy. "But you must be right. I'm not that unique."

Willis took Trudy's hands. "You are unique," he said, "but your problems aren't."

Tears suddenly filled Trudy's eyes. Willis took her in his arms. "We'll get through this, you'll see."

Willis decided not to tell Trudy about the plans for his trip to New York until she had settled in. It was the middle of the month when he broke the news. They were at the dining room table, finishing their coffee.

"Oh, Willis," said Trudy, clearly upset, "can't they send someone else?"

"I'm afraid not. I have the contacts, and the hotel reservations have already been made. It will only be for a few days."

"I don't want be left alone here in a new city with Andy." Her voice was shrill. "I hardly know anyone."

"Maybe Alice could drive up and stay with you," Willis suggested. "I'm sure she'd like the chance to visit. She could bring Joan along, and the cousins would get a chance to play together."

Trudy was relieved. "I'll call her tonight," she said.

It was agreed. Alice would arrive the day Willis left for New York and stay for a week. Trudy's mood improved once the arrangements were made. Alice was her favourite sister, and she began to look forward to the visit. Alice promised to bring pictures she'd received from Edith in Germany. Their sister had given birth to a boy named Roger in Breslau earlier that month.

Willis took the overnight train to New York. After checking in at the Algonquin Hotel, he stopped in at the Automat for a quick breakfast of bacon and eggs, then walked to the WOR studios on Broadway, two blocks from Times Square. He'd done his homework on the station's history. It had begun broadcasting from Newark, New Jersey, but in 1926 it opened studios in New York. A year later it carried the Columbia Broadcasting System's first programs. In 1929 WOR severed its connection with CBS and soon built a reputation as a popular independent station. In 1934 it founded its own network, the Mutual Broadcasting System, and added studios in the New Amsterdam Theatre. It broke new ground by developing the TransRadio wire service, which provided five fifteen-minute newscasts a day to its affiliated stations.

Willis had met representatives from Mutual when CKLW became its Detroit affiliate. During these meetings he had learned that the New York station had petitioned Washington for permission to expand its signal strength to 50,000 watts. That petition, with support from President Roosevelt, had been approved, and WOR had begun reaching a much wider audience in 1935 as one of a select group of stations in the United States licensed to operate at that signal strength. Because it had already been operating at that power for two years, it was the ideal site for Willis's research. The early technical problems that had been encountered during the conversion had been solved, and Willis was confident that the new CBC stations in Toronto and Verchères could benefit from an understanding of these difficulties and their solutions.

He was welcomed in the reception area by WOR's chief engineer, Ray Banks, who ushered him into a spacious office on the second floor of the building. A native New Yorker, he was a fast-talking, energetic man who radiated optimism.

"So our Canadian cousins have come to us to get their act together," said Banks. "It's about time."

Willis smiled. "We're always a little behind you Yanks. We let you make the big mistakes first. That way we avoid them."

"There's something to be said for that," said Banks, laughing good-naturedly at the jibe. "I'm pretty tied up today, but I'm going to introduce you to someone who can show you around. His name is Terry Wilson and he just graduated from Columbia. He probably understands a good deal more than I do about the technical side of broadcasting."

"As you know," said Willis, "our main interest is signal strength. Our two new stations in Canada are going to operate at 50,000 watts, like this one."

"Well, I'm sure our Columbia hotshot will have the answers, and then some." He picked up the telephone and pressed the intercom. "Gladys, has Terry shown up?" He paused to listen, looking across his desk at Willis. "Great, send him in."

In a moment a young man entered the office. Slight of build, he wore tan slacks, an open-necked shirt, and two-toned shoes. A cigarette hung from the corner of his mouth. Banks introduced Willis and the two men shook hands.

"Will you show our Canadian friend around?" the chief engineer asked.

"I'd be glad to," Terry said.

"Are you free for lunch?" Banks asked Willis.

"Sure."

"Where are you staying?"

"The Algonquin."

"Perfect. I'll have Gladys book us a table."

Willis spent the morning touring the WOR studios. His guide was a walking encyclopedia of information. Willis recognized the trait. He'd been that way after his time at Queen's. Much of what Wilson had to say was not new to Willis, but he let the man ramble on as he took in details of the station. He was struck by how new everything seemed, and impressed by the interior design of the studios. It was obvious that no expense had been spared.

"My primary interest," he explained when they had completed the tour, "is signal strength. I'd like to see the antennae you're using."

"For that we'll have to cross the Hudson," said Wilson. "Our main transmitter is in Newark." He looked at his watch. "But it's time for lunch, so I'll leave you with Ray and we can get together again this afternoon."

Willis and Banks had lunch in the Oak Room of the Algonquin. Banks told him this was a favourite spot for writers from the *New Yorker* magazine. He rattled off several names, but Willis hadn't heard of them.

His interest did perk up when Banks mentioned F. Scott Fitzgerald and Ernest Hemingway. Willis's taste ran more to non-fiction, but he had read *The Great Gatsby* and *The Sun Also Rises*.

Eventually they finished their lunch and walked back to the studios. Wilson was waiting for Willis in Banks's office.

"Now take care of our Canadian cousin," said Banks, "and make sure he stays out of trouble."

"I'll do my best," said Wilson with a smile.

On the street Wilson suggested they take the subway to New Jersey. "It's faster than fighting traffic," he explained. As the subway car passed under the East River he told Willis that the station's first programs had been broadcast from studios in Newark. "We maintain an office there. It services the main transmitter and the antenna. They're located just outside the town."

At the subway station in Newark Wilson hailed a cab and directed the driver to an address on Market Street.

"Wait here," he told the cabbie, then turned to Willis. "I'm just going inside to let the folks know we'll be checking out the transmitter."

Willis lit a cigarette. He could make out the skyline of Manhattan across the river.

Wilson soon returned and they drove out of the city. Soon a huge, towering antenna appeared before them and the taxi turned down a gravel road to a red brick building a few hundred yards from it.

For the remainder of the afternoon Willis toured the transmitter site with Wilson at his side. As he took notes he was thankful that his guide was willing to explain everything in detail. There was more to learn than he'd expected, and Wilson provided many tips that would prove valuable in the future.

When it was finally time to leave, Willis suggested a drink. Wilson called a cab and they returned to the Market Street offices, where he hurried inside to let them know they'd finished the tour. He was back moments later and they quickly adjourned to a nearby bar.

"Well, what did you think?" Wilson asked.

"You have one helluva station here," said Willis as he sipped his rye and ginger ale.

"Yes," said Wilson with obvious pride, "we do."

Willis spent three more days in New York. He went over blueprints and spec sheets. He noted the make and model numbers of certain pieces of equipment the CBC would need for its stations. A lot of this had already been ordered, but he knew there was still time for some of his recommendations to be acted upon.

He decided to take the day train back to Toronto and use the time on board to write a summary of his findings. He had learned a lot in New York, including details that could save the Corporation time and money in the construction phase of the new transmission facilities in Canada. And once the new stations in Toronto and Montreal were properly launched, his New York findings would help CBC technicians avoid the pitfalls WOR had encountered in its early days.

He had called Trudy each night. She seemed to be coping pretty well, thanks in a large part to Alice. He wondered if her condition was finally improving.

Chapter 52

Toronto and Montreal, 1937

When Willis opened the door to his home on Cranbrooke Avenue the first thing he heard was the sound of his one-year-old son crying. He put his Gladstone bag down in the hall and moved toward the sound. It suddenly stopped, and he heard Alice's soft voice.

"There, there, little fella. You were hungry, weren't you?" Willis stood at the door of the nursery and watched as his sister-in-law fed Andy his formula. Sensing his presence, she turned. "And look who's here!" she cooed. "It's your daddy back from New York."

"Where's Trudy?" Willis asked, keeping his voice low.

"She's napping," Alice whispered. "I'll finish here in a few minutes and put him down." She nodded to a second crib. "Joan's already asleep. Why don't you get yourself a beer from the icebox? We'll wake Trudy later and have dinner."

Willis uncapped an O'Keefe's Old Vienna and sat down at the kitchen table. He didn't bother with a glass, just drank from the bottle. The trip had tired him more than he realized. He took his beer into the living room and stretched out on the sofa. He pressed the cold bottle against his forehead. The moist condensation felt good, and he closed his eyes.

Trudy's voice roused him. "Willis, you're supposed to use a glass. You know I hate it when you drink from a bottle."

Willis sat up on the sofa. "That's some welcome home," he said. "I thought you were sleeping."

"Oh, and if I was I suppose that makes it okay to forget your manners." She crossed the room and handed him a glass.

"How've you been?" he asked, pouring what remained of his beer into the glass.

"I managed." Trudy gave him a perfunctory kiss on the cheek and then sat down on a chair across from him. "It was a good thing I had Alice here."

"How was Andy?"

"Good as gold. Even slept through the night."

"Guess he didn't miss me," said Willis, feigning disappointment.

"Maybe not, but I sure did. I hope this new job of yours isn't going to involve a lot of travel."

Willis knew he would eventually have to visit the Verchères site outside Montreal, but decided to keep this to himself for now. "I think I'll have my hands full here with the new transmitter in Hornby."

At this point Alice returned from the nursery. "They're both asleep," she said, smiling, and sat down beside Willis. "When would you like to eat?"

"Just as soon as I finish my beer. I'm famished."

At dinner Willis talked about his visit to New York, keeping the technical details to a minimum. But he couldn't keep the excitement from his voice when he described the WOR transmission tower in New Jersey. "It dwarfs anything I've seen."

"Taller than the Empire State building?" Alice asked.

"Oh, I doubt that. The antenna we're planning will be similar to the one in New York. It will certainly be higher than any of the buildings here in Toronto. And when we're finished, the signal strength will be ten times greater than what the CBC station has now."

They decided to move back to the living room for their coffee. When they were settled, Willis turned to Trudy.

"What did the psychiatrist say this week?"

"He said there were two options." Trudy glanced at her sister.

Alice picked up the account. "One is for Trudy to let some time pass, and use medication to deal with the anxiety. The doctor thinks that she should begin to feel better, now that you have work and are getting settled here in Toronto."

"And the other option?" Willis asked.

"It's more complicated," Alice continued. "Have you heard of psychoanalysis?"

"Isn't that something they came up with in Europe? I remember reading about it in a newspaper a few months ago."

"It's a treatment invented by Sigmund Freud, a psychiatrist from Austria.

It's catching on in the States. But it involves a long-term commitment. And several appointments a week."

"And what do you mean by long-term?"

"Hard to say. But it could be a year or two."

"Lord!" Willis looked at Trudy. "Appointments every week for that long?"

Trudy shook her head. "It's not the first option. Just something the doctor mentioned."

"I don't think psychoanalysis is the answer," said Alice. "And the psychiatrist didn't recommend it. He thinks the postpartum depression will play itself out."

"We can only hope," said Trudy, without much enthusiasm.

When Willis reached for Trudy in bed later than evening, she turned away. "Not tonight," she said. "Just hold me."

Willis did, and within a few minutes he was sleeping soundly.

The next morning after breakfast Alice bundled Joan into the back seat of her car. She hugged Trudy. "Hang in there, Sis," she said. "You'll get through this." She got behind the wheel and gave them a wave as she drove off. Willis and Trudy stood watching until the Buick turned north and was lost from view.

"I've got to get a move on," said Willis as they returned to the house. "The folks at the CBC will be expecting my report."

Construction at the Hornby site wasn't scheduled to begin for another two weeks, so there was still a chance for a last-minute review of the plans for the new transmitter. Willis soon found himself in a position of authority. He was the only in-house engineer in Toronto who had actually spent time at a 50,000-watt facility. The information he had gained would enable him to troubleshoot, even suggest shortcuts that would save construction time. For now, he was concentrating on the blueprints spread in front of him, with the Chief Construction and Design Engineer at his side.

The detailed plans were impressive. They called for the erection of an antenna 645 feet high, as well as a large building that would be constructed on the site to house the transmitter equipment and provide living quarters for the technicians, who would man the facility around the clock, seven days a week.

"I think we can safely say that the tower is the least of our worries,"

Harold Smith commented. "The company that's erecting it has a good deal of experience."

"I agree," said Willis, "but I am concerned about the underground wiring."

"I did some calculations while you were away," said the chief engineer. "Do you have any idea how many miles of wire we are dealing with?"

Willis offered a guess. "Must be thousands."

"More than ten thousand," answered Smith. "Probably closer to twenty."

Willis whistled in amazement.

"And every inch of every mile has to be treated carefully. They'll all be buried, of course, and that will offer some degree of protection. But whenever there's any interference we're going to have to dig down to find the source of the trouble."

"We'd better plan for that," said Willis, "and make sure we can monitor the system accurately. If there's a failure we'll want to be able to pinpoint its location precisely."

"My thoughts exactly. The last thing we want is to be digging in the dark."

"And if just one of those wires malfunctions we'll be in trouble," Willis added. "Dead air for our two million listeners."

"We do have a backup plan," said Smith. "In an emergency we can revert to the 5,000-watt signal we're currently using."

"I guess that will keep us on the air in Toronto," said Willis, "but our repeater stations will be out of luck."

Smith tapped the blueprints emphatically. "To make sure that never happens we've got to go over these plans with a fine-tooth comb."

It was a tedious task, and to make sure they didn't miss anything Willis would go over a portion of the plans first, then Harold would review his work. Occasionally, in order to alleviate the drudgery of this routine, they would reverse roles, with Willis checking Harold's work.

Ground was broken in June, and two temporary shacks were built, one for the construction workers and the other for CBC technicians. Willis and the chief engineer soon found they were spending most of their time at Hornby, supervising the installers as they began to lay the intricate network of wires in trenches that radiated from the centre to a wide circle extending to the far edges of the fifty-acre site.

From its base, a massive concrete slab, the tower gradually reached skyward. Watching the installers climbing the steel, Willis was reminded of the summer job he'd had with the hydro company servicing poles in Leamington. Heights hadn't bothered him then, but he wasn't sure he'd feel the same way now.

By the end of June the foundation had been poured for the transmitter building. It was to be a single-storey structure with an elaborate basement that would house the electronic equipment. On the main floor there would be offices and living quarters for the staff. As work progressed Willis and Harold fought a losing battle with the dust in their construction shack. Keeping the doors and windows closed helped, but the summer sun beating down turned their quarters into an oven, so when it was suggested in July that Willis visit the Verchères site outside Montreal he was happy to get away.

Trudy was less enthusiastic. "I don't see why you have to go," she complained. "Can't they send someone else?"

"No," Willis explained patiently, "they want someone who has seen a functioning fifty-thousand watt station. That was why I went to New York."

"How long will you be away?"

"Just two or three days."

"Well I can't very well ask Alice to come all the way from Detroit to help out."

"Isn't there someone else?" He wondered why Trudy couldn't manage on her own, but he knew she was still fragile and he didn't want to press the issue.

Trudy gave it some thought. "I suppose I could ask Bobbie Vale. We roomed together when I was studying at the Conservatory. She's living here now, but she works full-time and I hate to impose."

"Well, why not at least ask her? She might enjoy the chance to spend a couple of nights here." Willis smiled. "You two could relive some of those wild times you had together back in the roaring twenties."

As it turned out, Bobbie was receptive to the idea. She was two years younger than Trudy and had idolized her roommate when the two women were still in their teens.

Willis took the overnight train to Montreal and checked into the Windsor Hotel, where he had arranged to meet Alphonse Ouimet, the young engineer who would show him around the CBC's French-language operation, referred to as Radio-Canada.

"I haven't had breakfast," said Willis as the men stood talking in the hotel lobby. "Will you join me?"

"I'm afraid I've already eaten," said Ouimet, "but I can always use another cup of coffee."

In the hotel restaurant the Radio-Canada engineer gave Willis a progress report on the situation in Verchères. Most of it was familiar territory. Work on the tower was a little behind because of strong winds off the St. Lawrence River, but construction on the building was ahead of schedule. Willis was anxious to see it all for himself.

"Is this your first time in Montreal?" asked Ouimet.

"No," said Willis, "I was here after I graduated from Queen's. I worked at the Northern Electric plant in Ville St. Pierre for a year."

"Good," replied Ouimet. "I won't have to waste time showing you around the city. We can concentrate on Verchères."

They drove across the Victoria Bridge in Ouimet's 1935 Dodge sedan. As they headed along the south shore of the St. Lawrence River they exchanged details of their careers. Willis learned that Ouimet was a holdover from the Canadian Radio Broadcasting Commission and that he had been hired in 1934. He had graduated from McGill with a degree in electrical engineering. There was some good-natured banter about the Queen's-McGill rivalry, but when Willis learned that Ouimet had been involved in the construction of a television set in 1932, the two men clicked. Willis recounted his experience with General Electric in Schenectady, where he had first seen a practical demonstration of television. By the time they reached Verchères a friendship had taken root.

The two men spent the day at the site, talking to the engineers. Willis looked at the new transmitter building and recognized that it wouldn't be long before the Montreal staff would say goodbye to their cramped construction shack and move into their new offices.

"You're way ahead of us here," he noted. "I'm not sure why we're lagging behind in Hornby."

A Radio-Canada architect provided the answer. "The excavation for our foundation was a good deal easier," he explained. "At Hornby you're dealing with clay-based soil. Here, close to the river, there's more sand."

Ouimet drove Willis back to Montreal and the two men had dinner at Desjardins Restaurant. Since the menu was in French, Willis asked Ouimet to select for him. The young engineer chose coq-au-vin, and it didn't disappoint. After coffee and cognac they parted company, and Willis decided to walk back to his hotel rather than take a cab.

There had been no surprises for Willis in Verchères. Overall he was impressed with the skill and professionalism of the people he had encountered. He hadn't really known what he might find. There was a good deal of rivalry within the CBC, and some of his people in Toronto had been less than flattering in their assessment of the Radio-Canada engineers. Willis resolved to make sure this prejudice was removed. He understood that there would always be a turf war between the engineers, who wanted the best equipment, and the producers, who sought a bigger share of the budget for programs. But Willis felt it was important to create an alliance between the two engineering departments. It would give them more clout when it came time for budgetary allotments.

There were no messages waiting for him, and Willis chose to believe that Trudy was handling his absence without any difficulty. In the morning he boarded the train back to Toronto, once again using the travel time to prepare an up-to-date report, this time on his Radio-Canada findings.

Chapter 53

Toronto, 1937

Although there had been no messages from Trudy during Willis's stay in Montreal, his assumption that things had gone smoothly in Toronto proved false. When he arrived home Bobbie Dale met him at the door with Andy cradled in her arms.

"You're going to have to do something about Trudy," she said.

"Do something?" said Willis as he followed her into the living room.

"She's not herself." Bobbie shook her head sadly. "Not the fun-loving live wire I knew when we were rooming together."

"Where is she?"

"Sleeping," said Bobbie, then paused; "or else zonked out on those pills she's taking."

Willis felt a stab of fear. "But she's okay?"

"Yes . . . yes. I looked in a few minutes ago and her breathing is regular. But I'm worried."

Willis sat down with an exhausted sigh. "I'm worried too, Bobbie, but frankly I'm at a loss. I hoped she'd begin feeling better once we were settled in here, but that hasn't happened. You're right; she isn't her old self. Not even close."

"Well, I'm not a doctor, but I think there's something seriously wrong."

"She's seen a psychiatrist. He says it's just postpartum depression, made worse by the fact that we were out of work when Andy was born."

"Have you thought of getting a second opinion?"

"No," Willis admitted. "I thought the medication the doctor prescribed was going to help."

"I guess you could say it does," Bobbie said. "She certainly feels no pain when she's passed out."

"Passed out?" Willis wasn't sure he'd heard the words correctly.

"Yes, passed out," said Bobbie, exasperated. "We were eating dinner. Or at least I was. Trudy was just picking at her food. Then she suddenly got up from the table and rushed into the bathroom. I could hear her retching. I waited a few minutes and then knocked on the door to make sure she was all right."

"And?" Willis prompted.

"She said she was okay, and a few minutes later she came back to the table. When I asked her what had happened she said she had had an anxiety attack. I wasn't sure what that meant. She said the attacks come on suddenly, without warning. Apparently they terrify her; she feels as if something awful is about to happen. I tried to reason with her. I said she had nothing to be afraid of, that Andy was safely asleep and I was going to stay with her until you got back. She said that that was only part of the problem, and that the attacks happened more often when you are away."

"My work . . ." Willis started to explain.

"Oh, I know, it isn't your fault, you have to travel. But let me finish. As the evening wore on Trudy began slurring her words, as if she'd had too much to drink. And she hadn't touched a drop. That's when I realized it was the medication. I helped her up and into your bedroom. And she just passed out. I made her as comfortable as I could—took off her shoes and covered her with a blanket."

"Oh, Bobbie," said Willis, "I had no idea. Why didn't you call me?"

"I thought about it, but there wasn't much you could do. I decided I'd wait until morning and see how things were then." She glanced down at Andy, asleep in her arms. "Let me put this little guy to bed first and then I'll finish the story. By the way, I don't think Trudy should be using medicine to calm him down when he has trouble sleeping. It's one thing for her to use drugs, but they can't be good for him."

After Bobbie left the room, Willis lit a cigarette and leaned back on the sofa. He felt helpless. He was a practical man, used to solving problems with logic, but his wife's problems seemed to defy logic. Maybe Bobbie was right, maybe a second opinion was in order. And he agreed with her that it was time to wean Andy off his sleep medication. When she returned he looked up questioningly.

"I checked on Trudy as well, and they're both fine," she assured him. She sat down opposite Willis. "Where was I? Oh, yes. The next morning I fed Andy and changed his diapers. It was late when I finally heard Trudy stirring—about ten, I think. I knocked on the door and asked her if she needed anything. She said no, that she was going to take a shower. A few minutes later she came into the kitchen. She had a hangdog expression, said she was sorry for the trouble she'd caused. I told her it was no trouble, but that I was concerned. That's when she broke down and began to sob. Said her life wasn't worth living, that she wasn't much of a wife and wouldn't blame you if you left."

"I'd never leave . . ."

"Oh, I know, I know, Willis, and I tried to reassure her. I managed to get her to drink some orange juice. By then Andy was restless, so we got him into the carriage and went out for a walk. But I could see it was an effort for Trudy. When we got back she managed a few bites of a sandwich for lunch and then we put Andy down for a nap. She said she thought she could use a nap as well. This surprised me. I never knew Trudy to nap."

"That's true," Willis said. "She never likes to sleep during the day."

"Well, this time she did, and when I went into the bathroom I saw the prescription bottle open and realized she'd taken more medication. I counted the pills that were left, because I was worried . . ."

Willis felt a stab of anxiety. "And . . . ?"

"And I figured that what she'd taken hadn't been enough to pose a risk."

Willis wondered if the anxiety he was feeling was what Trudy experienced. But there was a reason for his anxiety; it was normal under the circumstances. What if he were to feel this way for no reason? That was a frightening thought. He began to understand for the first time the true nature of his wife's illness. A second opinion was definitely required. There had to be an answer.

A friend of Willis's at the CBC provided the name of another doctor. This time Willis accompanied Trudy to the consultation. He wasn't present for the physical examination, but he did take part in the discussion that followed. The second doctor confirmed the earlier diagnosis of postpartum depression, but found the anxiety attacks harder to fathom. He thought that the use of Sodium Amytal, carefully monitored, was an appropriate short-term solution.

"How long does postpartum depression usually last?" Willis asked.

"Hard to say," the doctor replied, "but your wife's case is unusual. It's been more than a year now since your son's birth. That's a long time for the condition to persist, particularly since you now have permanent employment."

"I seemed to be getting a little better," said Trudy, "but every time Willis has to travel I'm thrown into a tale-spin."

"Does your work involve much travel?" the doctor asked Willis.

"Not a lot," said Willis. "A few trips a year. Hard to predict, really."

"And that's when the anxiety attacks are at their worst?" the doctor asked Trudy.

"Yes. They still occur from time to time, even when Willis isn't away, but they're . . ." She searched for the word. "They're manageable. I simply take my prescription."

The doctor steepled his fingers. "These attacks are what trouble me the most," he said. "I suspect that the depression will lift in time, but you might want to consider another form of treatment for the anxiety."

"What other treatment?" Trudy asked.

"Psychoanalysis."

Trudy looked at Willis, and he shook his head.

The doctor picked up on the exchange. "Not for the faint of heart," he admitted. "It's a long, drawn-out process. And it's expensive."

The following Saturday Willis made a trip to the public library and was directed to the medical reference section. He spent the afternoon there, researching psychoanalysis. What troubled him most was the lack of hard scientific evidence supporting the practice. There were accounts of cures, but no one had managed a controlled study of the procedure. Ever the engineer, Willis was skeptical. And there was the cost. Willis was still helping support his parents. There was simply no way he could afford the long-term treatment psychoanalysis required.

At home, Andy was now walking and beginning to talk. Willis was enchanted, and Trudy's earlier indifference to the baby had given way to genuine affection. In fact there were times when Willis wondered if Trudy's constant mothering of the boy was a little extreme. But he kept these thoughts to himself, because Andy was now proving to be a tonic for his wife.

Willis stayed home one Saturday morning in April while Trudy took Andy to the local barbershop. Until now she had trimmed his curly locks herself, but with his first birthday approaching she decided it was time for his first real haircut.

"Willis," she called from the front door when she returned.

"What is it?"

"Come see your son."

Andy stood in the hall, smiling, a lollipop in his mouth.

"How did it go?"

"He was good as gold. Didn't cry or fuss."

"I guess he's no longer a baby," said Willis. "And the barber did a great job."

"That's not all," Trudy added excitedly. "He said our son has a double crown, something very rare."

"A double crown?"

"Most people have a single crown, where the hair grows in a circular pattern from the back of the head. But occasionally the barber does see a double crown. It's supposed to be a sign of superior intelligence."

Willis bent down to examine his son. "You're right," he said. "I can see the double swirl."

"I knew he was bright," Trudy concluded, "and this just confirms it."

Willis laughed. Privately, he didn't agree that the double crown was anything more than an oddity. But then he thought that it might distract her attention from the tiny crinkle in Andy's left ear. It hadn't filled out with time, as Trudy had hoped it would, and Willis knew it still troubled her, although he didn't share her concern.

There were no plans for Willis to travel in the immediate future, and he assured Trudy that with construction at Hornby in its final stages it was unlikely he'd be called away. Construction at the CBC site was behind schedule. They had hoped to have the new station on the air by September, but it was clear that this deadline would not be met. Even the new target, the beginning of October, proved unrealistic. Some of the equipment they had ordered had not been delivered on time, and there were unforeseen construction delays. But the cool fall weather did offer a respite from the oppressive summer heat in the cramped quarters of the engineers' shack that Willis shared with Harold Smith.

Willis was gradually gaining insight into the workings of the CBC. There were so many competing factions that it was often hard to tell who was winning. At the top were the executives at head office in Ottawa. They were locked in a perennial struggle with the federal government for more funding. There was also the English-French rivalry within the Corporation for the money that was available. The ruling Liberal party, under Prime Minister Mackenzie King, needed the support of the French-speaking voters in Quebec. If he were to be seen to favour the English-language branch of the CBC he risked alienating Quebec. Willis had never been actively involved in politics, and he was thankful that it wasn't a requirement for his work. His goal was simple, to make sure that the quality of the transmission signal from the Hornby site was on a par with the world's best. Once that was accomplished he had no doubt that CBC stations outside Montreal and Toronto would demand more powerful transmitters of their own.

The original press release from the CBC head office in Ottawa proudly proclaimed that the network would expand its programming from six hours a day to sixteen hours a day once the Hornby facilities were finished. However, this announcement had been made without prior consultation with the engineers. Harold Smith was quick to point out the perils of such a sudden shift, and it was scaled back. The general manager of the CBC, Gladstone Murray, quickly issued a second release that outlined a more modest plan. The goal would remain sixteen hours of daily programming, but it would be phased in over a number of weeks, an hour at a time.

In October the engineering staff was at last able to move out of the construction shack and into the almost-completed transmitter building. Technicians were still working on the wiring, but the offices were finished. On his first night in the new facility Willis stayed late, organizing his papers. It was dark when he left with Harold Smith. From the parking lot the two men looked back at the transmitter. The outside wall of the control room on the ground floor was built of translucent glass blocks, and in the darkness the interior lights bathed the main entrance in a soft, diffuse glow.

"Impressive," said Smith.

"It is," agreed Willis. He lit a cigarette. "Could be the set for a science-fiction movie." In the faint glow he could make out the ghostly structure of the tower looming into the night sky. To Willis, it was just about the most beautiful sight imaginable.

"See you tomorrow?" said Smith as he opened the door to his car.

"Tomorrow it is." Willis turned and walked in the direction of his own car.

New dates were set for the conversion to the higher frequencies in

Montreal and Toronto. For the first time, Willis was confident that these deadlines would be met. The Quebec transmitter was scheduled to begin broadcasting at fifty thousand watts on December 11. The Toronto station would follow, two weeks later, from Hornby. Its first program would begin at ten o'clock on Christmas morning. The federal Minister of Transport, C.D. Howe, would be on hand for both occasions, while the Acting Prime Minister, Ernest Lapointe, would join him in Verchères for the Quebec launch.

There were some grumblings within the CBC staff in Toronto about the French network being on the air first, but Willis wasn't among the complainers. With identical equipment at both sites, he reasoned that any technical problems encountered in the Montreal launch could be corrected at Hornby before the Toronto station signed on.

It was a hectic time for Willis, Smith, and the others. They continued to put in long hours to meet the new deadline. Trudy complained that Willis was never home in the evening. Relations between them came to a head at the beginning of December.

"I'm afraid our Christmas celebrations will have to be put on hold," said Willis.

"What do you mean?"

"I'm going to be working Christmas Day."

"Oh, Willis, you can't be serious."

"'Fraid so." Willis shrugged his shoulders apologetically.

"Can't they get someone else?" she asked. "Someone without a family?"

"No, we'll all be working that day."

"Last Christmas was bad enough, with you out of work," Trudy huffed. "Not only that but we were living with your folks. This was going to be Andy's first Christmas in a place of our own."

Willis raised his palms in a gesture of helplessness. "I'm sorry," he said.

"Well, I'm sorry too!" Trudy was shouting now. "Sorry you're putting the damn CBC ahead of your own family!" She stormed out of the living room, sobbing, and hurried down the hall into their bedroom. When the door slammed it woke Andy and he began to cry.

Willis went into the nursery and lifted his son from his crib. "There, there," he whispered. "Settle down, little fella, there's nothing for you to be crying about."

Andy buried his head in Willis's shoulder and began sucking his thumb. Willis walked him back and forth a few times in the living room until he was sure the boy was asleep again, then returned him to the crib. In the kitchen he made himself a ham sandwich, uncapped a beer, and went back into the living room. He switched on the radio, sat down on the sofa, and began to eat. He realized he'd once again forgotten to bring a glass for his beer. Ah, what the hell, he thought, and sipped directly from the bottle.

Chapter 54

Toronto and Leamington, 1937–38

By the end of first week in December all the essential work on the Hornby transmitter had been completed. Now it was time for testing. As the team worked its way through the procedures, Willis noticed with pride that much of the equipment they were using had been manufactured by Northern Electric. It brought back fond memories of his time in Montreal.

As predicted, some technical difficulties were encountered when Montreal signed on. This enabled the Toronto engineers to take steps to ensure they wouldn't appear again on Christmas morning, when the Hornby launch was scheduled.

On the big day Willis rose early, while it was still dark, and drove out to Hornby. There had been a light sprinkling of snow overnight, and Willis knew that Trudy would be pleased about that. They'd have a white Christmas. She had calmed down when he had assured her that he would be home early, in plenty of time for dinner. They had bought a tree from a man selling them in the parking lot of a service station. She had been busy all week decorating the tree, hanging mistletoe, baking cookies, and wrapping gifts.

At the Hornby site Willis watched as the dignitaries assembled. He'd met some of them, including Gladstone Murray, his Quebec deputy August Frigon, and Ernie Bushnell, who was in charge of English programming. Some members of the CBC's board of directors were on hand—the chairman, Leonard Brockington, the feminist Nellie McClung, from Victoria, Colonel Wilfrid Bovey from McGill, and Allan Plaunt, who was widely acknowledged to be the driving force behind the creation of the CBC. These were names Willis had heard often, and now, at last, he could put faces to them. The highest-ranking federal politician present was C.D.

Howe, and he was joined by the local member of parliament, the mayor of Hornby, and a host of other lesser luminaries.

In the main control room, which was crowded with people, Howe approached a bank of glowing dials. Promptly at ten o'clock he pressed a switch, and the sound of the Toronto Symphony Orchestra playing "God Save the King" flooded the room. Everyone stood. When the music ended there was an enthusiastic cheer, and congratulations were extended all around. The new station was on the air.

Willis and Harold Smith sighed with relief. "I guess all those long hours we put in paid off," said Smith with satisfaction.

Willis scanned the smiling faces. "Every second was worth it."

A reception followed at which Ernie Bushnell introduced Willis to C.D. Howe.

"Well done, young man," the politician intoned, and shook his hand.

After a few more introductions and the lapse of a suitable period of time, Willis quietly slipped away. He tuned the car radio to the CBC and felt a surge of accomplishment as the music played.

It was almost noon by the time he crossed the Toronto city limits. He'd soon be home with Trudy and Andy, in plenty of time to open the presents and enjoy a Christmas dinner.

When Willis opened the door he was overcome by the aroma of roasting chicken. Andy toddled to greet him, and he swept his young son up in his arms. "Merry Christmas," he said, squeezing him gently.

Trudy appeared from the kitchen, smiling. "You kept your promise." She looked at her watch. "I didn't expect you this soon. How did it go?"

Willis put Andy down and kissed Trudy lightly on the lips. "Like clockwork," he said. "All the bigwigs were there. I even got a chance to meet C.D. Howe."

"My," Trudy teased, "hobnobbing with the elite! Are you still talking to us common folk?"

"Not only talking," said Willis with a wink as he caressed his wife's behind.

"Willis," she said in mock outrage, "not in front of your son!"

"I don't think he noticed," Willis said. "He seems preoccupied with something else." They looked down and saw Andy tugging away at the wrapping of a Christmas present he'd managed to drag from beneath the tree.

The rest of the day unfolded better than Willis had dared to hope.

Trudy seemed decidedly upbeat, thrilled with the gifts Willis had chosen for her. Andy watched as they took down the stockings that hung from the mantelpiece. "Let's see what Santa brought you," Trudy exclaimed excitedly.

Willis watched as his wife helped his young son unwrap his gifts. He was sure Andy was still too young to appreciate the significance of the day, but it was clear that Trudy was enjoying every minute. By mid-afternoon they'd put their son down for a nap, and Willis gave Trudy a full account of the launch.

She was particularly interested in his impression of Nellie McClung, the only woman on the CBC's board of directors.

"She's quite the character," Willis said; "very outspoken."

"Did you know she wants the United Church to allow women to become ministers?"

"No, I didn't" Willis admitted. "I bet that wouldn't go over very well with your father."

Trudy laughed. "Oh, I'd love to see it. He thinks women belong in the home, looking after the children."

As the afternoon wore on Trudy busied herself in the kitchen while Willis relaxed with an eggnog, liberally laced with rum, and listened to Christmas carols on the CBC. Andy woke from his nap as Trudy was serving dinner, and Willis strapped him into his high chair.

"I wonder what he makes of all this," mused Trudy.

"Given his age he probably won't remember it," said Willis thoughtfully, "but he certainly had fun with all that wrapping paper."

"And he seems to like the mashed potatoes," said Trudy as she spooned another mouthful into her son's open mouth.

When Willis reached across the bed for Trudy that evening she didn't shy away from his touch. They made gentle love, and for the first time in more than a year Willis sensed that his wife was enjoying it. He hoped this was an indication that she was actually getting better.

On the last day of 1937 Trudy and Willis drove to Leamington, with Andy in the back seat, to celebrate the arrival of the New Year with Anna Maria and Clyde. It had been eight months since Anna Maria had last seen Andy.

"Goodness gracious," she said, "how the lad has grown!"

Andy hid his face in the folds of Trudy's coat.

"He's a little shy," said Trudy. "Give him some time to get used to things."

Willis crossed the room to his father, who sat in his accustomed place in his beloved rocking chair. He reached down and placed his hand on Clyde's shoulder. "How've you been, Dad?"

"Oh, I have my good days and my bad days," said Clyde as he began filling his pipe with tobacco from a humidor. "We listened to your Christmas Day broadcast from Toronto, and it was clear as a bell. You must be very proud."

"Well, it took a team to bring it off."

"Willis," said Trudy in mock exasperation, "you're too modest." She looked at Anna Maria. "He was one of the leaders of that team. And he put in very long hours."

"He always was a hard worker," Clyde commented.

By now Andy was exploring the room.

"Come here," said Willis. He opened his arms and his son toddled into them. "Give me your hand, Dad." He took Clyde's hand and placed the palm on Andy's head.

"Well, now, your grandmother was right. You are a big boy."

Andy looked at his sightless grandfather with open curiosity. He reached up and took Clyde's pipe. Willis rescued it, but before the child had time to protest he lifted him off the floor and placed him in his grandmother's waiting arms. She gave him a gentle hug and began singing quietly in Swedish, moving in a slow waltz around the room.

The Littles' welcome to the New Year had become a tradition—listening to the live broadcast of Guy Lombardo and the Royal Canadians from the Roosevelt Grill in New York City. Their radio was tuned to a Detroit station.

"It's hard to believe the CBC isn't carrying this," Willis complained. "After all, Lombardo is a Canadian."

"Do you know how the band got its name?" Trudy asked.

"No," said Willis.

"Their agent originally wanted the musicians to dress as Royal Canadian Mounted Police, complete with red tunics. Lombardo refused, but he did agree that they needed a way of distinguishing themselves from other orchestras in the States. He was proud of his Canadian heritage, so that's the name he chose."

"How many Canadians are in the band?" Willis asked.

"In the beginning they were all from Canada," said Trudy. "The

Lombardo brothers grew up in London, Ontario. But they've been in the States so long I'm sure they have a good many American musicians."

"At least their leader is still a Canadian," Willis countered.

"Yes," agreed Trudy with an edge to her voice. "Too bad Canada doesn't value its musicians the way the Americans do. We have to make it in the States before anyone in Canada takes notice."

A musical number finished and an announcer, speaking from Times Square in New York, began what had become an annual tradition, the countdown to the New Year.

"Three . . . two . . . one . . .," he intoned. A rousing cheer went up and the Royal Canadians broke into their signature tune, "Auld Lang Syne."

"Happy New Year," said Willis, raising a glass of eggnog. The others echoed his words and sipped their drinks.

"Mmm," Clyde murmured with obvious relish. Because of his diabetes, this was the only time each year he was permitted alcohol.

New Year's Day of 1938 fell on a Saturday, and since they were in the area for the weekend, Trudy decided it was time for her parents to see their new grandson. Essex, where Blanche and J.B. had settled, was a half-hour drive from Leamington. Blanche had written Trudy, begging her to come. As much as Trudy hated the idea having to deal with her father, she didn't want to deprive her mother of the opportunity to see Andy. She left after lunch on Saturday, leaving Willis behind with his folks. He knew that his father-in-law still deemed him an unsuitable match for Trudy, and didn't want to needlessly complicate things.

When she returned to Leamington she reported that, much to her relief, J.B. hadn't been present for her visit. "Mom wouldn't admit it, but I think she'd banished the old boy."

Willis smiled at the idea.

"Andy was good as gold," said Trudy with pride. "Didn't cry or fuss at all. And of course Blanche was delighted to finally see him."

It was clear to Willis that, as disappointed as Trudy was with her parents, she still felt the need to stay connected to them. This visit had been an important first step, and he was pleased it had gone well.

Later that evening, after Trudy had settled Andy and had gone off to bed herself, Anna Maria commented to Willis that her daughter-in-law's spirits seemed to have improved.

"Some of the old life is coming back," she suggested.

"Oh, I hope so. She's been through a lot."

"What does her doctor say?"

"It takes time. No two cases are the same." Willis decided against mentioning psychoanalysis. It was hard enough for him to understand, and he was sure it would only confuse Anna Maria.

On Sunday morning, after breakfast, they left for Toronto. As they drove off, Willis glanced in the rear-view mirror and caught sight of his mother waving goodbye. He acknowledged her wave with a double beep of his horn.

Willis would later characterize 1938 as the year when he was able to spend more time at home. The long hours at Hornby were a thing of the past. Most of the network development was happening away from Toronto. Willis did put in time refining plans for new, more powerful transmitters in the Maritime and Prairie provinces, but whenever travel was involved he steered the work to others. One unexpected result of this decision was a promotion. Early in the New Year he was named the CBC's Chief Engineer for the province of Ontario.

At home, Trudy was still subject to anxiety attacks, but the depression had eased. With Willis travelling less and working regular hours, their lives had taken on a degree of predictability.

His promotion meant an increase in salary, and Willis was finally able to fulfill a promise he'd made to himself in high school. He had learned rudimentary woodworking skills in the Industrial Arts class, and he liked to build things, so he had resolved that someday he would have his own workshop. There had been no opportunity in Windsor, because they lived in an apartment, but their Toronto home had a basement, so it wasn't long before Willis began acquiring power tools.

"You're spending too much on those . . . whatever they are," Trudy complained one evening as he assembled a new jig saw.

"Just wait," Willis replied. "Once I get properly set up I'll be able to build us all kinds of things. You'll see; we'll save money in the long run. Andy will soon be out of his crib, and I intend to make him a proper bed."

Trudy held her tongue. At least she had Willis home evenings, even if he was in the basement tinkering with what she'd come to think of as his latest toys. But the sounds of the various power tools frightened her, and she worried he might injure himself.

Andy's bed was his first project, and even Trudy had to admit that it was impressive, as good as anything available in stores. Then came a coffee table for their living room. Willis had decided to keep this project a secret. The carefully selected wood was the finest-grained maple available. He worked slowly, fashioning a delicate pattern of parallel grooves around the edges of the tabletop. The legs were glued into place and secured by clamps. When the glue dried, Willis sanded the table down, then applied a weak bleach solution to tone down the colour of the maple. It was sealed with several coats of wax.

One evening after Trudy had gone to bed he brought the finished coffee table upstairs. In the morning he feigned sleep, making sure Trudy rose before he did.

There was a shriek of delight from the living room.

"Willis," she shouted excitedly, "it's gorgeous!"

When he joined her, she was running her hands over the smooth surface.

"You like it?"

"Like it? I love it! And the colour is so subtle. Amazing!" She turned and embraced him enthusiastically. "You're a genius."

"Do I get a reward?" he asked mischievously.

"Tonight," she promised with a knowing smile.

Trudy never again complained about the money Willis spent to outfit his workshop. She even began to take a certain comfort from the sound of the power tools. The noise meant that her husband was at home and not working overtime. Occasionally she would bring Andy down to watch Willis. At first the noise frightened the child, but over time he became accustomed to it. He watched with fascination as the shavings and sawdust flew from his father's creations.

Trudy noticed that whenever Willis was concentrating on a woodworking project, he would bite down gently on his tongue, keeping its tip barely visible between his lips at one corner of his mouth.

"Why do you do that?" she asked him one evening in the basement.

"Do what?" Willis looked up, puzzled.

"You bite down on your tongue when the work is challenging."

"I do?"

"Yes."

"I wasn't aware I was doing that. Does it bother you?"

"Not really, I guess. I just think it looks a little silly, that's all."

"Well, if it helps me concentrate I can't see the harm."

"Do you do it at work?"

"I don't know." Willis shook his head. "No one's ever mentioned it."

"Well, I hope you don't," Trudy continued. "It's something a kid might do, but not a grown man."

"I'll bear that in mind," said Willis, a little testily.

Chapter 55

Toronto, 1938

At the CBC the hours of daily programming, most of it consisting of musical shows, steadily increased. A live noontime production, a mixture of comedy and song called *The Happy Gang*, was proving especially popular. Each program began with the sound of a knock at a door. An announcer called out, "Who's there?" and a chorus of voices replied, "It's the Happy Gang!" "Well, come o-o-o-o-o-n in!" called the announcer, and then the theme music followed. Another hit was *Treasure Trail*, whose listeners could write in and win prizes.

In an attempt to achieve a wider audience in the evenings, the CBC increased the number of shows imported from the United States. *Kraft Music Hall*, hosted by Bing Crosby, was Trudy's favourite, and for comedy the CBC offered Jack Benny, Fibber McGee and Molly, and Edgar Bergen and Charlie McCarthy. *Lux Radio Theatre*, an hour-long program that featured adaptations of popular movies, was broadcast each week. Nationalists complained, but the advertising revenue that these American programs generated was proving essential.

Farm broadcasts were introduced. They were made up of a mixture of agricultural information, weather reports, and features on interesting people from farming communities across Canada.

Mary Grannon, a schoolteacher from Fredericton who wrote stories for children, had become a fixture on the New Brunswick station, reading her own works. In 1938 she was brought to Toronto and given network exposure. Her program, *Just Mary*, was one of the first CBC shows designed for children. Parents across the country were delighted.

When Willis came home with the news that he'd met Mary Grannon, Trudy couldn't contain her excitement.

"What's she like?"

"Very sweet," said Willis. "Down-to-earth. I mentioned that we had a young son and we were looking forward to the day when he could enjoy her program."

"It won't be long," Trudy promised. "I'm going to buy one of her books. We should start reading to Andy at bedtime."

"Isn't it a little early for that?" Willis asked. "He won't understand anything."

"Oh, I know, but I think it's important that we read to him anyway. Even if he can't follow the stories, he'll get used to the idea. And I'm sure he'll be lulled into a peaceful sleep listening the sound of our voices."

"I guess it can't hurt," Willis acknowledged.

Reading to Andy soon became an evening ritual. Trudy still sang lullabies to him, but she began to add bedtime stories. Her son seemed as content listening to her spoken words as he was to her singing. Willis and Trudy took turns.

One night Andy interrupted his father's story to say, "Song, Daddy."

"What?" His son's words were limited and Willis wasn't sure he'd heard correctly.

"Song?" It seemed to be a question this time.

Willis understood. "Song?" He smiled down at his son. He placed his finger on his own chest. "Song . . .," he said, then shook his head and said, "Daddy, no." Then he pointed in the direction of the living room, where Trudy was listening to the radio. "Song . . .," he repeated, then, nodding, said, "Mommy, yes."

Andy looked up at his father. Willis wondered if his shortcoming would disappoint his son. After a moment, Andy pointed to the book in Willis's hands. The signal was clear. He understood. Daddy didn't sing, but he did read. Willis continued with the story. Although Andy was alert at the beginning, his eyelids began to flutter, then closed as Willis read on. When he was asleep, Willis closed the book, but he remained by the bedside, enchanted by the look of his son dreaming peacefully.

While the year 1938 saw the restoration of a degree of domestic harmony to Willis and his young family, forces outside his home and his work were anything but harmonious. Hitler was menacing Europe, a civil war raged in Spain, and Japanese troops had invaded China. Add to this the sobering fact that recovery from the Great Depression seemed to be stalled, and you had what Willis characterized as a "decidedly uncomfortable" situation.

He read that, as a consequence of Hitler's oppression of the Jews, Sigmund Freud had fled his home in Vienna for sanctuary in London. And in Italy, psychiatrists had begun using something called electroshock therapy on schizophrenics. Although Willis thought that science would eventually produce a cure for mental illness, he found this idea barbaric.

There was some positive news. In London, John Baird demonstrated colour television for the first time, while in the United States General Electric introduced the fluorescent light bulb. The American millionaire Howard Hughes flew his twin-engine Lockheed airplane around the world in three days and nineteen hours, a record.

At home, Trudy and Willis took Andy to see Walt Disney's first full-length animated film, *Snow White and the Seven Dwarfs*. Their son soon fell asleep, but Trudy was enchanted. As they walked to the car after the show, Willis asked Trudy if one of the songs had reminded her of anything.

"Which song?" Trudy asked, genuinely curious.

"'Some Day My Prince Will Come.'"

Trudy snickered. "Funny, but I don't think of you as a prince."

"You did once," Willis teased. "I seem to recall you referring to me as Sir Willis Little Lancelot."

Trudy was puzzled for a moment, and then it came to her. Their conversation in Windsor, when she'd just moved there from Wheatley. She'd been keeping her relationship with Willis a secret from her parents.

"Lancelot wasn't a prince," she said, "he was a knight. You're still my knight in shining armour."

"I guess I can live with that," said Willis with a laugh.

There were rumours floating about the CBC that King George VI and Queen Elizabeth might be visiting Canada. Willis knew that most Canadians outside Quebec were fiercely loyal to Britain and devoted fans of the monarchy. He didn't share this view; he was largely indifferent to British royalty. But he realized that if the King and Queen did come to Canada the CBC would cover every minute of their tour. When Edward VIII, the previous king, had abdicated in 1936 to marry the American divorcee Wallis Simpson, the fledgling network had carried his speech, and millions of Canadians had listened in. In 1937 the CBC had broadcast the coronation of his successor, George VI. That coverage lasted seventeen consecutive hours, most of it originating with the BBC.

The summer of 1939 was considered to be the most likely date for the royal tour. Willis realized that if this plan materialized it would mean that his own nine-to-five existence would suffer a major interruption, but he kept this information to himself, not wanting to worry Trudy.

English programming on the CBC steadily increased as the year wore on, but at Radio-Canada things really took off. With the increased signal strength from Verchères, the French-language network now reached virtually every home in Quebec. Soon there was an explosion of talent among French Canadians, and music, drama, comedies, soap operas, and educational programs flooded the airwaves. Willis had remained friends with the Montreal engineer Alphonse Ouimet, who kept him informed about the French network's phenomenal growth as he climbed its ranks. There was an English-language CBC station in Montreal, but its programming was dwarfed by the city's Radio-Canada counterpart.

One evening in December Willis had worked late, and light snow was falling as he left the CBC offices. Maybe they'd have the white Christmas Trudy loved. He hoped so. He took the streetcar to his Cranbrooke Avenue stop, got off, and began walking the two blocks west to his home. As he approached the house he noticed a figure huddled on the front steps, illuminated only by the faint glow of a streetlight. He wondered who would be sitting there at this hour. As he drew closer he realized it was too small to be an adult, and he quickened his step. When he recognized his young son he ran up to him. Andy looked up at his father, tears streaming down his face.

"I'm sorry, Daddy," he sobbed.

"Sorry?" Willis was confused.

"I had to go, I couldn't wait."

Willis scooped up his son and realized the boy had wet himself. "That's all right son, it's not your fault." Anger replaced pity as he climbed the stairs to the porch and untethered Andy from the rope that was attached to the small harness he wore over his snowsuit. He could feel his son shivering. "We'll get you inside and warmed up, out of those wet clothes."

He tried the front door and found it locked. Fear pushed aside his anger as he fumbled awkwardly for his keys. He managed to get the door open and stepped into the front hall. There were no lights on, so he had to feel his way into the living room. He found a switch and, still carrying Andy, hurried down the hall to the main bedroom. He could just make out the form of Trudy huddled on the bed. He moved quickly to her side. As he did, she sighed and turned to the wall. Willis felt relief surge through him. He had feared the worst.

Willis backed out of the room and removed the wet snowsuit from his son, who was still shivering. He wrapped him in a towel and carried him into the bathroom. He turned on the hot water and rubbed Andy vigorously as he waited for the tub to fill.

"What happened?" he asked.

His son's halting words soon made it clear what had happened. Trudy had put Andy outside that afternoon. This in itself was not unusual. Safely tethered, he was free to play in the front yard, and could come back inside whenever he grew tired or cold. But this time something had gone wrong. When he had pushed on the front door it wouldn't open. He banged on it with his tiny fists and called for his mother. He could not reach the doorbell. After some time had passed and no one came, he decided he would wait for his father, who usually returned before supper. But then it grew dark. Snow began to fall, and still his father failed to appear. He began to panic. In search of shelter he crawled under the front porch. A couple of times he called out as people passed, but they didn't hear him. Soon he had to relieve himself, but his snowsuit made this impossible. He finally let go, and felt the brief warmth it provided, but he also felt ashamed, because he was too old to wet himself. He eventually crawled out from his hiding place and banged again on the front door. When no one answered, he sat back down on the steps. That was when Willis finally appeared.

Safe in the warm bath, Andy was soon playing happily with a toy duck, his troubles apparently forgotten. Willis looked in on Trudy again and found her sleeping soundly. He returned to the bathroom, washed his son thoroughly, then lifted him from the water and dried him off. Once Andy was in his pyjamas Willis took him into the kitchen and sat him down in his high chair. He warmed up the boy's favourite food and then played a game with each spoonful.

"Open wide," Willis would say. "Plane coming in."

Andy obediently complied and soon finished his super. Willis made himself a sandwich. After that, it was time for the bedtime ritual. He tucked Andy in and began to read one of Mary Grannon's stories. Gradually his son's eyes closed.

Willis stayed by Andy's side as he reviewed the situation. Trudy was no drinker, so it seemed clear that she had overdosed on Sodium Amytal. Had it been deliberate? Willis didn't think so, because there had been no note. She must have had one of her anxiety attacks and simply taken too much medication. When he rose to leave the room, Andy's voice stopped him.

"Daddy, sleep with me tonight?"

"Sure," said Willis, and knew at that instant what he had to do. "I'll change into my pyjamas and be back in a minute." He went into the bedroom, quietly undressed, and put on his pyjamas. In the bathroom he washed himself and brushed his teeth. Back with his son, he slipped beneath the covers. Fortunately he had recently built a full-sized bed for Andy.

Soon he was cradling his son's warm body close to his own.

"Love you, Daddy," Andy murmured.

"I love you, too," Willis whispered.

In the morning Trudy awoke first. When she realized that she was alone in bed, she panicked. Where was Willis? She tried to reconstruct the previous day's events, but the last thing she could recall was getting Andy ready for his playtime outside. When this thought hit her she sat bolt upright in bed, panic racing through her. She struggled to her feet, head pounding and nausea gripping her stomach. Panting, she padded to her son's room. When she saw Willis and Andy curled up together, sleeping soundly, relief replaced her fear. It was not long, however, before shame took over, and Trudy retreated to the bathroom, where she threw up again and again, her head bent over the toilet.

The sounds of her dry retching roused Willis. He got up and stood in the open bathroom door, silently observing his wife. When she finally finished, she rose unsteadily to the sink.

"Are you all right?" Willis asked, his voice tight.

Trudy was startled. "Oh, my god," she rasped, "I didn't see you." She turned on the cold water and splashed it on her face.

Willis remained silent.

Trudy sipped some water and rinsed out her mouth. "I don't know what happened," she said, and began to sob.

Willis took her in his arms and held her until the crying stopped. Then he led her into the kitchen.

"Tell me Andy is all right," she implored.

"He's okay now," said Willis, "but he was pretty terrified by the time I got home." Trying to keep the anger from his voice he explained what had happened.

"I can't believe it," said Trudy. "I just can't believe it," she repeated, and began to cry again.

Willis waited patiently until Trudy regained her composure.

"Will you ever be able to forgive me?" she whispered, her cheeks still wet with tears.

"It isn't a question of forgiving you," he said. "We've just got to make sure it never happens again."

Chapter 56

Toronto, 1938–39

The woman's name was Mrs. Fairchild. She was in her fifties, a widow with two grown children. She was the third candidate Trudy had interviewed, and there was something reassuring about her. She was plump, had a twinkle in her eye, and showed none of the nervousness that she saw in the other women.

"So you'd be available as a full-time housekeeper, beginning after Christmas?"

"I would," Mrs. Fairchild replied.

"Well," said Trudy, "I'd like you to meet my husband before we agree on anything."

"Of course, and your young son."

"Right. He's napping just now and I'd rather not wake him."

"Sleep is precious," Mrs. Fairchild agreed.

Mrs. Fairchild returned that evening to meet Willis and Andy. In the interim Trudy had checked out her references. She had previously been employed as a companion to an elderly woman in failing health. The woman had recently died, and her son had nothing but praise for Mrs. Fairchild.

"What do you think?" Trudy asked Willis after the prospective housekeeper had left.

"I liked her," said Willis, "and so, apparently, did Andy."

"Yes, he did," said Trudy with a smile. "And that surprised me. He can be shy with strangers."

"Did you ask her if she'd be willing to stay overnight whenever I'm out of town?" He hadn't mentioned the royal tour, but it now seemed a certainty.

"She said as long as she has advance notice. Seems she has a cat."

"Why don't we suggest a month's trial. Let her start in January and take it from there. See how things work out."

"Good." Trudy felt she had little choice. She wasn't sure how she'd like having someone around full-time, but the episode with Andy in November had convinced her she should at least try.

Willis and Trudy decided to spend Christmas at home in Toronto. Once the gifts were opened, Willis brought out his folding Kodak camera, set up some photofloods, and snapped off a roll of film. A few days later he picked up the developed pictures at the drugstore, and they drove on to Leamington to ring in the New Year with Anna Maria and Clyde.

On New Year's day Trudy decided she'd take the opportunity to visit her mother again in Essex. She wrapped Andy in blankets for the drive and headed off after breakfast. She wondered if her father would be home this time. He hadn't yet seen his grandson.

It was late afternoon when she returned.

"How did that go?" Willis asked.

Trudy shrugged. "You'd think nothing had happened. He was polite with me and distant with Andy, and he didn't mention you. He was so formal with me I might have been one of his parishioners. After a few minutes he excused himself and went upstairs to his study. Blanche was excited to see us, though. She remarked that Andy had grown a good deal in the year since she last saw him."

"Maybe next year I'll tag along. It'll be interesting to see if the old boy still resents me."

"Oh, I'm sure he does, but he's such a hypocrite you'll never know. Now that you're working for the CBC he might be more polite to you. It's a status thing."

"I'll let you decide if you want me there," said Willis.

"We might see them sooner if we decide to spend our summer vacation on Lake Erie. Alice and Al are going to rent a cottage and they've invited us to join them."

"Sounds good," said Willis, but he wondered if he would even get a summer vacation with the prospect of a royal tour looming. Trudy had read about it, but she hadn't put two and two together, and didn't realize it would mean a lot of travelling for Willis.

Back in Toronto Mrs. Fairchild proved a godsend. Trudy's concerns about sharing her home with a stranger were quickly dispelled. The woman was cheerful and accommodating. Andy took to her immediately, and the affection was mutual. In fact there were times when Trudy felt a tinge of jealousy at how willingly her young son let himself be cuddled in the arms of the new housekeeper.

Willis took the streetcar to work so that Trudy could have the use of the car. With Mrs. Fairchild looking after their son, he hoped she'd get out more. She visited Bobbie Dale, her old university friend, and drove to museums and the occasional lunchtime concert. At other times she just roamed the city, enjoying her newfound freedom. She still suffered from anxiety attacks, but with a housekeeper around and medication available she worried less about them. She wasn't the Trudy that Willis had married, full of fun and always the life of the party, but then Willis had changed too. They were both parents now, and that had made a difference.

Andy, for his part, was thriving. He did have a setback that spring, though, when he developed a painful case of gingivitis.

"It's an infection of the gums," the pediatrician explained; "it's often called 'trench mouth.'"

Trudy shuddered. "Ugh! What an awful name. It is serious?"

"It can be, if it's left untreated. It was called trench mouth because the problem was common among the troops fighting in the Great War. But we've caught it early, and he should be fine in a month or so."

"A month?" Trudy was shocked. "I thought you said it wasn't serious."

"It isn't, although I suppose you could say it's a serious challenge to the parent who's administering the treatment."

"What does that involve?"

"You'll have to swab his gums with hydrogen peroxide three times a day and get him to rinse his mouth with sodium perborate. I'll write you a prescription."

"Speaking of prescriptions, I'd like more of Andy's phenobarbital."

The doctor shook his head. "No, I think it's time you cut back on that. He's well past the stage where it's appropriate."

"But it works so well," Trudy protested.

"I'm sure it does, but we want him to get to sleep naturally, without medication."

"I suppose so," Trudy agreed, a little reluctantly. "As for the trench mouth, have you any idea how he might have caught it?"

"Hard to say. Maybe a cigarette butt picked up from the street."

Andy's low-grade fever disappeared a few days after treatment was begun. Administering the perborate was a tiresome daily procedure that, thankfully, Mrs. Fairchild quickly mastered, but it left him looking as if he'd just eaten a serving of blueberry pie. Willis had to stifle a laugh when he first saw his young son in this condition.

Trudy gradually reduced Andy's sleep medication, so that within a month she was no longer using it. Willis was relieved. "It may mean the occasional restless night," he said, "but we'll just have to put up with it."

Plans for the month-long royal tour of Canada were formally announced early in the New Year. Since it would be the first time a reigning monarch had visited Canada, King George was determined to see all nine provinces. (It would be ten years before Newfoundland joined the Confederation, and it was decided that the remote Northwest Territories and the Yukon would not be included.) The royal couple would arrive at Quebec City by ship and cross the country by train, taking the Canadian Pacific Railway for the journey to Vancouver and returning to Halifax on the Canadian National Railway. The Prime Minister was an enthusiastic monarchist, and planned to accompany them. With a federal election scheduled for the following year, Mackenzie King shrewdly realized that being seen with the popular King and Queen would boost his chances for victory.

The CBC's radio coverage of the tour would be the biggest challenge ever undertaken by the fledgling network. The government voted a supplemental budget of fifty thousand dollars to make it possible.

It was a Monday morning in March when Willis broke the news to Trudy that he would be travelling.

"I thought you promised me that the Montreal trip would be your last," Trudy protested.

"I didn't see this coming back then," he said.

"How long will you be gone?"

"Hard to say." Willis decided there was nothing to be gained by minimizing his commitment. "At least six weeks."

"You're kidding," said Trudy, incredulous. "And you'll be leaving us here alone all that time?"

"'Fraid so. But you won't be alone. You'll have Mrs. Fairchild."

"But six weeks, Willis. I can handle a few days here and a few days there, but six weeks is . . ."—she searched for the right word—". . . well, it's simply unacceptable."

"I'm afraid I have no choice."

"But you've never cared a damn for royalty. Can't they get someone else?"

"Honey," said Willis, "you know I've never liked the idea that we Canadians should be paying homage to a king who lives thousands of miles away, but this is a huge technical opportunity for me. It's what I dreamed about when I was an undergraduate at Queen's."

"I don't think the CBC understands the strain this will put on our family."

"I'm not alone. There will be a hundred people travelling with me. Lots of other guys with families."

"And lots of women, too, I expect."

"Very few, Trudy."

"I don't like it," said Trudy petulantly.

"Why not look at it from the bright side. The tour will be finished by July. We can take the vacation we planned and stay with Alice and Al."

Trudy was about to reply when the doorbell rang. It was Mrs. Fairchild reporting for work. Willis took the opportunity to slip out. He hoped Trudy would discuss things with her helper. She could move in—even bring her cat. Andy would like that.

Chapter 57

Across Canada, and Washington, D.C., 1939

Andy did, indeed, like the cat, but the affection wasn't mutual. Named Kitkat, it was a neutered five-year-old female tabby and was, in the beginning at least, reclusive. Andy would squeal with delight at the sight of the cat, but when he toddled in its direction Kitkat would scamper safely out of his reach. In time this would change. Andy was taught to wait until the animal was settled comfortably in Mrs. Fairchild's lap and then approach it slowly. He learned to pet it gently and was occasionally rewarded with a resonant purr. Trudy, less enchanted, complained about the smell of the litter box. Kitkat was an indoor pet, but her claws were intact, and they soon took a toll on the furniture. Trudy was less concerned about this than about what those claws might do to her curious son.

The royal tour was scheduled to begin on the fifteenth of May, when the King and Queen would disembark from the *Empress of Australia* in Quebec City. But Willis's work began much earlier. He crisscrossed the country, checking out the various facilities along the route. The main cities on the itinerary posed no problem. Their stations were already well equipped, and he was sure that they could meet the technical requirements for broadcasting the royal tour. But the CBC also wanted to carry accounts from smaller towns along the route, particularly in the prairie provinces and Northern Ontario.

"There's no getting around it," Willis explained to a meeting of producers that included Ernie Bushnell, the man with overall responsibility for the CBC coverage, "we'll have to rely on local stations."

"But are they equipped to handle our needs?" Bushnell asked.

"Some are, others aren't."

"So what's the plan?"

"The way I see it, as long as we limit these broadcasts to towns that are within a hundred miles of our larger stations we should be able to provide decent coverage. We'll need mobiles at these sites. That way we can relay the signal from these remotes to more powerful transmitters, and they, in turn, can pass them along the network to the rest of the country."

"How many of these remote sites are feasible?" Bushnell asked.

"At least fifty," Willis replied. "And we're going to need at least three new vehicles outfitted with mobile transmission equipment."

"Well, gentlemen, it sounds as if we all have our work cut out for us," said Bushnell with a smile. "It's going to be one helluva challenge. But if we can bring it off we will be making broadcasting history."

Bushnell must have made a convincing case to federal officials about the need for more money if the CBC was going to broadcast from remote locations, because the government quickly approved an additional $100,000 to buy and equip three mobile vans. Willis helped with their design and proposed a plan for the tour coverage. Three teams of CBC technicians would drive across the country, leapfrogging from site to site. One team would get everything ready at a specific location for the arrival of the producers and announcers travelling with the royal couple, so that they could broadcast their account of that part of the tour. Meanwhile, another unit of technicians would be setting up at the next venue, and a third would be heading to a yet another stop. Twelve announcers were assigned to the tour, and they all had to be coached on the finer points of royal etiquette. In addition, there was the problem of the King's shyness and his tendency to stutter. Bob Bowman, one of the CBC's most experienced announcers, came up with a suggestion.

"Whenever the King is expected to speak, why not put an announcer equipped with a live mike within clear sight of him. Then if he senses any real hesitancy, or if the King stumbles, he can fill in with commentary."

"Brilliant idea," said Bushnell. "And you're just the man for that role."

"Me and my big mouth," said Bowman with a smile.

"Can we manage that?" He directed the question at Willis.

"Shouldn't be a problem," said Willis, with more confidence than he felt. Just what we need, he thought to himself. Bad enough having to deal with a shy, hesitant king, but why did he have to have a speech impediment too? Willis didn't share these thoughts with the other members of the team. He had decided from the outset to keep his negative views on the monarchy to himself. Only his closest friends knew the truth.

Fortunately his role was the technical co-ordination of the broadcasts, not loyalty to the King. His task would require a degree of diplomacy, but not in relation to royalty. It would be called into play in his dealings with personnel from the various regions, from the French network, and from the BBC.

Trudy coped better than Willis had expected while he was away on his survey. In fact she was somewhat chastened over her earlier protests about his travel.

"I guess I had no idea how important the royal tour was going to be. The newspapers have been filled with stories about the King and Queen and their travel plans. One even carried a big spread on the CBC's coverage."

"I knew it would be a major event," said Willis, "but I admit I'm surprised at the excitement it's generating. I don't get it. Canada is perfectly able to function as an independent country. Why do people feel such loyalty to a king? He wasn't even elected, he simply inherited his position."

Trudy decided to shift the topic to safer ground. "What part of Canada impressed you the most?"

"Well, as you know, I'd never been out west, or even to the east coast for that matter." He paused, thinking about the question. "For sheer grandeur, I guess the Rockies are hard to beat."

"And the people?"

"I found Maritimers the most friendly. They go out of their way to make you feel welcome. Prairie farmers are friendly too, but in a different way. They're quieter, a little on the shy side. But loyalty to the crown seemed to be strongest in British Columbia."

"Do you think you'll get a chance to actually meet the King?"

"I don't know, and frankly, I don't really care. You know my feelings on that subject. I suppose it's possible. On the final day of the tour there'll be a small reception for the folks involved in the broadcasts."

"I do hope you meet him. Dad's a big fan of the royal family. To have a son-in-law actually meet the King . . . why, he might even begin to respect you."

Willis changed the subject. "Has Andy made friends with the cat?"

"When he can find him," said Trudy with a smile. "Kitkat manages to keep out of Andy's way most of the time."

One evening Trudy went to bed early, giving Willis a chance to speak to Mrs. Fairchild on her own.

"Trudy seems to be holding up pretty well," he remarked.

"Yes, she is, for the most part. There have been a few down days, but on the whole I think she's managing remarkably well."

"I appreciate all the support you're providing. It's great to know I won't have to worry about Trudy during the tour."

"In spite of the difficulties that come from your being away, I think that deep down your wife is very proud of you."

The King and Queen arrived in Canada on the seventeenth of May, two days late. Bad weather and an unusual number of icebergs in the Atlantic had delayed their crossing. It was not an auspicious beginning for the people in charge of broadcasting the tour, who had to revamp the entire schedule. In all, there were ninety-one stops across the country, and it would require long hours of work by the CBC teams to ensure that everything was ready at each new destination.

In Quebec City they encountered their first problem. One of the CBC reporters, J. Frank Willis, was positioned in the belfry of a Roman Catholic church, one hundred fifty feet above the city streets, from which he had a splendid view of the port and its surroundings. He could provide listeners with a unique bird's-eye account of the arrival of the *Empress of Australia*. Tests were carried out and his voice came through clearly.

On the ground, Charles Jennings reported the excitement as the ship came into view. When the crowd began to cheer wildly and he could no longer see the *Empress*, he said, "We take you now to J. Frank Willis, situated high above the city, broadcasting from the spire of the St. Roch Church."

In the studio, several blocks west of the church, producer René St. Pierre cued the announcer. "Come in, Frank," he said over the intercom, and signalled the soundman to switch to the St. Roch remote.

Then the unexpected happened. The studio was suddenly filled with the deafening sound of bells.

"Jesus," muttered the harried St. Pierre, "I can't make out a word Frank's saying." He looked to Willis, hoping for a solution.

From the studio doorway Willis gave the producer the cut signal, a slice of the hand across his throat. A switch was thrown and the bells were silenced.

"Cue Jennings," the producer shouted.

Charles, a seasoned pro, took over: "As you know doubt heard, our correspondent found himself competing with the bells. It was no contest. In fact I can barely hear myself speaking. Every church in Quebec City seems to have joined in a magnificent tribute to Their Royal Majesties. Let's listen in."

The producer, who by now had regained his composure, cued the soundman. Once again the studio was filled with the sound of bells, but this time they were no longer deafening.

The rest of the arrival broadcast went smoothly. At the post-production meeting, though, tempers flared. "You should have anticipated this," an angry Frank Willis sputtered.

"I'm sorry," said St. Pierre, "but mistakes occur. We'll simply have to learn from them."

"They better not happen again," Frank replied, still seething, "not when I'm on the air."

"Just be thankful you had Charlie to bail you out," Willis countered, his own temper flaring.

At that point Ernie Bushnell interceded. "Gentlemen, gentlemen, simmer down," he said. "We have a lot of work ahead of us. We have to expect problems along the way, but we need to handle them as professionals."

After the official welcome ceremony at the port, the King and Queen were driven to a formal luncheon in a specially equipped convertible McLaughlin Buick touring sedan. Two of these had been custom-made for the tour.

At the luncheon the Prime Minister officially welcomed the royal couple: "Today as never before, the crown has become the centre of our national lives. In coming from the old land to the new, you have not left one home, but come to another."

It was a trouble-free broadcast, and the CBC technical team was able to relax as they boarded the train for Montreal.

There were actually two trains. The first, called The Pilot, accommodated railway crews, police, and the journalists covering the tour. It preceded the royal train. From the outside the two trains appeared almost identical. Both were painted bright royal blue trimmed with gold, and decorated with the royal coat of arms. They were pulled by massive Hudson locomotives, also painted blue.

It was on the inside that the two trains differed. The last two cars of the royal train were luxurious, outfitted with all the modern conveniences.

There were two comfortable sitting rooms, an area for dining, and a tastefully appointed bedroom with an adjoining bathroom.

"I guess this is one time when the phrase 'fit for a king' is appropriate," Willis joked.

There was also a business car with an office for the King's use, a rail operations office, and a general train office. The remaining cars had been altered to provide a dispensary for the royal surgeon, wardrobe rooms with steam pressing tables, a barbershop, and even a mini-post office. There were two baggage cars. One had living accommodations for diner crews, extra food storage, and a telephone switchboard that was hooked up at each stop, while the second carried luggage, floodlights, spare parts, including touch-up paint, and even ropes for crowd control.

From his window in the Pilot train Willis saw thousands of people lining the route, waving Union Jacks and seeking a glimpse of the royal couple. The train moved along at a leisurely forty miles an hour, and at each brief whistle stop he heard rousing cheers indicating that the King and Queen had stepped out onto the rear balcony of their car. This surprised Willis. He hadn't expected this show of affection from the people of Quebec. Why did they feel loyalty for this foreign monarch who didn't even speak their language?

In Montreal Their Majesties were driven in the second McLaughlin Buick from the railway station to the Windsor Hotel. It was there that Camillien Houde, the rotund mayor of Montreal, hosted an elaborate banquet. Willis knew that Houde could pose a problem. He was on record as a supporter of the Nazi government in Germany. This was a popular position with many French Canadians, who did not want to see Canada drawn into another war. The English in the city were enthusiastic supporters of the monarchy, but Willis had wondered about the French-speaking population. These doubts were quickly dispelled. French Canadians in Montreal seemed genuinely intrigued by the King and Queen, and turned out in vast numbers for a glimpse of British royalty.

There was a minor embarrassment at the dinner hosted by the Mayor.

"I wish to extend the warmest welcome to Your Majesties," he said, "and for including Montreal on your tour. I thank you from the bottom of my heart, and"—he turned to his wife—"from my wife's bottom too."

Willis groaned. "I hope that slip of the tongue was accidental," he remarked, "but with Houde you can never really be sure."

Ottawa was the next city on the agenda, and in many ways it was the most important. There was a lavish reception at Rideau Hall, the residence of the governor general, the King's official representative in Canada. John Buchan, Lord Tweedsmuir, held that office at the time, and he hosted the event.

Prime Minister Mackenzie King had long cherished the hope that the first visit to Canada by a reigning British monarch would occur while he was in office, and he did everything in his power to ensure that the royal couple received the warmest of welcomes. There was a state dinner at the Chateau Laurier Hotel, where the King and Queen were staying. A tour of the Central Experimental Farm in an open landau afforded thousands of ordinary people a chance to see the royal couple. The Queen laid the cornerstone for the new Supreme Court building, and later the King addressed the Senate, stressing Canada's importance within the British Empire.

The most moving event of the Ottawa visit was the official dedication of Canada's National War Memorial. Six thousand Canadian veterans of the Great War watched the King lay a wreath at the foot of the imposing monument, after which the King and Queen moved freely among the veterans, stopping to thank them individually for their contribution to the victory over Germany.

Charles Jennings was part of the cheering crowd, and his commentary betrayed his excitement: "The King and the Queen are walking this way, toward us. They're not more than six feet away now. Can you hear the King and Queen? I don't suppose you can. They're chatting with people. The King, smiling, is talking to a man in a royal blue beret. Her Majesty is speaking to a blind veteran now. He can't see her, of course, but inclines his head so that he might hear her better."

"Amazing," said Harold Smith to Willis. "They're so unpretentious. I thought they'd be much more formal."

"I wasn't sure what to expect," said Willis, "but I have to admit they do seem at ease talking to those veterans."

As the train pulled out of Ottawa for Toronto there was a massive sigh of relief from the CBC personnel who had gathered in the capital to cover the many events. It had been a complicated logistical undertaking, but they had brought if off.

The turnout in Toronto was the largest in the city's history, but there were fewer events than in Ottawa. Willis was busy, but he did manage a phone call to Trudy from Woodbine Racetrack, where the royal couple were watching the annual running of the King's Plate. "It's a zoo here," he told her.

"A zoo?" Trudy teased. "I thought you said it was a horse race."

"Oh, it's a race all right. But for us this is just another leg in a marathon. I can't believe the crowds. It seems that everyone in the city expects to see them."

"I know," she said. "Most of the people on our street are there at the track. Even Mrs. Fairchild. I gave her the day off. But Andy and I aren't going anywhere."

"A wise decision," said Willis.

From Toronto the tour skirted the Great Lakes, stopping in the twin cities of Fort William and Port Arthur before moving on to Manitoba. It rained for most of the time the royal couple were in Winnipeg. The premier, Joseph Bracken, apologized: "We are sorry the weather isn't cooperating for Your Majesties, but we are in the midst of a major drought and this rain is desperately needed by our farming communities."

Saskatchewan was next; crowds gathered in Regina to see the King and Queen. By now the CBC technicians had fallen into a routine. The team-members knew their roles; they were able to organize, install, and dismantle equipment in a highly efficient manner and with almost military precision.

One the most colourful events of the tour had been arranged for Calgary. Members of several Indian tribes—Piegans, Bloods, Blackfeet, Stoneys, and Sarcees—appeared in their ceremonial garb. The voice of commentator Ted Briggs, with its faint traces of a British accent, seemed the perfectly appropriate instrument for the description of this first meeting between royalty and the chiefs of Canada's aboriginal people: "The Indians, sitting on their horses, looking magnificent with their buckskin heavily embroidered, are waiting patiently."

The arrival of the royal couple was signalled by the beat of tom-toms. "Their Majesties have just come into sight, escorted by a mounted guard of the Lord Strathcona Horse." When the royal limousine stopped, the Indians laid down buffalo robes to serve as a carpet. When the King and Queen got out of the car, another announcer, Frank Witters, took over the commentary: "We have a sound-effects mike hidden over where a ceremony will take place, so we'll be quiet sometimes so you can hear for yourself."

No words were picked up, but the chants of the Indians could be heard clearly as they staged a ceremonial dance for the visiting couple. Then Briggs resumed: "The Great Duck Chief, senior among all the chiefs of the tribes assembled here this afternoon, is crowning the King. His Majesty is being made an honorary chief . . . Chief Sitting Albino . . . which means Great White Buffalo, sitting there, over all of us."

His voice was suddenly lost in a round of cheering, clapping, and whooping. After this subsided, Briggs continued, "The amazing part about this particular scene was that the Indians actually broke into cheering. You all know their reputation for being quiet, stolid sort of men, but this time they really let themselves go. Most of the chiefs have their full war paint. The medicine man has an extraordinary headgear made of weasel skins and small pieces of otters' tails. The mounted police are moving the crowds back so that Their Majesties may have a clear passage to their car."

At this point the people in the stands broke into a rousing version of "God Save the King." As the last words of the anthem faded, the royal couple drove off, and Briggs concluded, "This is certainly going to go down in history as one of the most unprecedented events in the visit of Their Majesties to Canada. It is a spectacle that nobody who witnessed it today will ever forget."

In the trip through the Rocky Mountains the royal couple and their entourage were captivated by the scenery.

"Remarkable!" the King was overheard to say as he gazed at the towering peaks.

"It's breathtaking," added the Queen.

Vancouver was next. The city boasted a large community of British expatriates, including retired civil servants and military officers. When it rained, the King remarked that it had been thoughtful of the tour organizers to arrange some "proper English weather" to remind him of home.

The royal couple then boarded a yacht for the crossing to Vancouver Island, to visit Victoria, the capital of British Columbia. King George addressed a huge crowd, as well as millions of radio listeners in Canada, the United States, and Great Britain. He spoke hesitantly, pausing between sentences. Bob Bowman was ready to take over should it turn out to be necessary. "The Queen and I have crossed Canada from east to west, from ocean to ocean. When I remember that here, on the shores of the Pacific, I am as far from Ottawa as Ottawa is from London, I realize the vastness of this country."

He thanked the Canadian people for the hospitality they had shown at every stop along the way. He specifically mentioned how much he had enjoyed the chance to meet the "original inhabitants of this land." In the speech there was only a vague reference to the looming threat posed by Germany: "Some day the peoples of the world will come to realize that prosperity lies in cooperation and not in conflict. On the dawn of that brighter day, I look to Canada to play an increasingly important role."

In order to provide the royal couple with some different images of the country, the royal train was moved from the Canadian Pacific Railway line to that of Canadian National for its return trip. On board the train, Bowman drew Willis aside. "That went well," he said. "I was ready, but I was glad my services weren't required. And I've noticed something."

"What's that?" said Willis.

"The King isn't the one to focus on."

"No?"

"No, I watch the Queen. She's the best barometer of how the King is doing. I've seen her prompt him quietly when he begins to fumble for a word."

"That's interesting," said Willis, smiling. "Maybe we should have hired her instead of you."

"I'd be more than happy to relinquish my role," said Bowman with a chuckle.

Willis remembered how easily the Queen had mingled with the veterans in Ottawa—and now this, a quality he'd seen before, the simple act of a wife helping out her husband. Trudy had filled that role for him on many occasions. He realized with surprise that his feelings about royalty were beginning to thaw, at least where the Queen was concerned.

Once through the Rockies the train stopped in Jasper, and then it was on to Edmonton, and another record crowd.

In Saskatchewan a major technical challenge occurred in the small prairie town of Melville. It was supposed to be a brief stop, and as a result Willis seriously underestimated the number of people who would turn out. They began arriving in unprecedented numbers the day before the event, and when Willis drove out from Regina he found the road jammed with cars. Most were from Saskatchewan, but to his surprise there were also many American licence plates.

At one point the traffic came to a halt, and Willis got out of the car to stretch his legs. A man approached him. "Got a light?" he asked. There was no mistaking the accent. This stranger was from the States.

"Sure." Willis produced his Ronson lighter and lit the man's cigarette.

"Some traffic jam." The man inhaled and blew out a puff of smoke. "Didn't expect this, not out here in the country." It wasn't a complaint, just an observation.

"Where you from?" Willis asked.

"Billings, Montana."

"How long did it take you to get here?"

"Started out yesterday afternoon. Drove all night."

"That's some trip. I'm impressed"

"Not every day you get the chance to see a king."

"I guess not," Willis admitted. "But he's not your king."

"Nope, that's true. But a king is a king. Lots of folks from Billings drove up to see for themselves."

"So it seems."

The line of cars had begun to move. "Thanks for the light," the man said.

"You're welcome."

The drive from Regina to Melville, which would normally take a little over two hours, took Willis twice that, and during the trip he had time to wonder about the appeal royalty held for so many different folks. He could understand it among people descended from English stock, but French Canadians and now Americans were caught up in the excitement too. Maybe he was missing something.

His apprehension grew as he neared Melville and saw the size of the crowds that were assembling. It was going to be a major challenge to cover the event according to the usual plan. He envisioned a nightmare of difficulties. How could he keep the hundreds of feet of wires and cables intact in such a sea of spectators?

When he got to Melville he made an emergency call to the station in Regina.

"We need volunteers—secretaries, accountants, executives, members of the cleaning staff—anyone willing to lend a hand. They don't have to know anything about the technical side of radio. I want to position these people all along our network of cables and wires. With any luck they can keep the crowds from trampling them."

More than thirty volunteers arrived the next morning. They were quickly briefed and then sent out to protect the wiring. Willis looked out over the gathering throng, later estimated to number more than sixty thousand, and crossed his fingers.

There was some pushing and shoving as the crowd tried to get closer to the railway station. Willis noticed a big man in a plaid shirt arguing with one of the volunteers, a tiny older woman who was supporting herself with a cane.

He moved quickly to intercede. "What seems to be the problem?" he asked.

"This lady here won't let me by," the big man blustered. "Who is she to give me orders?"

"She's a member of the CBC's broadcast team," Willis explained, deliberately keeping his voice low. He pointed to ground behind the woman. "We have to keep people away from our wires and cables."

"Who says?" The man was clearly ready for trouble.

"I say," Willis answered. Then, raising his voice a little, he added, "And so do the millions of Canadians who are at home listening to their radios. You step on those cables and they won't be able to hear a thing."

"Shit!" the man muttered, but the fire had gone out of his voice.

"If you move down the line a little, there's a place to cross safely."

The man shrugged and moved off.

"Thank you," said the tiny woman. "He was beginning to be a real pain in the butt."

Willis smiled. "I'm sure you could have handled him on your own, but I was close by and thought I'd put in my two cents worth."

"Glad you did. I was about ready to give him a whack with this." She raised her cane defiantly.

The train finally pulled into the station, and a roar went up as the royal couple emerged on the platform at the rear of the last car.

Charles Jennings described the scene to radio listeners across the country. His diction was flawless, as was the whole broadcast. Willis's rag-tag team of volunteers managed to succeed in holding the lines.

"That was a close call," said Willis with a sigh after Jennings had signed off.

"Too close," agreed the announcer.

Another highlight of the return trip was a stop in Sioux Lookout, in Northern Ontario. It was the only occasion when Willis wasn't part of the team. The long hours and the lack of sleep had finally caught up with him, and he felt a stab of pain in his back while he was lifting a suitcase in his hotel room. He'd been somewhat troubled in the past with back spasms, which he attributed to a soccer injury he'd sustained at Queen's, but this time the pain was excruciating. He could barely make it across the room to his bed. When Harold Smith knocked on his door a few minutes later, Willis staggered up to let him in.

"I'm afraid you're on your own this time," Willis muttered through clenched teeth. "My back has given out."

"You're not coming?"

"No, I don't think so. Do you have any Aspirin?"

"Sure, back in my room," said Smith, and started for the door.

"Harold . . ."

His friend turned.

"You don't happen to have anything to drink, do you?"

"No, but I can probably arrange something from the bar downstairs. What's your poison?"

Willis managed a smile. "Rye would go a long way."

The rest of the day he drifted in and out of sleep. For once he was a listener rather than a participant in the tour. He had a small radio and it was tuned to the CBC. Charles Jennings was handling the commentary: "Here comes the royal train now. The locomotive is just passing me. Then we have one, two, three, four, five, six, seven, eight, nine, ten, eleven, twelve—an even dozen coaches. All royal blue, trimmed in gold."

At this point, wild cheering filled the air. Then, faintly, in the background, there was the sound of a local band playing "O Canada." When the hubbub subsided, Jennings continued, "Why, you might well ask, is the royal couple stopping in Sioux Lookout, a remote mining town? Well, we understand His Majesty is supremely interested in Canada's natural resources. I would say nowhere in Canada can the royal party capture a better picture of pioneer mining enterprise than here, and Sioux Lookout is a key point for at least six rich mining areas."

There was another burst of deafening applause when the train stopped, and finally the moment came that the crowd had been anticipating: "The King and Queen have just appeared on the platform at the rear of the coach, the

Queen dressed in white, the King is wearing a double-breasted business suit. His Majesty waves; Her Majesty waves. Flash bulbs are going off."

Willis drifted off to sleep. The trickiest part of the remote broadcast was always the arrival, because that was when it was hardest to anticipate what might happen. The more formal ceremonies were more predictable.

When Smith returned to the hotel that evening he explained at the reception desk that he was concerned about his friend's health, and persuaded the clerk him to provide him with a key. He hurried up to Willis's room, quietly opened the door, and then gasped. Willis lay stretched on his back on the floor, the bottle of rye beside him. Smith was relieved to see that it was almost full. He knelt down beside his friend and discovered that he was snoring softly. This was encouraging. He gently prodded a shoulder.

Willis stirred and opened his eyes. "It sounded great," he said, his voice a whisper.

"What?" Smith was confused.

"The broadcast. I listened on the radio."

"Oh that. It was fine, no problems. But what are you doing on the floor?"

"It's the only place I can get relief from the pain. The bed is too soft."

"Have you eaten?"

"Room service brought me a sandwich."

"When?"

"I dunno. Noon, I guess."

"How about some supper?"

Willis thought this over. "Maybe a bowl of soup. Nothing heavy."

Smith called down to room service and placed an order. While he waited, he managed to help his friend up and into the bathroom. Then he pushed the bed against the wall, spread a blanket on the carpeted floor, and put down a pillow.

Willis emerged from the bathroom and chuckled. "I see you've rearranged things." He winced as he lowered himself onto one of the straight-backed chairs.

"You should really go to the town hospital," Smith suggested, "and have your back properly checked out."

"Let's wait till morning. If I'm not better by then I'll consider it."

There was a knock on the door. Smith took the dinner tray and signed a chit, then sat down opposite Willis. The two men ate in silence. Willis even managed some bread with his soup.

"Tell me about the broadcast," he said.

"Not much to tell. The King is interested in mining, as you know. The remote worked fine, no real challenges. Another day and we head back to Toronto."

In the morning Willis told Smith he was feeling better. "I think all I need is a couple of day's rest."

"You sure?"

"I'm sure. We're going on to Toronto anyway, so I'll take some time off when we get there. For now I'll just relax here in my room."

Willis was true to his word. He lay flat on the floor most of the day, only getting up for meals and to use the bathroom. His back was still painful, but it wasn't as crippling as it had been the previous day.

Smith checked in on him in that evening. "How you doing?"

"Getting there," said Willis, grimacing. "How'd today's broadcast go?"

"Pretty routine. It's become so predictable most of us could do it in our sleep."

"Speaking of sleep, I think I'll try the bed tonight. It was hard to get comfortable on the floor. Can you help me with my luggage tomorrow?"

"Sure thing."

When Willis woke up in the morning he got out of bed gingerly. He had slept well, thanks to a nightcap of two liberal shots of rye. His back was still stiff, so he was grateful when Smith arrived to carry his bags. In the taxi, on the way to the station, he outlined his plan.

"I've called my wife. I didn't mention my back, just that I needed a break. She'll pick me up when we get into Toronto. I should be fine by the time the tour heads to Washington for the King's meeting with Roosevelt."

A long train ride wasn't the best treatment for Willis's back problem. He made sure he didn't sit for too long. From time to time he rose and walked carefully up and down the corridor of the swaying railway carriage. Aspirins helped, but he felt his back beginning to tighten.

At Union Station he found a redcap to help with his luggage and limped along behind him.

Trudy was parked outside, waiting. When she spotted Willis she could tell that something wasn't right, because normally he carried his own bags. She got out of the car and ran to meet him.

"What's wrong?" she asked, fearing the worst.

"Oh, it's nothing serious." He took Trudy in his arms and held her in a tentative embrace. "Just my back. It's been acting up again."

"Do you need help?"

"No, just some rest."

"Well, let's get you home. You look exhausted."

"Oh, I am, I am."

Willis's conviction that rest would resolve his problem proved to be correct. His back pain gradually subsided, and on the eighth of June he was able to rejoin the royal tour. He was still stiff, though, and realized he would have to be careful or risk a relapse. He promised Trudy he would see a doctor once the tour was over.

The royal train crossed from Canada into the United States at Niagara Falls. Americans seemed as captivated by the British royalty as Canadians had been. All along the route to Washington they lined the roads, hoping for a glimpse of the King and Queen. To accommodate them the train moved slowly, and there were a number of whistle stops.

In Washington President Franklin Delano Roosevelt and his wife, Eleanor, welcomed the royal couple at Union Station, then walked them to an open limousine that took them to the White House. Cheering crowds, three and four deep, lined the motorcade route. They were rewarded with a clear view of their president and the King as the car moved slowly through the streets to 1600 Pennsylvania Avenue.

The royal visit to Washington was the most lavish event the city had ever seen. There was an afternoon reception at the British embassy and a formal evening of dining and musical entertainment at the White House.

The CBC and the BBC had pooled their resources to provide radio coverage during the Canadian tour. Willis had been pleasantly surprised at how well they had worked together; the Brits had been particularly helpful when it came to matters of protocol. But in Washington the Americans incensed the BBC team because they wanted to call the shots. "Damn Yanks," one of the BBC's producers complained to Willis, "you'd think he was their king, not ours."

"They can be a pain the ass," Willis conceded. Privately he had anticipated this clash. He had worked with a number of American radio producers during his years at CKLW, when the station served both Windsor and Detroit, and he knew they could be difficult. But during his trip to New York the folks at WOR had been very helpful; they had been more than willing to show off their powerful new transmitter. The station's chief engineer, Roy Banks, happened to be in Washington to help co-ordinate the Mutual network's radio coverage of the King's visit, and he ran into Willis in the lobby of the hotel where both men were staying.

"I wondered if you'd be here," he said with a smile. "Our Canadian cousin."

"I'm flattered you remember me."

"Well, you made a pretty favourable impression in New York. Not all the visiting firemen we hosted knew much about radio. You were an exception."

"Maybe you can help me with something," said Willis. "Can I buy you a drink?"

"Sure."

When they were settled into the hotel bar Banks sipped his Scotch and looked across the table at Willis. "Shoot," he said.

"Well, some of the Brits covering the tour are upset. Their noses are a little out of joint. They feel they're being patronized by you Americans."

Banks raised his palms in mock surprise. "Us? Patronizing?" Then he laughed. "I guess we're used to running our own show. It's not every day we have to share our turf with outsiders."

"Maybe you could say a word to some of your buddies who are in charge?"

"I'll see what I can do," Banks promised, then asked, "Do you want to order some supper? I'm starved."

"Sure," said Willis as he lit a cigarette.

Over coffee Banks explained why the American networks were putting so much effort into their coverage.

"Did you know it was Roosevelt's idea to invite the King and Queen to visit Washington?"

"I thought it was a mutual arrangement."

"No, it was FDR's suggestion. When he learned about the King's tour of Canada he pushed for a meeting in the States. And it was a risky move. This is the first time a reigning British monarch has set foot on American soil since the U.S. declared its independence from England in 1776. I wouldn't say that all Yanks dislike the Brits, but there is still a residue of animosity, dating back to the Revolution. And now that war with

Germany seems inevitable in Europe, Roosevelt knows that people in the United States want no part of it." Banks leaned forward and lowered his voice. "Privately, the President doesn't share this isolationist view, but he can't say so publicly. It would be political suicide. But he understands that the British will certainly need some sort of American help in any future conflict."

"So there's a lot more at stake here than just pomp and circumstance?"

"A good deal more. This visit to Washington is crucial. Roosevelt hopes that when they see the royal couple in person Americans will become more sympathetic to the threat the British are facing. Then there's his own wish to foster a closer personal relationship with England."

"Maybe if I explained this to the BBC folks they might see things in a different light."

"I'd rather you didn't," Banks said. "Some of this is off the record. Let me see if I can smooth things over. I know we Yanks tend to be a little too pushy at times, but in this case I think you can see why."

On their second day in Washington, the King and Queen took in the sights of D.C. They boarded the presidential yacht and sailed up the Potomac River to George Washington's Mount Vernon; then they visited the Lincoln Memorial and the Arlington Cemetery, where they laid a wreath at the Tomb of the Unknown Soldier.

To comment on the significance of the visit, the CBC commissioned Gregory Clark of *The Toronto Star*: "Today will probably go down in history, so mark it well in your memory. A King visiting his lost domain, a people risen to greatness and power welcoming with warmth and joy the wearer of a crown they rejected with blood and fire. A growing friendship has been consummated. More than a handclasp binds us now. A long chapter ends and a new page turns. Yes, we're working our way out by friendship, by mutual regard; by understanding and patience we are driving to their rest the ghosts of yesterday."

After two days in Washington, the royal couple went with the Roosevelts to their country home in Hyde Park, ninety miles north of New York City, where the tone of the visit changed from formal to informal. This event was designed to illustrate to the American people that although the King and Queen were royalty, they also enjoyed the simpler things in life.

In contrast to the elaborate state dinner at the White House, the meal at Hyde Park was casual, with two families engaged in simple conversation, unfettered by formalities. Even more relaxed was the following day's event, an old-fashioned American-style picnic. The King and Queen of England were served hot dogs on the front porch of the cottage. Although the press paid a lot of attention to the hot dogs, the menu also included more delicate fare, such as smoked turkey with cranberry jelly, a green salad, and strawberry shortcake. Reading the menu over later, Willis smiled to himself. He noted that these hot dogs weren't quite the standard American fare; they were specially prepared from the finest aged and honey-cured Virginia ham.

The royal couple were clearly delighted to have the chance to get to know the president and his wife in an informal setting. In a letter to her mother-in-law, the Queen wrote, "They are such a charming and united family."

The departure from Hyde Park was emotional; a large crowd gathered at the small train station to wish the royal couple luck. They seemed to sense that the King and Queen would be returning to an uncertain world, with Britain on the verge of war.

As they headed north on the Pilot train, Ernie Bushnell sat down beside Willis with a sigh. "I'm glad we're finished here," he said. "How's your back holding up?"

"Pretty well. Just have to be careful, that's all."

"Well, we're in the home stretch now. Just the Maritimes and then we'll be through. It's been one helluva ride."

Willis nodded in agreement. "It has," he said.

"What did you think of FDR?"

"Impressive," said Willis.

"He's one wily old fox," Bushnell chuckled. "Makes our Prime Minister look like an amateur. Mackenzie King can't hold a candle to him."

"How so?"

"The PM behaves like a smitten schoolboy around the royal couple. Roosevelt was respectful, but he clearly regarded himself as their equal. I'd give anything to know what went on behind closed doors at Hyde Park. The President invited the King to Washington for a purpose. He knows Britain is going to need American help if and when the war with Germany begins."

"But FDR can't be seen to be taking orders from the King."

"No, I think it's the other way around."

"What do you mean?"

"I think the King accepted the invitation because he wanted reassurance from the President."

"Interesting thought," said Willis. He reached into his pocket for his cigarettes and offered the pack to Bushnell. "Smoke?"

"No, thanks," he said as he rose from the seat. "What I need now is a good stiff drink."

As Bushnell moved away down the corridor, Willis lit up. He looked at the rich farmland stretching into the distance. Could be Canada out there, he mused. Americans aren't all that different.

The train crossed the border into Canada at Delson, then made its way to Quebec City. From there it was on to the Maritimes, whistle-stopping through New Brunswick with brief stays in Fredericton and Saint John. A coastal steamer took them to Prince Edward Island, of particular significance because it was in Charlottetown, the province's capital, that the Fathers of Confederation had convened to unite Canada into a single country in 1867.

The royal couple's final destination was Halifax. As promised, a reception was held for the journalists and technicians who had covered the tour. Each member of the CBC team was presented with a gift from the royal couple, a sterling silver cigarette box with a tiny map of Canada engraved on its cover, showing the route of the tour.

Willis stood in line, waiting to be introduced. The King paused when he reached him, and asked, "How is the back, young man?"

Willis was momentarily speechless, but he recovered in time to reply, "It's fine, Your Majesty."

"Good," said the King, and moved on.

Willis was stunned. How had the King heard about his problem? Surely he had more pressing concerns. At that moment his understanding of the monarchy changed. The last shreds of the resentment against royalty that he had harboured for so long vanished. He no longer saw the King and Queen as mere symbols, residing in a distant castle across the sea. They were real people, he realized, but burdened with a responsibility he couldn't begin to imagine. He didn't envy them.

Finally, the couple boarded the *Empress of Britain*, sister ship of the *Empress of Australia*, the liner that had brought them to Canada. Willis, having gained experience with short wave during his work on the Hornby transmitter, suggested that they place a commentator on board one of the Canadian destroyers escorting the *Empress*. His account could be relayed from the ship to the station in Halifax for inclusion in the final broadcast, scheduled to be heard across the country. The connection was made, and it was a technical triumph. The same could not be said for the commentary. Ted Briggs' choice of words when he described the scene left something to be desired: "The Queen, I think I told you, is wearing powder-blue. And now as she moves away and juts her bow out into the sun . . . we can make out a great deal of her green boot topping."

Willis groaned. "We began our coverage of the tour on a sour note, the St. Roch bell fiasco. Now we end it this way. I just hope no one mentions it to Her Majesty."

"Amen to that," said Ernie Bushnell.

The royal couple had one more stop—Newfoundland—before returning to England. The island was still a British colony at that time; it would not become a Canadian province until a decade later.

Much to Willis's satisfaction, there had been no major technical foul-ups during the tour. Given the complexity of the undertaking, he considered this a minor miracle. When he returned home to Toronto he was given a week off. For once he was happy to simply relax—no woodworking projects, just sleeping in, puttering around the house, and getting reacquainted with Andy, who was now talking a blue streak. Mrs. Fairchild had returned to her own home, along with her cat. Andy didn't seem to miss Kitkat, but there was now talk of his getting a pet of his own. Both Willis and Trudy thought a dog might be a better choice than a cat.

When Willis showed Trudy his engraved gift from the royal couple she was delighted.

"It's perfect," she said. "What a wonderful souvenir!" She kept prompting Willis to tell her more about Their Majesties, and he was glad to oblige. No detail was too trivial. She had kept a scrapbook of pictures from the papers, and Willis was able to supply additional commentary.

At the CBC, Willis learned that his annual salary had been increased to $2,700 as a reward for his work on the tour. Newspaper reviews of the Corporation's coverage had been unanimous in their praise. There had been letters from the BBC expressing thanks for the quality of the technical assistance that had been provided, and even a note from President Roosevelt expressing his personal gratitude. Willis would look back on this event as a coming-of-age accomplishment for the CBC. It was a heady moment for the Corporation—the first occasion on which Canadians fully embraced the new network.

Maybe, Trudy thought to herself, Willis's role in this event would finally impress J.B. She decided it was time to reintroduce Willis to him.

Alice had written to say that she and Al had rented a summer cottage in Port Dover and wanted Willis and Trudy to spend some time there. Claire and Ed Pritchard had been invited as well. They now had a young son, Michael, another cousin for Andy. Alice was suggesting a family get-together, one that would include J.B. and Blanche. It would be an ideal opportunity to put Willis back in the Champion family picture.

Chapter 58

Port Dover and Leamington, 1939

One day in July they settled Andy into the back seat of the car and drove the seventy miles to Port Dover. Alice had described the cottage, she had even sent pictures, but these pictures didn't do it justice. It was more a home than a cottage, with four bedrooms and a large screened-in porch that overlooked Lake Erie.

"What do you think?" Alice asked after they'd had a quick tour.

"It's amazing," said Trudy. "Much larger than I expected."

"Plenty of room for everyone. You and Willis can have your own room, and so can Claire and Ed when they arrive. We could even accommodate Blanche and J.B. if they decide to stay over."

The words sent a stab of fear through Trudy.

"I don't think that's likely, though," Alice added, sensing her sister's apprehension.

It would be a week before the arrival of the Pritchards. Each morning after breakfast Trudy and Alice carried Andy and Joan down a steep path to the sandy beach. The two children played happily at the water's edge while the sisters caught up on the latest family news.

"What do you hear from Edith?" Trudy asked.

"We're all worried about her," said Alice. "With Hitler on the rampage in Europe and the talk of war, we've been trying to convince her to come home."

"Any luck?"

"I'm afraid not. She doesn't believe the Germans want another war."

"I hope she's right," said Trudy. "It's the last thing we need. The Depression is bad enough."

"Speaking of depression," said Alice, resting her hand lightly on Trudy's arm, "how've you been?"

"It comes and goes," said Trudy, then she shook her head. "It's the anxiety attacks that really trouble me."

"The medication doesn't work?"

"It does, but I don't have any advance warning, so by the time the attack hits it's too late. I take the pills and they take a half-hour to work. Then I'm zonked out."

"I guess what you need is something you could take ahead of time to prevent the attacks. I'll ask at the hospital."

"That would be nice," said Trudy. "By the way, since we're on the topic, how's Dad?"

"Oh, like you, I guess—good days and bad. But I think there's been a gradual improvement."

Back in the cottage Willis and Al read the newspapers and listened to the radio. As an American, Al had been puzzled by the enthusiastic reception the King and Queen had received in Washington.

"You'd think they were our king and queen," he said good-naturedly.

"I must confess it surprised me too," Willis admitted. "But your FDR really put on a show."

"Don't get me started on him."

"Not a fan?"

"I'm a third-generation Republican, and proud of it."

"What do you Americans make of all this talk of war?" Willis asked.

"Not our war," said Al with a smile. "You might get dragged into it, but we won't. Hitler's not our problem."

The children were put down for a nap after lunch, and when they woke up it was the men's turn to supervise them. If the weather was fine they might return to the beach or take a walk. On cooler days, or when the wind whipped the waves into a rough surf, they would take the children to the town pier and watch commercial fishermen coming in with their catch. If there were fresh perch, they'd buy some and take them back to the cottage, where Alice would pan-fry them for supper.

When Claire and Ed Pritchard arrived, Michael, their son, who was a year younger than Andy and Joan, tended to complicate things. He wanted to do everything they did, but he wasn't quite up to the challenge. Claire had to make sure he didn't get ahead of himself. Willis found Ed Pritchard a little distant. He was the complete opposite of Al Kingerley. While Al was tall and charming, a true salesman, Ed was short and quiet, even taciturn. But the three men managed to get along.

The stage was set for the arrival of J.B. and Blanche. Doris, the eldest of the Champion offspring, would drive them from Essex in time for lunch. When the plans were in place Trudy began to fret. She was determined that Willis would gain the approval of her family, but she was beginning to have trouble sleeping. She had increased her bedtime dose of Sodium Amytal, and when she woke up late on the morning of her parents' visit she was groggy from the medication.

At the breakfast table Alice could see that her sister was apprehensive. "Relax, Trudy," she said as she poured her a glass of orange juice. "Everything is going to be just fine."

"You're sure they know Willis is here?"

"Yes, they do. I talked to Blanche last night."

Willis, Ed, and Al agreed to take the children to the beach so that the wives could prepare lunch. This gave Trudy something to do and helped distract her from what lay ahead. They set the table on the screened porch. At least the weather had cooperated. It was a fine sunny day.

Later, the men brought the children back from the beach, and they were bathed and outfitted in clean clothes. Shortly after noon the rumble of a car making its way along the lane from the main road to the cottage could be heard.

It pulled into view and came to a stop. Doris emerged smiling, and Trudy, Alice, and Claire crossed the lawn to greet her. Their husbands accompanied them. Then the back door opened and Blanche got out, followed by J.B.

Doris looked around. "What a place you've got here," she exclaimed.

"We like it," said Alice. "Plenty of room for everyone."

Suddenly the screen door swung open and the three children ran to meet their grandparents, followed by Willis and Al. Blanche bent down to kiss them as J.B. watched impassively.

Al stepped forward and gave Blanche a hug. "Hope the drive wasn't too tiring," he said, and turned to J.B. "How've you been?"

J.B. shrugged.

Blanche turned to Willis. "And it's been a while since we've seen you, young man. I hear you've been busy. You must tell us all about it."

Willis bent down to give Blanche a kiss on the cheek.

The moment had arrived. Willis managed a smile and extended his hand to his father-in-law. "Good to see you, sir."

J.B. looked at Willis. Trudy held her breath. After an uncomfortable pause the two men shook hands.

Alice broke the spell. "Let's get inside, out of this heat." They moved into the cottage. "If you'd like to wash up, the bathroom is through the hall. When you're ready we can have lunch."

When their parents were out of sight, Alice gave Trudy a thumbs-up.

At lunch Blanche quizzed Willis. "So you actually met the King?"

"Briefly," said Willis.

"What's he like?"

"Surprisingly shy. But the Queen was very friendly and down-to-earth."

"We listened to the tour on the radio. As we did I kept thinking that you were out there somewhere, making it all happen."

"I had a lot of help," Willis said modestly.

"Did you see the gift the royal couple gave Willis?" Alice prompted.

"A gift?" said J.B., raising an eyebrow.

"Trudy, why don't you show them?"

Trudy rose from the table and went back into the living room. When she returned she was carrying the sterling silver cigarette box. She handed it across the table to Blanche.

"My, my! This is beautiful! And there's a map engraved on it!"

"It shows the route the tour followed," Willis explained.

She handed the box to her husband. "Look at this," she said, then turned to Trudy. "You must be very proud of Willis."

"Oh, I am, Mother, very proud."

J.B. was examining the box. "Sterling silver," he commented. "Very impressive." He handed the box back to Trudy. "Very impressive," he repeated.

The rest of the lunch was uneventful. Much to the relief of Trudy, her parents turned down the invitation to stay overnight. "Doris has to get back," Blanche explained.

When their car drove out of sight Trudy gave a sigh and fell back on the sofa beside Willis. "I'm so glad that's over." She patted her husband on the knee. "But maybe Dad is finally coming around."

"I guess he is," Willis admitted. "But you know, he doesn't look well. He's lost weight since the wedding."

"He's been through a lot," Claire explained. "First of all there's Ethel;

he was never comfortable there. Then there was the move to Essex, and the leave of absence from the church, and eventually his retirement."

"Now he has plenty of time to listen to the CBC," said Willis in an attempt to introduce a little humour into the conversation.

"You know," said Trudy, "he's a devoted fan, although he probably wouldn't admit it. The CBC provides classical music and Dad loves that, particularly Bach."

"And those Saturday afternoon broadcasts of the opera from New York," said Alice with a smile. "He never misses those."

"He certainly liked that cigarette box," Trudy mused.

"I think we can safely say you've broken the ice, Willis," said Alice. "No more hiding away when there are family affairs."

Willis smiled. "I suppose so." But in truth he hadn't minded the estrangement. It had been hard on Trudy, but not on him. In fact he wouldn't have cared if he never saw his father-in-law again. He couldn't put his finger on the reason he felt this way. It could be Trudy's obvious dislike for her father, of course, but Willis thought there was more to it than that. He'd always had a feeling that the man was hiding something.

Willis and Trudy decided to wind up their vacation with a visit to Leamington, so that they could spend some time with Anna Maria and Clyde. It was out of their way, but Willis was worried that if war did break out, gasoline would most certainly be rationed, and that would make it more difficult to get to see his parents.

The afternoon after their arrival Anna Maria and Willis were sitting in the kitchen. Willis had been out most of the morning helping a friend repair a faulty radio. Andy was in his grandmother's lap, nibbling a donut. Upstairs they could hear the sound of Trudy's footsteps as she moved about.

Willis smiled at his son. "It's a nice afternoon. I guess I should take him for a walk."

Anna Maria shifted her grandson a little, relieving the pain in her arthritic hip. "I spoke to Vernon this morning," she said. "That's Vernon McGwire, one of the locomotive engineers who works the Heinz siding. He said things are pretty quiet these days. He wondered if you'd want to show Andy the engine."

"That'd be great," said Willis. "Do you think he'd let us up into the cab?"

"I'm sure he would."

Anna Maria knew most of the train engineers by name. Over the years she'd provided them with many a cool drink of well water during breaks in the shunting routine. They'd stop their big locomotives opposite the house, jump down from the cab, and take the chipped enamel water jugs from her. Later in the day, on a return run, the jug would be handed back.

"What time will he be coming through?" Willis asked, glancing at the alarm clock on the shelf above the stove.

"In an hour or so," said Anna Maria.

Willis had Andy ready to go when the big locomotive shunted onto the siding that ran parallel to Clyde and Anna Maria's backyard. His son had been playing with some blocks on the front porch. The noise of the trains had frightened him at first, but he'd always been a safe distance from them, and in time that fear had disappeared. He'd even learned to wave at the engineers, and was usually rewarded by a wave back.

Willis picked Andy up and headed down the steps as Vernon descended from the cab of the locomotive. Anna Maria, on the porch, stood watching, and Clyde was beside her in his rocking chair.

As they approached the train Andy grew apprehensive. The huge black locomotive, hissing steam, seemed alive to him. He started to cry. Willis, embarrassed, sought to reassure his son.

"It's okay, Andy, there's nothing to fuss about. We're just going to meet Vernon. You know Vernon—he waves to you." Willis wasn't paying much attention to the crying, which was getting louder and more frantic the closer they moved to the locomotive.

Andy buried his head in his father's shoulder, but couldn't hide from the sound. He began to struggle, kicking his feet.

"There, there, you're not going to let a train scare you, are you?" Willis spoke in a soft, reassuring voice.

Vernon, not quite sure how to handle this, approached awkwardly.

"Hello, little fella," he said tentatively.

They were beside the locomotive now. The smell of the oil, the heat, and the hissing of the steam were too much for Andy. By this time he was sobbing non-stop.

"Maybe he'll calm down if we get him in the cab," Vernon suggested. He turned and mounted the ladder.

When Vernon was in the cab Willis lifted his son up toward the engineer. At this point Andy became hysterical, kicking and screaming. One of his kicks struck his father's cheek, and Willis staggered back, almost dropping the boy. When he recovered his balance he tried to restrain Andy and stop the struggling by clasping him close to his chest.

"Maybe today isn't the day," said Vernon from the cab.

"I guess not," said Willis, still trying to control Andy. "Perhaps another time, Vernon. Thanks anyway."

Willis was disappointed. As they made their way back to the house, he admonished his son. "I'm surprised at you, a big boy, three years old and frightened by a train."

Andy's crying subsided into whimpers. But Trudy had heard the fuss and had come down from the bedroom. She crossed the lawn and took Andy from her husband's arms.

"There, there now . . . it's all right . . . it's all right," she murmured. "You're with Mummy now; it's going to be okay."

She looked at Willis. "That was a damn fool idea. Anyone with half a brain could see a child would be frightened by something like that."

Willis looked past Trudy to the porch, where Anna Maria stood watching and his father sat listening. He shrugged his shoulders. "I think part of his problem is that you're too protective," he said as he climbed the stairs and went into the kitchen, letting the screen door slam behind him.

"And you . . .," said Trudy, looking angrily at Anna Maria, "you should have put a stop to this nonsense before it began." She followed Willis into the house, holding her son tightly in her arms.

It was a long time before Andy waved at passing trains again.

Chapter 59

Toronto, 1939–41

When Hitler's troops invaded Poland, on the first of September, 1939, Britain immediately declared war on Germany. At the CBC this was no great shock. The Corporation's foreign correspondents had been predicting it for months. Official word of Canada's declaration of war, which happened ten days later, reached the CBC newsroom on a Sunday afternoon. The network was carrying an NBC program, *Music for Moderns*. A studio technician faded the music and cued the announcer, Austin Willis: "We interrupt this regularly schedule program to bring you a news bulletin from Canadian Press. Canada has officially declared war on Germany. I repeat, Canada has declared war on Germany. That was a news bulletin from Canadian Press. We now re-join the regularly scheduled program."

The fact that this significant development was announced in such a bland manner prompted *The Financial Post* to comment the next day that the CBC had failed Canadians, that it had provided "no sense of the gravity of the moment."

Willis heard the news at home. He had just finished reading the Sunday paper and was listening to the radio. Trudy was feeding Andy.

"There'll be hell to pay tomorrow," he said angrily. "We should have had a program ready for this. We knew it was only a matter of time. Or at the very least we should have cut away from the NBC show and played a suitable selection of classical music."

"I can't believe it," said Trudy, shaking her head. "Another war. And there's Edith stuck in Germany with a two-year-old son. She wouldn't listen to reason, wouldn't get out last year when she had the chance. I just wish there was something we could do."

"We've been on this story all week," said Willis. "The CBC carried the

BBC broadcast of the King's message to his subjects, as well as accounts of the fighting in Poland. We covered our own parliamentary debate on the war, and just yesterday there was a speech by the Prime Minister. We prepared our listeners for the worst, but then when it came to one of the most important moments in history, we dropped the ball."

Willis decided this wasn't the appropriate time to mention his own intentions. He was determined to enlist. The next day at lunch he went to one of the hastily opened recruitment centres. It was crowded, but a harried clerk handed him forms to fill out. He sat down and did the paper work, then stood in line, waiting his turn.

A young officer looked up and Willis handed him his application. The man read it through, taking his time. Then he looked up.

"It says here that you are thirty-four?" said the officer, indicating the forms.

"Yes, that's right."

"And you have a wife and a young son?"

"I do."

"Any other dependents?"

"I help my parents out. They live in Leamington. Dad is blind, has diabetes."

The officer nodded, wrote something down, and looked up again.

"Any previous military experience?"

"No."

"You weren't in the cadets?"

Willis shook his head.

The officer slid the application into a file folder. "We'll keep this," he said. "I appreciate that you'd like to serve, but I'm afraid you don't meet our requirements, at least not for now. We're looking for younger men without children. Perhaps if the war drags on we may eventually need someone with your qualifications, but not at present."

It was clear to Willis that the interview was over. "Well, thank you," he said, and turned away. He left the office feeling a mixture of sadness and relief. As much as he would have liked the chance to fight for his country, he realized that risking his life would be foolhardy, particularly with a fragile wife and a young son to worry about.

At the moment he had another problem. The owner of the house on

Cranbrooke Avenue had died suddenly, and his heirs had decided to put it up for sale. They offered it to Willis and Trudy at a reasonable price.

Trudy had mixed feelings. She had put down roots, and had made friends in the neighbourhood. "I'm not sure I can take another move," she said. "And then Andy would have to find new playmates." But she also knew that even at a decent price, the purchase of a home would put a strain on their finances.

"With a war on, I don't think the timing is right," said Willis. "My future isn't that clear. I've really just begun at the CBC. I could be transferred at any time."

Fortunately, a solution soon presented itself. The bottom half of a duplex was available just down Cranbrooke from their bungalow. The rent was slightly less than what they had been paying, and there was even an extra bedroom. They could move in at the end of the month. They signed a lease, and Trudy breathed a sigh of relief. Her only concern now was privacy: there would be people living above them. She hoped they would prove to be friendly.

It was out of the question for Willis, with his bad back, to try to move the furniture and his heavy woodworking tools himself, so he called a local transport company. On the first of October a truck arrived with three men, two of whom were burly and the third small and wiry. As Willis and Trudy watched, the team quickly and efficiently transferred their belongings to 133 Cranbrooke Avenue.

At the CBC everyone was caught up in the challenge of broadcasting the news of the war to Canadians. The Corporation had some first-class announcers, including Austin Willis, Lorne Greene, and Earl Cameron, but the news department was dependent on Canadian Press for all its information. There were editors, but they were essentially re-write men, restricted to the facts provided by CP. This arrangement had been a sore point with CBC journalists from the day the Corporation was established. In one camp were those who did not envision the CBC as a newsgathering organization. They were content to make do with token newscasts from Canadian Press wire copy. They had had their way until the war broke out. Opposing them were a group who saw a wider role for the news department. Ernie Bushnell put it this way: "What we get from CP is not good enough—stale news poorly written and often poorly chosen. They do not know how to write for oral delivery; they do not know how to build a radio newscast, giving it shape just as a musical production has a shape."

Fortunately the CBC correspondents who covered the war were under no such restraints. Canadians soon became familiar with the names of Matthew Halton, Bob Bowman, and Peter Stursberg as these men sent back first-hand reports of the fighting in Europe. These accounts were relayed by "transatlantic beam" from London.

Art Holmes was the engineer who made many of these broadcasts possible. He drove the CBC's recording van around London throughout the Blitz. Nicknamed Big Betsy, it was one of the three mobiles that had been built for the royal tour, Thanks to it, Canadians were able to hear air-raid sirens warning people of impending attacks, the actual sounds of bombs exploding, and the wail of fire engines and ambulances.

At home, a war of words was taking place between politicians and CBC executives. The federal government's position was that in wartime it should be the one to decide what should and should not be broadcast. Opposition parties complained that the CBC was becoming a tool for spreading Liberal propaganda. Newspapers picked up on this theme, and editorials questioning the CBC's integrity began to appear. Willis was aware of this infighting, but fortunately it hadn't permeated the engineering department. If anything, the war provided a justification for the expansion and further improvement of the studios and transmission facilities across the country.

Early in 1940 Willis received notice that his name had been selected for jury duty. He turned up on the appointed morning at the courthouse, fully expecting to be excused because of the nature of his work. But jurors were in short supply, and he was told that his job at the CBC couldn't be deemed essential to the war effort. He would be required to serve. This news did not go down favourably with his bosses at the CBC, and they protested, but in vain. So for three weeks in February Willis was forced to hear a criminal trial involving a bank robbery. Fortunately, the evidence against the defendants was compelling, and a verdict of guilty was reached quickly. However, when he returned to work he found a new man in charge of the Toronto engineering department. Royce Burton was a tall, taciturn individual with a perpetually suspicious nature. Willis wasn't impressed.

If he had felt free of the political infighting between the government and the CBC brass in the past, he soon realized that things had changed. The new appointee had connections within the Liberal party. He was an engineer, to be sure, but a mechanical engineer with what Willis deemed to be a decidedly limited understanding of radio. To make matters worse, it was clear that he regarded the fact that Willis had spent time on jury duty as a dereliction of his CBC responsibilities. He implied that if Willis had bent the truth a little he could have avoided it.

"I don't bend the truth for anyone," said Willis angrily.

"As you wish," Burton replied, his voice clipped and dismissive. "But I expect a certain degree of ingenuity from my staff."

Willis soon found himself out of the loop, frequently assigned to low-level duty at the Hornby transmitter. When he learned from his annual performance review that Burton had denied him his expected salary increase and deemed his work "unsatisfactory at this time," Willis began looking for the possibility of a change. He found it in a posting for a job in Saskatchewan, as the chief engineer at the network's repeater transmitter in Watrous, sixty miles southeast of Saskatoon.

Trudy was upset when he told her he had applied for the job. "Where the hell is Watrous?" she wailed.

Willis got out the map, but assured her that they wouldn't actually be living in Watrous, a tiny prairie farming town. "We'll live in Saskatoon," he explained, "and I'll commute."

"Saskatoon?" she said, her voice dripping with sarcasm. "Oh, that's just great. I can hardly wait."

"It won't be so bad," Willis protested.

"You and your principles," said Trudy with an exasperated sigh. "They cost you your job at CKLW, and now this."

"I can't help it, Trudy. As long as Burton is in charge I'll have no hope of advancement, and my salary will be frozen."

"In Saskatchewan more than your salary will be frozen. Do you have any idea how cold it gets out there in the winter?"

As it happened, they didn't have to worry about the prairie weather. At the last minute the plan changed. An urgent request for Willis's services had come to the CBC from the National Research Council in Ottawa. With his experience he was an ideal candidate for the position of supervising engineer in what was cryptically described as "top secret war work."

Willis was delighted. This opportunity meant that he'd be playing an active part in the war effort; he'd no longer be relegated to the sidelines. It would also give him the chance to meet Dr. Frederick Banting, the scientist who had had discovered insulin and thus saved his father's life. Banting was a fixture at the Council. Of course Trudy was relieved.

"I'll take Ottawa over Watrous any day."

The following week, an RCMP inspector was dispatched to interview the ideal candidate in his home. "We have to make sure you aren't a security risk," he had explained to Willis over the phone.

Inspector Wilde was in his forties, a squat, powerful man with a military moustache and a face that gave nothing away. He spent the better part of an evening going over a file he'd brought with him. It seemed, from his questions, that the man already knew a great deal about Willis. Most of the questions were simple and straightforward. Then the inspector surprised Willis.

"How well do you know George Ketiladze?"

"We were together at Queen's." Willis wondered what his old friend had to do with this.

"Have you stayed in touch?"

"Not for quite a while. He was teaching at Queen's, but left to work in advertising. It's been a couple of years . . . I think he's in Montreal."

The inspector folded the file into his briefcase. "Can we ask you to keep clear of him?"

"Why?" Willis asked.

"I'm afraid I'm not at liberty to say. But if he does seek you out, we'd like to know about it."

"Is he in some kind of trouble?"

"Sorry," said the inspector, but it was clear the word wasn't an apology. He got up to leave and shook Willis's hand. "We'll see you in Ottawa," he said, and departed.

The CBC's director of personnel drew up an agreement that Willis would be "loaned" to the NRC. But he'd be reinstated within the CBC, his seniority intact, once his duties in Ottawa were finished.

"Thank God for small mercies," Trudy exclaimed when she heard the news.

A representative from the National Research Council called the next day and, after explaining that because of the war there was a housing shortage in Ottawa, asked, "Would you and your family be willing to stay at a hotel until we find you a permanent place?"

"Sure," said Willis, glancing at Trudy.

"We'll make reservations at the Chateau Laurier," he said, "and perhaps you can arrange to have your furnishings put into storage for the time being. We'll reimburse you, of course, for whatever expenses this may involve. And when we find you a home to rent, we'll handle the cost of moving."

"What was all that about?" Trudy asked as Willis hung up the phone.

He explained the details.

"Wow, the Chateau Laurier! They must really want you." Trudy was even more pleased when she learned that her husband's annual salary at the National Research Council would be $3,240. It was, Willis noted, even more than he would have been paid at the CBC if he hadn't fallen out of favour with Burton.

VII
Ottawa

Chapter 60

Ottawa, 1941

Willis and Trudy drove to Ottawa early in May, with Andy playing happily in the back seat. He always behaved well in the car, and on this trip he even managed a nap. They checked into the Chateau Laurier, and after they had moved their bags into their room Trudy called Willis to the window.

"This is some view!" she said excitedly. From their south-facing fifth-floor room they could see the Rideau Canal and the small boats that were moored beside it. "We seem to be right in the heart of town!"

"We are," said Willis, who had become acquainted with Ottawa during the royal tour. "That's the War Memorial over there, to the right. The King dedicated it two years ago, and the CBC covered that. And the Parliament Buildings are just up the street."

"Tell the people at the Council to take their time finding us a house," she teased. "I think I could get used to living here."

Willis found that he could get to work by walking along Sussex Drive from the hotel. The National Research Council was an impressive building on the south bank of the Ottawa River with eight free-standing sandstone Doric columns on each side of the main entrance. It resembled Buckingham Palace, although it was on a smaller scale.

Willis paused outside the building and experienced a moment of regret that he would never have a chance to thank Dr. Frederick Banting, the man whose discovery of insulin had saved his father's life. The NRC scientist had been killed in a plane crash in Newfoundland a few months earlier. Willis would quickly learn how huge a loss it had been. Banting had been one of the driving forces behind the Research Council's war effort.

Willis's first day at the Council was taken up with interviews. Some were security-related, covering ground already discussed with Inspector

Wilde. Others, of a more technical nature, tested Willis's knowledge of electronics. He was introduced to John Henderson, the co-ordinator of radio research, a tall, rangy man with a mop of unruly white hair.

"How much do you know about short-wave radio?" Henderson asked.

Willis explained that he had helped install a short-wave transmitter in the CBC's Hornby facility. "It was tested, but never used," he said.

"Why not?"

"It was seen as insurance, something to be used in an emergency. But we did use short-wave during the King's visit to Canada in 1939. When he was leaving from Halifax one of our reporters relayed his commentary via short-wave from an escort destroyer."

"How about radar?"

"Isn't that what the British are using to locate enemy bombers crossing the English Channel?"

"Yes, the Brits have a network of radar stations all along the coast. Radar can pick up ships as well as planes. These stations have been rushed into production, because the British are afraid the Germans may try to launch an invasion."

"Where do we come into the picture?"

"Have you heard of the cavity magnetron?"

Willis shook his head.

"It was a major breakthrough in the evolution of radar. It greatly extends the range and accuracy of the microwave beams. We've been asked to design and produce mobile radar stations using the cavity magnetron."

Willis felt a surge of excitement. "Sounds challenging," he said.

"Oh, it is," said Henderson, "but it's a challenge we're ready to face."

As he walked back to the hotel Willis realized that all those interviews had taken their toll. He felt tired, but at the same time energized. At the end of the day he'd been assured that the screening was almost complete; he'd soon receive his assignment. He wondered exactly what he would be doing. When he asked, he was given only vague, unsatisfactory answers. It was clear that secrecy was paramount at the Council.

Ottawa was no longer the quiet capital he had surveyed for the tour in 1939. It was still dominated by civil servants, but the war had brought in a new group of professional people from across the country. Willis felt that there was a sense of urgency in the city, something that had been lacking the last time he walked the downtown streets.

Trudy seemed to be enjoying herself as well. "We visited the Parliament Buildings today," she said excitedly, "and we went to the top of the Peace Tower. The views were spectacular."

"How about you, young fellow?" Willis asked his son.

"I liked the Mounties," said Andy. "They all had shiny leather boots and red jackets."

"Do you know what their motto is?"

"Motto?"

"It's a saying, a slogan."

Andy shook his head.

"They say that they always get their man."

Andy looked questioningly at his father, seeking to make sense of this.

"They're policemen," Willis continued, "and when someone does something wrong, it's their job to catch the bad guy. They're very good at catching them, so the slogan 'They always get their man' means that they always capture the bad guys."

"Do they all dress that way, with those shiny boots and funny hats?"

Willis laughed. "No, most of them dress the way I do when I go to work. The red uniforms are what they call 'ceremonial.' They only wear them when they're on duty in public."

The NRC personnel department found other temporary accommodations for the family in June. It was a cottage on Britannia Bay, a small community on the Ottawa River just west of the city. Trudy was sorry to have to give up the luxury and convenience of the hotel, but the idea of living in a cottage for the summer was appealing.

On Saturday morning they packed their things and made their way to their new neighbourhood. They had to step carefully along a makeshift stone path to reach the cottage. No one had warned them that the spring floodwaters had yet to recede. Inside the cottage, a dank, musty smell greeted them, a clear indication that it hadn't been in use since the previous summer. Trudy looked around in dismay. "My God, Willis, this place is a mess," she said, cradling Andy protectively in her arms.

Willis tried to be upbeat, "We can clean it up," he said. "Air it out." He reached for an overhead light, but nothing happened when he flicked the switch.

"Oh, great!" Trudy moaned. "Stuck out here in the dark ages." She moved into the kitchen. Beside the sink was an old fashioned pump. "I haven't seen one of these since we were kids in New Brunswick."

Willis tried the pump. It squeaked in protest.

Trudy was close to tears. "No lights, no water!"

"Just needs priming," Willis explained. He went outside, filled a bucket from the river, and returned. He poured water into the pump and tried the handle. Still nothing. He added more water. This time there was less squeaking. He added still more water, and eventually he was rewarded. The pump produced a rusty trickle.

"Ugh!" said Trudy in disgust.

"Give it time," Willis said, and continued to pump. Eventually the water ran clear. "That's one problem solved." He went back outside and found the electrical switch box at the back of the cottage. He turned the power on and heard Trudy's voice from inside the house.

"Lord, let there be light," she called, relief evident in her voice.

They spent the better part of the day cleaning the cottage, airing it out, and making an inventory of what they would need. It was bad enough that there was no hot water, but when Trudy realized that there was no toilet she lost control.

"This is impossible!" she shouted. "Do they really expect us to live here under these conditions?" Andy, who had been playing contentedly on the kitchen floor, looked up at his mother in alarm.

Willis was becoming fed up with her complaints. "Trudy, in case you hadn't noticed, there's a war on. Everyone has to make sacrifices. This is just temporary, until they find us something else."

Willis needed the car to get to work, and this increased his wife's sense of isolation. There was no Mrs. Fairchild in Britannia Bay, and Trudy's depression began to take hold once again.

"I can't take this anymore," Trudy sobbed one night in July when Willis returned from work. "I'm losing what little peace of mind I'd gained while we were living in Toronto. When I looked at the Ottawa River today I was tempted to just walk into the water and keep on walking."

"Don't even think such thoughts," said Willis. "We'll eventually get you the help you need. I'm sure of it. This situation is just temporary, it's just a rough patch."

"It's too rough."

"I have an idea," said Willis. "Would it help if you had the car?"

"I suppose so," Trudy managed. "At least I could escape from this . . . prison. But how would you get to work?"

"You could drive me to the streetcar stop in the morning and pick me up again in the evening. I don't have any real use for the car while I'm at work. It sits in the parking lot all day."

"That might help," Trudy conceded.

"Let's give it a try," said Willis.

This arrangement proved to be the tonic Trudy needed to make it through the summer. She didn't flee the cottage often, but just having the car there and knowing she could escape any time she wanted provided a measure of relief. Andy was clearly enjoying himself, paddling happily at the river's edge with some new playmates. The outhouse remained a sore point, though. Trudy never used it without registering a protest.

At the Research Council Willis had been assigned to the Radio Branch, where radar was the top priority. Radar had proven effective against planes and ships, and had helped give the Allies victory over the Luftwaffe in 1940 during the Battle of Britain. Now it was to be employed at sea in the Battle of the Atlantic, where it was hoped that it could be used to detect German submarines. Nazi subs were sinking an alarming number of freighters carrying troops and supplies across the Atlantic. Passive sonar was being used—it operated by lowering microphones into the water to detect the sound of submarine engines—but it had limitations. When depth charges were deployed, the location signal was interrupted and accuracy was compromised.

"Using radar principles," Henderson explained, "we transmit signals under the water. When they hit a submarine they bounce back. This helps us determine the location of a sub more precisely, and no problems arise when depth charges are used. Instead of passive sonar, we have active sonar."

"Impressive," said Willis.

"So now we have to design radar equipment for use on ships."

"Sounds like an expensive proposition," said Willis.

Henderson smiled. "It is, but fortunately Santa Claus came through."

"Santa Claus?" Willis wondered if he was being teased.

"A group of wealthy Canadian businessmen realized that the government was dragging its heels at a time when we desperately needed more money, so they donated one hundred million dollars for war research. It came as a complete surprise to us, and we took to calling it the Santa Claus fund."

Willis gave a low whistle. "Some gift!" he said.

"It was a godsend. It tided us over until the government got its act together. Now, with an increased budget, we're able to expand into new areas. Today it's less about money and more a about finding properly trained personnel. That's where men like you come in."

"I'm flattered," Willis said.

As he was about to leave Henderson's office Willis remarked, "It must have been quite a shock to lose Banting."

"It was an absolute tragedy. We couldn't believe it when his plane went down. He was an inspiration to all of us here. Everyone loved him."

In September the personnel people at the NRC finally found Willis and Trudy a real home, the bottom half of a duplex at 46 Western Avenue. While it was across town from his offices, the streetcar connections were direct. Trudy would once again have the use of the car, although with gas rationing in effect, her trips were restricted.

They enrolled Andy in kindergarten, and Trudy feared there'd be a scene when she dropped him off on his first day. He surprised her, though. "He just waved goodbye and walked into the classroom without even looking back," she told Willis over dinner. "I checked with the teacher later and she assured me he'd had a fine time; he had fit right in."

"That's good," said Willis. He thought Trudy worried too much about their young son.

They had decided against having more children for the time being, because of the frightening bout of postpartum depression she had experienced. When Trudy expressed some concerns about raising an only child, Willis pointed out that he'd been an only child. "I survived," he said good-naturedly. "I'd say I turned out pretty well."

Trudy responded with an arched eyebrow.

Now that they were installed in a proper house it was time to make good on an earlier promise to Andy. He could have a dog. After some discussion they agreed to get a cocker spaniel. Trudy found an ad in the paper, and on the first weekend in October the family drove to see the litter. Fascinated and awestruck, Andy stared down at the tiny puppies.

"Which one would you like?" the owner asked. She was a middle-aged woman, plump and engaging.

Andy instantly pointed to a tan pup. "That one," he said emphatically.

"I think you've made a fine choice. Come back in a few weeks and you can pick her up."

Andy looked at his parents questioningly. "Can't we take her now?"

Trudy bent down to his level. "No, she's still a baby. She needs her mommy for a while."

Andy's eyes brimmed with tears.

"What would you like to call her?" asked Trudy, cajoling the child, hoping to forestall a round of full-scale crying.

It worked. "I'd like to call her Taffy," Andy said.

"Taffy it is, then," said Willis.

For the next two weeks Andy couldn't stop talking about their new pet.

When the big day finally arrived, however, they received a shock. The tan pup had been sold.

"I'm sorry," said the owner. "My husband didn't realize she'd been your choice."

The remaining pups were all black. Andy looked at his parents, clearly disappointed.

"If you have your heart set on a tan pup we can make sure you have one from the next litter."

Willis reached down and picked up one of the tiny black spaniels. He held it out to Andy, who took it tentatively in his hands.

"He's so warm," he said, burying his nose in the fur. "And he smells so good."

"It's up to you," Willis said. "We can wait a few months or take this one home."

It was no contest. Andy cradled the pup in his arms as they drove home.

"What shall we call her?" Trudy asked.

"Taffy," Andy said without hesitation.

Trudy looked at Willis questioningly. He shrugged. Then he reached across and stroked the new pet. "Welcome to the family, Taffy," he said, doing his best to contain a smile.

Taffy wasn't the only addition to the family. Recalling how Mrs. Fairchild had helped Trudy cope in Toronto, Willis sought the services of a housekeeper. They soon hired Mrs. Elliott, a petite woman about the same age as Trudy whose husband was away in the navy. Trudy was used to the matronly Mrs. Fairchild, so it took her a few weeks to get comfortable with a younger woman. But Andy was no longer a baby, and Mrs. Elliot took on a larger role in his care than Mrs. Fairchild had. She became part housekeeper and part nanny. Andy, for his part, adored her.

Willis began to search for a doctor for Trudy. He steadfastly believed they'd eventually find someone who could restore Trudy's health to the state it was in before the birth of their son. Some of Willis's colleagues at the National Research Council had medical backgrounds, and they provided him with the names of Ottawa physicians, including a pediatrician to tend to Andy's needs. Appointments were made, and Trudy began making the rounds of the recommended doctors. Unfortunately, none of them had anything to say that Trudy hadn't heard before in Toronto, so she continued to rely on Sodium Amytal to treat her anxiety. Gradually the combination of a proper home and the services of a housekeeper helped Trudy emerge from the depression brought on by their summer in the primitive conditions of the Britannia Bay cottage.

Willis soon set up his power tools in a basement workshop and began making new furniture for their Western Avenue home. One of his first tasks was to fashion a ramp, so that Taffy could have access to the fenced-in yard at the back of the house. A small balcony off the kitchen overlooked the yard. Willis found that if he removed one of the vertical slats in the railing there would be enough space for the spaniel to squeeze through. He attached strips of wood to the ramp surface to provide her with purchase on her way up and down. It took him a few days to train the pup, but she was soon negotiating the ramp with remarkable speed. They still took her for walks, but when time didn't permit this, they'd simply open the door to the balcony and she would bound out. The only negative aspect of this arrangement was the weekly chore of picking up her deposits from the yard.

Chapter 61

Ottawa, 1941–42

Alice and Al Kingerley drove from Detroit to Ottawa in October for a first-hand look at Trudy and Willis's new home. Gasoline wasn't rationed in the United States, so they took an indirect route, travelling east across Michigan and through upstate New York, crossing the border south of Ottawa. They brought Joan along with them, and soon Andy was proudly displaying Taffy to his cousin. The encounter produced the inevitable request from Joan "Mom, why can't we have a dog?" she asked plaintively.

"We'll see," Alice answered, glancing at her husband.

Al shook his head. "Not until you're old enough to look after it."

"I look after Taffy," Andy offered helpfully.

"I'm sure you do," said Alice, then turned to her daughter. "We can talk about this when we're back home."

Joan understood from the tone of her mother's voice that the subject was closed, at least temporarily.

The talk among the adults was about the war. They had finally made contact with Edith in Germany, through the Red Cross. The details were sketchy, but at least she was safe for the time being.

Trudy asked about her brothers. "Wendell has joined the army," said Alice, "but George is now an American citizen, so thankfully he is out of it."

"For now," said Willis cryptically.

"What do you mean?" said Al indignantly. "There's no way the United States is going to be drawn into a European war."

"Perhaps." Willis conceded the point, but he knew from the nature of his work at the National Research Council that Americans were highly active behind the scenes. This was confidential information that he couldn't share with the Kingerleys.

"How have you been?" Alice asked Trudy when the two sisters were alone. "Do the doctors here have any suggestions?"

"No," said Trudy, shaking her head sadly. "I'm afraid Ottawa is a little behind the times when it comes to medicine. I actually think I was better off in Toronto, although no one there had a solution either."

"Well," said Alice, "the whole subject of postpartum depression is just beginning to be understood in the United States. I'm sure there'll be developments soon, with all the research that's being carried out. No one knew how serious the problem was until recently; happily, that's changing."

"I hope so," said Trudy. "Willis is optimistic, though. He keeps my spirits up."

Later that evening, after Trudy had gone to bed, Alice joined Willis in the living room. "How are you holding up through all of this?" she asked.

"Oh, it's not easy, but I'm managing."

"It must be hard for you to concentrate at work, with Trudy always on your mind."

"It's always there, but over time I've learned to compartmentalize things. I do the best I can for her, but I don't let it interfere with my job. It might sound callous, but for me work is an escape. I lose myself in whatever I'm doing. I rationalize it this way: I can't very well stay home with Trudy; the year I was unemployed was a disaster. I have to earn a living, and she needs the financial security my job provides. So while I'm at work I'm doing the best I can for her, in my own way. Does that sound selfish?"

"No," said Alice, "not at all. It sounds practical . . . pragmatic. You're doing the best you can for her. Other men faced with your situation might have walked out."

Willis shook his head. "I could never do that. Never."

"I know," said Alice. "And Trudy is lucky to have such a loyal husband."

Since they were in Ottawa, Alice asked to visit the Britannia cottage. Trudy had written about it in her letters and her sister was curious. Could it really be as primitive as it seemed from Trudy's description?

"Willis can take you there," said Trudy with a shiver. "I'll stay here and look after the children."

They decided to take Al's Buick for the trip. The cottage was closed up for the winter, but Willis had a key, and he showed them around. "To be frank, I think we were lucky to have it. Housing is a real problem in Ottawa with the war on. Andy loved it, but Trudy felt isolated. I can understand her feeling that way, but we were only there a couple of months."

Alice took in the surroundings and immediately understood the reasons for her sister's complaints. With no hot water and an outhouse for a toilet she would have been uncomfortable there too.

Back at Western Avenue she shared her impressions with Trudy. "I'm amazed you stuck it out," she told her sister. "I think you're a lot stronger than you sometimes think."

"It wasn't easy," Trudy said, "but I could see that Andy was enjoying it, and that helped a lot."

Alice smiled. "Kids are certainly adaptable," she said. "And speaking of kids, Doris's boy is talking about joining the air force."

"Don? My little nephew?" Trudy was amazed. "Is he old enough?"

"He'll be eighteen this year."

"The war news isn't good," said Trudy, shaking her head. "And now, with Wendell already in the army and the prospect of Don joining up, it's becoming a lot more personal."

"You must be thankful Willis didn't enlist."

"I'm sure he gave it some thought. But thank goodness he found a better way to serve. He can't talk about his Research Council work, of course, because it's all top secret, but I know it's completely focused on war projects."

"We're so lucky in the States," said Alice. "I'd hate to think of our boys off fighting in some foreign country."

Two months later the Japanese bombed Pearl Harbor and the world changed. The United States was suddenly thrust into the war. Alice phoned, in a state of shock.

"We've just got through the Depression," she said, "and now this."

Trudy found herself in the unusual role of comforter of her older sister. "I'm so sorry," she said.

Willis took a more philosophical view of the development. "With all the resources the U.S. has, I think the balance has shifted. We need to have the Americans fully engaged if we're going to win."

Although Willis had been hired because of his background in radio, he was working almost full-time on radar. Now, with the Americans involved and Nazi subs preying on U.S. ships, there was increased pressure from Washington. Willis was also looking into methods of intercepting enemy radio transmissions. The Council had developed an efficient means of recording these signals, but since most of them were encoded the problem became one of deciphering the messages.

"What's the point?" an exasperated technician asked Willis. "It's all gobbledegook to me."

"The more gobbledegook we come up with, the more likely we'll find a key."

A team of code-breakers in England was able to do just that, with help from Polish scientists who had fled to England to escape the war. The machine the Germans had used to generate the code was called Enigma. Once the Allies had the means of deciphering Enigma's messages, it was up to intelligence experts to make use of the information. But discretion was necessary. They didn't want the Germans to realize they had broken the code for fear it would be changed.

In 1942 Willis was asked to report to the NRC's personnel office. He had no idea what to expect. When he arrived, he was surprised to find that Inspector Wilde of the RCMP was there. The two men shook hands and Wilde motioned to Willis to sit down.

"What's this about?"

"A minor detail, Willis," the policeman answered.

Willis looked over at the personnel director, but he wasn't volunteering anything.

"We've noticed that there's something going on in the evenings in your basement. You've curtained off the recessed windows, and we've picked up the sounds of machinery."

Willis was stunned. Then he broke into a grin. "I have a woodworking shop in the basement," he explained. "Lately I've been working on a couple of Christmas presents."

Inspector Wilde opened a notebook and began to write.

"I'm currently making a Mickey Mouse rocking horse for my son. You're welcome to come home with me and see for yourself." He decided to attempt a little humour. "But only if you promise to keep it a secret from my boy."

For the first time Wilde smiled. "I see," he said, and finished writing.

"You might remember that in our interview in Toronto I mentioned that my hobby was woodworking," Willis added. "I had power tools in the basement there, as well."

"I think that answers our questions satisfactorily," said Wilde, and put away his notebook. "Consider the Mickey Mouse project a secret. We won't bother you any more about this."

Willis rose and headed for the door, but the Inspector called him back. "Perhaps you might leave those basement curtains open in the future," he said with a smile.

Christmas in Ottawa was a subdued occasion for Willis and Trudy. The shock of Pearl Harbor was still resonating throughout the world, and Germany appeared to be winning the war in Europe. With Canadian casualties mounting, it didn't seem an appropriate time for excessive celebration. Andy's expectations, however, like those of any five-year-old, were undiminished by worldly circumstances.

They'd taken him to see Santa at a downtown department store, and his stockings were hung by the fireplace on Christmas Eve. Willis had refined his photoflood reflectors, and he took lots of pictures. No one told Andy that the fireplace in the Western Avenue home was purely decorative. Santa was expected. Cookies and milk were set out for him, and in the morning they were gone.

The Mickey Mouse rocking horse proved a big hit. Willis wondered if the Mounties had managed to glimpse it through the now parted basement curtains. Not wanting to worry Trudy unnecessarily, he hadn't told her that they were spying on him. He just hoped they'd be discreet. The sight of mysterious strangers peering in their windows was the last thing he wanted his fragile wife to encounter.

Willis was still a loyal listener to the CBC. Every evening at ten he would tune in to the CBC News Roundup and listen to Lorne Greene's account of the latest developments in what was now, with the United States involved against Japan, truly a World War. Over time, Greene became synonymous with bad news, and pundits soon dubbed him "Canada's Voice of Doom."

In the week between Christmas and New Year's Trudy brought up a touchy subject.

"Don't you think we should teach Andy a bedtime prayer?"

"Why? I thought we'd agreed to let him make up his own mind about religion when he's older."

"I know, but with everyone talking about the war it might provide him with some comfort."

"He doesn't seem worried to me."

"But he hears people talk."

"You don't pray," said Willis.

"No."

"And I don't pray."

"No."

"We don't even attend church."

"We don't, but so many people do. The parents of most of Andy's friends go every Sunday."

"Is this about them or is it about Andy?"

"It can't hurt," Trudy protested. "His cousins . . . Joan . . . Mike . . . they say their prayers."

"Trudy, I'm an atheist. I'm not sure what you are. Your father is an SOB and he's a minister. I thought he'd turned you off religion."

"He did. I'm not suggesting the Lord's Prayer, just something simple, a few words before he goes to sleep."

"After we read him his bedtime story? That would certainly confuse him." Willis had let a note of sarcasm creep into his voice.

"He just could ask God to bless Mommy, Daddy, Taffy, his grandparents . . ." She let her words trail off.

Willis tired of the topic. "Why don't we ask him in the morning? See what he thinks of the idea." Willis couldn't imagine his five-year-old son being excited at the prospect of memorizing what he regarded as a meaningless bit of mumbo-jumbo.

But Andy surprised him. "I guess I could," he ventured. "Most of my friends say their prayers every night."

"Okay," said Willis, defeated. "But this is your mother's idea. She'll have to teach you."

The prayer Trudy found was short and, to her mind, simple.

Now I lay me down to sleep.
I pray the Lord my soul to keep.
If I should die before I wake
I pray the Lord my soul to take.

A couple of nights later, when it was Willis's turn to read Andy a bedtime story, he chose one by Mary Grannon, CBC's "Just Mary."

When he finished reading, he closed the book and bent to kiss his son goodnight.

"What about my prayer?" Andy asked.

"What prayer?"

"The one Mommy taught me."

"You've learned it already?"

"Sure, it was easy." Andy started to get out of bed.

"What are you doing?" Willis asked.

"I'm going kneel to say my prayer," Andy explained.

"You don't have to," Willis suggested. "You can stay tucked in and say it that way."

"Are you sure?"

"Yes, I'm sure," said Willis, keeping the resentment from his voice.

Andy sighed and repeated the prayer. He followed it up with, "God bless Mommy, Daddy, Taffy, Grandma and Grandpa Little, Grandma and Grandpa Champion, Cousin Mike, Cousin Joan, Aunt Alice and Uncle Al, Uncle Wendell . . ." As the list continued, Willis saw his son's eyes grow heavy. He was asleep before he managed to reach "Amen."

Chapter 62

Ottawa, 1942

The following evening it was Trudy's turn to read to Andy, and Willis waited until she returned to the living room to mention something that had been bothering him all day.

"When you suggested the idea of a bedtime prayer," he said, "you said that you thought, with all the talk of war in the air, it might be reassuring."

"Yes?" Trudy wasn't sure where this was going. She thought she'd won the argument.

"I don't think that 'If I should die before I wake, I pray the Lord my soul to take' is very reassuring."

"Oh, Willis," said Trudy, exasperated, "they're just words. I doubt if he has even given it a second thought."

"Well, that could be true, but there will come a time when he does. Are you prepared for that?"

"Let's cross that bridge when we get to it."

Willis could see that changing the prayer now would likely prompt his son to ask some questions. And that might bring the idea of death into sharper focus. He decided to let the matter rest.

"Andy said you let him say the prayer in bed, without kneeling down?" Trudy said.

"Damn right," said Willis, and now there was an edge in his voice. "You can get him to kneel when it's your turn, but that won't happen when I'm putting him to bed."

"Okay, okay," said Trudy, conceding the point.

Willis mulled the matter over the next day at work. He realized that his wife was more concerned with appearances than she was with the question of religion. She was worried about what the parents of Andy's new friends might think of her if they learned that her son didn't say his prayers. No doubt the day would come when she would send Andy off to Sunday school "because all the other kids are going." Willis didn't like this need to conform to the collectivity; it wasn't part of his character.

He had other concerns. He thought Trudy was being overprotective. At the slightest sniffle she'd keep Andy home from kindergarten. She discouraged friendships she thought were inappropriate, especially with boys she considered "too rough" for her son. Whenever a nightmare woke him he would be welcomed into their bed. Willis wasn't sure what to do about these tendencies. He wasn't home enough to enforce his own code of behaviour, but on weekends, when he had the chance to be alone with his son, he allowed him more freedom. He didn't want Andy to become a "mama's boy," and resolved to do everything he could to make sure this didn't happen.

At work, every day was a new experience. Mobile radar stations had been designed, prototypes had been manufactured and tested, and full-scale production was now underway. There were whispers about chemical and germ warfare. The development of higher-yield explosives was beyond Willis's area of expertise, but he knew that something big was going on in the Council's Montreal facility. When he learned that uranium ore was being mined in Chalk River, Willis was almost certain that the research involved splitting the atom.

The Council had been set up during the First World War, and consequently it was perfectly structured for advanced research when the Second World War broke out. In the beginning the NRC co-operated mainly with British scientists, but once the Americans had entered the fray they became trusted partners as well. There were new, bright people coming in every week, and Willis found the experience stimulating. It had the feel of an advanced graduate program, except that the goal was winning the war, not gaining a degree. Even the building itself had the look of something you might find on a university campus.

In April they celebrated Andy's sixth birthday with a small party. Trudy invited a few of her son's classmates and Gerald Kenyon, a neighbourhood boy a year older than Andy who had become her son's best friend. As usual, Trudy outdid herself for this festive occasion. The dining room was decorated with streamers, and every guest had a little party hat. She organized some games—pin the tail on the donkey, blind man's bluff, musical chairs. Then it was time for the cake. As Trudy brought it in from the kitchen she began to sing, and the children joined in:

> *Happy Birthday to you!*
> *Happy Birthday to you!*
> *Happy Birthday, dear Andy,*
> *Happy Birthday to you!*

"Now, make a wish," she said to her son, placing the cake down in front of him. "And blow out all the candles in one big breath."

Andy did as he was told, drawing in the air and blowing as hard as he could. Willis snapped a picture. All six candles were extinguished.

Trudy led a round of clapping, then put the knife carefully in her son's hand. "Make another wish and then cut the cake." Again Andy did as he was instructed. Trudy took the knife from him, cut individual portions, and handed plates to the children. As they ate, Willis folded up his camera.

Suddenly one of the children put his hand to his mouth and went pale. Willis thought at first that he'd found one of the pennies Trudy had baked into the cake as a surprise, but it was soon clear that this was something else. When the child vomited violently on the table everyone else froze in fright. Trudy rushed to his side. The boy was flushed and perspiring.

Trudy instructed Willis to take the children into the living room.

A call to the child's mother revealed the nature of the problem. The youngster was allergic to nuts. Trudy had been told about this condition and had made sure no nuts were on the menu, and she had warned the people at the pastry shop where she had bought the cake as well. However, she learned later that they had misplaced her instructions and had included ground walnuts in the cake. Fortunately the youngster had only managed a single bite. By the time his mother arrived to pick him up he seemed fully recovered, although he was still a little shaken.

"I'm dreadfully sorry," said Trudy.

"I'm sorry it spoiled your party," the woman replied. "This allergy is such a nuisance."

After the children had left and the table was cleared, Willis and Trudy sat down on the living room sofa with Andy cuddled between them. They explained allergies to him in simple terms.

"I'm just relieved you don't have that problem," said Trudy, hugging her son.

At the foot of Western Avenue a path wound through a wooded area down to the bank of the Ottawa River. Andy was too young to explore this area on his own, but Willis would take him there on weekends. At certain times of the year this shallow region of the Ottawa River was a repository for trees harvested from the Gatineau hills, in Quebec. This vast expanse of floating logs, retained by chains, was a tempting playground for children. As long as the logs were tightly packed there was little risk, and Andy and Willis watched as several older boys hopped from log to log, playing tag.

"Let's go out there, Daddy," Andy begged. "It looks like fun."

"Okay, but we aren't going very far," Willis warned. "Take my hand."

They made their way carefully from log to log, never venturing more than a few yards from the shore. Occasionally the logs would bob slightly under their weight, and Andy would shriek with a mixture of fear and delight.

Safely back on the shore, Willis unpacked their lunch and they sat in the shade eating sandwiches and watching teenagers roaming over the logs.

"The real danger comes when it's time to move the logs," Willis explained to his son. "There's a pulp and paper mill across the river. Men use poles to separate some of the logs from the main group and float them downstream to the mill. The other logs shift and bounce around until they settle back in place. It's these bobbing logs that are dangerous to play on. One slip and you could find yourself in the water and trapped under the wood."

"That's scary," Andy commented, shivering at the thought.

"Very scary. But there's some danger even when the logs aren't moving. I want you to promise me you'll never come down here alone."

"I promise," said Andy. "Mom doesn't want me going through the woods anyway."

"And she's right about that," Willis agreed. "But let's keep this little adventure on the logs today a secret." There were times, he thought to himself, when Trudy's advice to their young son made sense. He pointed across the river. "Next week we'll drive out to that island and I'll show you the rapids."

Andy spent the week in anticipation. He told his friends at kindergarten about the trip he would be taking. He could see that they were envious.

The following Saturday Willis backed the car out of the garage and Andy hopped in beside him. Trudy had once again packed a lunch, a peanut butter and honey sandwich for her son, and ham and cheese for Willis. They drove west along Wellington Street, then turned right onto the Driveway. They soon crossed a bridge, and Andy glimpsed the expanse of logs from his window. From this elevation there seemed to be even more of them. At the entrance to Bate Island they turned right onto a paved road that led to the island's north shore, facing the Gatineau hills. Andy jumped from the car and ran to the edge of the river, peering out across the water.

"Where are they?" he asked. "I don't see them."

Willis joined his son at the water's edge. "Why, they're right there," he said, and pointed to the swirling waters.

"Are they swimming?" Andy asked, squinting.

"Swimming?" Puzzled, Willis looked down at his son.

"The rabbits you told me about," said Andy. "I don't see any rabbits."

Willis stifled a chuckle. "I said 'rapids,' Andy," he said, articulating the word carefully, "not 'rabbits.'"

"But you said they were white, tumbling over each other," his son protested, looking up at his father.

"And so they are." Willis pointed to the churning waters.

"Oh." Andy looked back at the rapids, clearly disappointed.

"Close your eyes," Willis advised him. "You can hear the rush of the water."

Andy did, but it wasn't any consolation. He'd envisioned hundreds of white rabbits, hopping madly about in all directions.

"Can we go back to the logs?" he asked.

"Sure," said Willis, "but let's eat our lunch first."

Andy learned one more lesson that day. On their way through the woods he noticed what looked like a white balloon hanging limply from a branch of one of the trees.

"Look, Daddy," he said, pointing it out. "And there are more balloons over there on the ground."

Willis looked at the used condoms and was momentarily stumped. "I see them," he said, trying to buy some time.

"Can we collect them and blow them up?"

"No, Andy, definitely not. They're dirty; they've been out in the woods too long."

"We could wash them," Andy suggested.

"No," said Willis, and then came up with a diversion. "They're all white, anyway. Boring. When we drive home we'll stop at the store and buy some new balloons—coloured balloons—red, blue, yellow . . ." He didn't finish the sentence.

"Okay," said Andy, mollified.

Chapter 63

Ottawa, 1942

It was an exciting summer for Andy. His uncle Wendell was briefly stationed at the Camp Petawawa army base near Ottawa. On his last visit before shipping out for Europe, he promised to bring Andy a souvenir from the fighting when the war was over.

"What would you like?" he asked.

"A gun," his nephew replied without hesitation. Andy had been begging his parents unsuccessfully for a BB gun.

"What kind of a gun?"

Trudy rolled her eyes in exasperation.

"A German gun."

"Ah," said Wendell, smiling. "How about a Luger?"

"A Luger?" Andy was confused.

"It's a famous German pistol," Wendell explained.

"A Luger would be great," replied Andy excitedly.

Willis interrupted the conversation."Let's take a picture," he suggested.

Wendell, decked out in his uniform, stood next to Andy on the front lawn.

"At ease, young man," Wendell advised, and assumed the position, with his legs spread apart and his hands behind his back. Andy followed suit and Willis snapped the picture.

"You'd make a great soldier," said Wendell.

That evening, after her brother had left, Trudy looked at Willis. "Andy has no idea how serious war really is."

"Probably just as well," Willis commented, looking up from the newspaper.

"Do you think he realizes his Uncle Wendell might never return?"

"I doubt it. In Andy's mind the good guys always win. In spite of the bad news from Europe he seems pretty confident."

"Let's keep it that way," said Trudy.

Later that summer Andy's cousin Don, the son of Doris Elseley, Trudy's sister, began training with the Royal Canadian Air Force. The RCAF base was in Trenton, within driving distance of Ottawa, and Don was soon a frequent visitor. Andy noticed that the uniform his cousin was wearing was different from his uncle's.

"Wendell is in the army," Don explained. "The army wears khaki."

"Khaki?"

"It's a light shade of brown. Since he'll be fighting on the ground it sort of blends in."

"Your uniform is blue," Andy noted.

"That's because I'll be up in the sky."

"And the sky is blue!" said Andy, excited at having made the connection.

With Don posted in Trenton, it wasn't long before Doris decided to visit. Doris was the oldest of the Champion children and Trudy's least favourite sister, but there was little she could do when Doris wrote to say she was coming to see her son.

"I can understand her visiting," said Trudy. "I just wish she'd stay at a hotel."

"Come on, Trudy," Willis chided, "it will only be for a weekend. You can put up with her for a couple of days."

The visit was a decidedly uneasy one. Doris complained about Trudy's not having been to Essex to see her parents, to which Trudy countered that travel was difficult with a war on. Fortunately, a good deal of Doris's time was spent with Don, and it was he who showed her around Ottawa.

"I'm surprised you didn't enlist," Doris said to Willis on Saturday evening after dinner.

Trudy's anger flared. "Willis is doing important war work," she said indignantly. "He's a lot more value to his country working here at the National Research Council than he would be dodging bullets in Europe."

"I was a little too old for the draft," said Willis, as calmly as he could, "especially since I have two dependents."

"I just pray Donald gets through the war in one piece," said Doris.

Trudy decided to change the subject. "Why don't you put Andy to bed," she suggested to her sister.

When Doris left the room, Trudy looked at Willis. "See what I mean?" she said. "What nerve, asking why you weren't in uniform!"

"Let it go, Trudy. She's on edge. Can you imagine how you'd feel if Andy was about to fight in a war?"

"I suppose," Trudy conceded.

When Doris returned she was smiling. "I'm glad to see you're raising your young son as a good Christian," she said. "He was very cute, down his knees saying his prayers."

Willis was relieved to hear that his son had opted for the kneeling version of his prayers on this occasion.

"He's probably ready for something a little more formal," Doris suggested. "They're never too young for the Lord's Prayer."

By the fall, both Wendell and Don were overseas, Wendell in the infantry and Don as an RCAF fighter pilot.

"I hate to think of it," Trudy remarked, "but wouldn't it be awful if Edith or Roger were killed by somebody from their own family!"

"The chances of that happening are pretty remote," said Willis. "Last we heard they were in Breslau. Don's missions are over France, and Wendell will likely be deployed somewhere in Belgium."

"I know, but it's hard to get the thought out of my mind."

One evening Willis decided that it was time his son understood that war wasn't a game, but serious business. He told Trudy he was taking Andy for a drive by the river, but instead headed downtown. It was raining, and the lights on Wellington Street were reflected from the black, wet pavement. It was cold and blustery, but the car heater was on and Andy was snug and warm in the front seat.

After turning the corner of Wellington and Elgin, Willis pulled the car over and parked, leaving the motor running and the heater on. He pointed to the left.

"That's Canada's War Memorial," he said. "It's a monument that was put there to help us remember the soldiers who died fighting in the last war. Their names are all engraved on the base of the statue."

Andy could make out lifelike shapes of men in fighting poses on the cenotaph.

"Someday the names of soldiers who die in this war will be written there."

Through the beaded droplets of rain on the car window they stared at the statues of the charging soldiers, their guns and bayonets raised, frozen in time.

"Will Uncle Wendell's name be there if he gets shot?"

Willis didn't answer, just looked out through the rain. Then he turned. "I suppose it would," he said. "We just have to hope that doesn't happen."

"What about Cousin Don?"

"His name could be added, too, if he is shot down."

"I'm glad you're not over there, fighting."

"Oh, I wanted to go, but they wouldn't let me."

"Who wouldn't let you?"

"The government. They decided I'd be more use here in Ottawa, working on war projects. That's why we moved to Ottawa. I work on things your cousin uses when he flies his plane."

"I'm going to ask Santa for one of those little Dinky Toy fighter planes for Christmas."

Willis glanced at his son. "I wouldn't ask this year, Andy." He looked back out into the night. "You see, Dinky Toys are made in England, and with the war on, the British aren't making them anymore." Willis spoke as if he were talking not just to his son, but to the world out there beyond the car.

Trudy was furious when she learned Willis had taken Andy to see the War Memorial. "Why in God's name would you do a thing like that?"

"I just thought it was time to introduce him to some of the reality of the war. Until now he's seen it as a glorified game without any serious consequences, a game of cops and robbers, with bad guys and good guys."

"So now he'll go to sleep every night worrying about Wendell and Don? Is that what you want?'

"No, Trudy," said Willis, trying to remain patient, "I don't think it will carry over that far."

"And you were the one who complained that the prayer talked about 'dying before I wake.' That was about death and dying."

"That was different. That was personal. That suggested that he might die. Seeing the War Memorial is more . . . abstract."

"Abstract?" said Trudy, her voice becoming shrill. "You think that soldiers dying in Europe and now in the Far East is abstract?" Trudy burst into tears and ran into the bedroom, slamming the door behind her.

That fall there was another minor domestic crisis. Andy had attended kindergarten in a school east of Western Avenue, but when it was time for Grade One he was assigned to another school, which was in the other direction. It had a troubled reputation, and Andy would have to cross several busy streets to get there. Trudy complained to the school board, but was told that students living on the west side of Western Avenue had to go to the school in question regardless of where they attended kindergarten.

"You mean to tell me that if we moved across the street he'd be attending another school?"

"That's right."

"That's ridiculous," said Trudy. "Can't you make an exception?"

"I'm afraid not," said the clerk. "If we make an exception for you, others would want the same privilege."

Defeated and fuming, Trudy returned home. That night she poured out her frustration upon Willis. He listened patiently, keeping to himself the thought that she was overreacting. The new school might even be good for Andy. It might toughen him up.

Chapter 64

Ottawa, 1942–43

Christmas in Ottawa was, again, a quiet affair. Thanks to the Red Cross, another letter from Edith had arrived. It was heavily censored, but it reported that she and Roger were still safe. None of Trudy and Willis's friends was in a mood for celebration. Many had brothers overseas, others had uncles and cousins. Andy noticed small Canadian flags hanging in the front windows of homes. Willis explained that this was because those families had sons serving in the war.

One day Andy came home from school with the news that in a window down the street a black flag had replaced the Red Ensign.

"What does that mean?" he asked.

Trudy looked to Willis to provide the answer.

"I'm afraid it means they've lost someone."

"Lost?"

"Yes. A soldier from that family has been killed, or is missing in action."

"Missing?"

Willis sighed. "If a plane is shot down and the crew doesn't manage to bail out, they assume the worst. Those airmen are listed as missing in action."

"Oh."

Willis watched as his son tried to take in the meaning of this.

"Could that happen to Cousin Don?"

"It could," Willis said, "but we've been told that the heaviest Air Force losses came last year, during what they called 'The Battle of Britain.' We won that fight, and now Don's main job is escorting bombers on raids over Germany."

"Isn't that dangerous?"

"Yes, it is, but sometimes when a plane is hit the pilot has time to bail out. He's usually captured by the Germans and put in a prisoner-of-war camp."

"What about Uncle Wendell?"

"He's safe. He's still in England, waiting. At some point our troops will cross the English Channel again and begin fighting the Germans, probably in France."

"Again?" Andy asked.

"Well," said Willis, "they tried it once before but they weren't very successful." He didn't want to explain Dunkirk to his young son.

The CBC News Roundup at ten became an evening ritual for Willis and Trudy. They had listened, terrified, a year earlier, when Matthew Halton reported from London as Nazi bombers pounded the city night after night. The sounds of air raid sirens, fire engines and ambulances, and explosions brought the war right into their living room. But the British, inspired by the words of Prime Minister Winston Churchill, had survived the Blitz, and now the probability of an invasion of England seemed remote. The Soviet Union had entered the fray, thus opening a second front against the Germans, this one to the east.

The sea battle against German submarines was also turning in the Allies' favour. Thanks to refined radar and the ability to decipher the enemy's coded transmissions, Allied depth charges were finding their mark. As a result, more and more Allied ships were making their way across the previously treacherous waters of the North Atlantic.

For Andy's seventh birthday Willis managed a minor miracle. Because of the war, very few new bicycles were being manufactured, but Willis knew that the Canada Cycle and Motor Company, known throughout Canada as CCM, had planned to manufacture a small two-wheeler, designed for younger children. The company had abandoned this program when it shifted production to accommodate the war effort, but Willis learned through an old Queen's classmate in Toronto that a number of prototypes of these smaller bikes had been made. He persevered until he managed to track down one that was gathering dust in a CCM factory outside Toronto. Calling in a favour, he implored his old friend to see if he could buy it. The friend was successful, although the bike cost Willis twice what he would have paid for a new, full-sized bike in 1939. He didn't ask, but suspected that some of the money had been used as a bribe.

The bicycle was too large to wrap, so Willis propped it against the fireplace in the living room. Andy's other gifts had been placed in his bedroom after he went to sleep, so that when he woke early, as he always did on his birthday, he could open his presents without waking Willis and Trudy. The plan worked for a while, but then their young son became bored and padded into the living room. His shriek of delight woke Trudy and Willis with a start. Then, realizing what had happened, they smiled at each other, got up from the bed, and joined him in the living room.

When they entered, Andy flung himself into his father's arms. "Dad!" he shouted, "My own bike! And it's the perfect size!" Then, disentangling himself from Willis, he hugged his mother. "Thank you, thank you, thank you!" he said.

"It was really all your father's doing," Trudy said modestly. "He has friends in high places."

After breakfast they took the bike out to the sidewalk and Andy climbed aboard. Willis held onto the seat and began to jog along behind him. It didn't take long before he was able to let go and his son began cruising on his own. However, the moment Andy looked back and realized his father was no longer helping, his front wheel wobbled and he fell to the pavement. Fortunately he hadn't been going very fast, and wasn't hurt. He got back on the bike, Willis supported him again, and they resumed the drill. This time Willis promised to jog along behind Andy so that he could catch him and prevent another fall. This worked, and within half an hour Andy was riding confidently on his own.

Trudy had been watching from the front porch. When Willis joined her, she smiled. "Another Wheeler," she said, and they laughed.

There were strict rules, of course, and Trudy extracted a solemn promise from her son that he would stay on the sidewalks when riding alone, and would get off the bike whenever he had to cross a street.

Willis and Trudy both had bikes of their own, but Trudy was reluctant to use hers. Willis, on the other hand, true to his roots, loved to cycle. Now, with Andy as his willing companion, he set out to explore Ottawa. On weekends they'd pack picnic lunches and head out on excursions.

One of their favourite destinations was the Experimental Farm. Established by the federal government before the turn of the century to test new seed crops and innovative agricultural methods, it was originally located on the outskirts of Ottawa. Eventually, as the city's population grew, it began to be surrounded by new homes and buildings, but this fully functioning farm, with its fields and barns, remained untouched. There were sheep and cows, tractors, and row after row of crops. The quiet farm roads, safe from city traffic, were ideal for biking.

Andy and Willis liked to picnic by the side of these roads, with the cows grazing peacefully just a few yards away.

"We get our milk from cows," Willis explained.

"How?" Andy asked, chewing on a peanut butter and honey sandwich.

Willis pointed to a nearby cow." Those things hanging down are udders. When farmers squeeze the udders they produce milk, which they collect in pails. Then they heat the milk, pasteurize it to make sure it is safe, and bottle it."

Andy understood his father's explanation, but had a hard time relating these huge animals to the cold glasses of milk he consumed every day.

Later, Willis explained how sheep were sheared and their wool used to make socks and sweaters.

Andy marvelled at how much his father knew. He wondered if he'd ever be as smart as his dad.

"Is there anything you don't know?" he asked.

"Oh, lots of things," Willis replied modestly.

Other favourite destinations were the Ottawa River and the Rideau Canal, which, now that they both had bicycles, Willis and Andy were able to explore. The Canal linked the city with Lake Ontario. Andy was fascinated by the system of locks that allowed boats to be raised and lowered.

"How far does the Canal go?" he asked.

"All the way to Kingston," Willis said, "where I went to university. Maybe someday you'll go there as well."

"Can we bike to Kingston?"

Willis smiled and shook his head. "That's a little too far for us, I'm afraid."

School finished in June, and soon Andy was biking every day. One evening in July when Willis returned home from work he could tell by the look on Trudy's face that something was wrong.

"Someone has stolen Andy's bike," she said.

"How did that happen?" asked Willis. "Wasn't it locked?"

"I'll let him tell you."

Andy emerged from his bedroom, rubbing tears from his cheeks. He'd gone to the corner store for candy, he said, and had left the bike without locking it. He'd only be inside for a minute and didn't think it was necessary. But when he came out it was gone.

Willis sighed. "Well, let's have dinner and then I'll see what we can do."

It was a quiet meal. When Willis finished his coffee he told Andy to get into the car and he'd join him.

"Shouldn't we call the police?" Trudy asked.

"Maybe, but first let me see if I can track it down."

Willis drove to the candy store and parked. "You wait here," he told his son, then went inside. A few minutes later he returned to the car and they drove off.

"Where are we going?" Andy asked.

"I have a hunch," said Willis. "Mr. Sheehan, the man in the store, thinks he knows who took your bike."

"Who?"

"I'm not going to name names until I'm sure," Willis replied.

They drove out past the city limits into the country. After they'd gone a short distance on a dirt road, Willis pulled the car over and parked. In the distance, up a lane, Andy could make out a dilapidated farmhouse. His father strode confidently to the front door and knocked. A woman answered and Willis was soon ushered inside. Minutes passed, and then, as Andy watched, his father emerged from behind the house carrying his beloved CCM two-wheeler. Willis had been told by parents in the farmhouse that their boy hadn't stolen the bike—he just "borrowed" it. Willis didn't believe them, and warned them that if it happened again he'd report the matter to the police.

Andy jumped from the car and ran to meet Willis.

"You found it, you found it!" he called excitedly.

"Yes, I did," said Willis grimly as he wheeled the bike to his son. "And from now on I want you to promise me you will lock it, no matter where you leave it."

"I will, I will, I promise." Andy mounted the bike and rode it back down the lane to the car. Willis followed, and soon they were driving back home. On that day Andy decided that his father not only knew everything there was to know but could solve just about any problem.

Chapter 65

Ottawa, 1943

As the year progressed, there were a number of developments that illustrated the fundamental differences between Trudy's approach to raising their son and Willis's. In particular, Willis felt that Trudy, with her overprotective attitude, tried to shield Andy unnecessarily from some of life's inevitable experiences.

One Saturday afternoon Andy came into the house wiping tears from his face.

"What's the matter?" said Trudy.

"Tom pushed me down," said her son with a sniffle, "and when I tried to get up, he pushed me down again."

"Willis?" Trudy called to her husband, who was reading his newspaper in the living room.

"What is it?" he responded, looking up.

"I think this is something you should handle." She gently directed Andy to the living room.

"What's the matter?" Willis put down the newspaper and looked at his son.

Andy repeated his story, and Willis put his hand on his son's shoulder. "Tom is being a bully," said Willis, "and there's only one way to deal with a bully."

His son looked questioningly at his father.

"Go back out there and clean his clock."

Andy looked confused. "I don't think he has a clock," he said.

Willis smiled. "I mean you can't let a bully push you around. You have to push back."

"But Tommy is older and bigger," Andy protested.

"Doesn't matter," Willis said. "You give him a good punch and he'll back off."

Andy looked at his father, anticipating more in the way of advice.

"Now wash your face and go back outside," said Willis, and resumed reading his paper.

When Andy had left, Trudy came back into the living room. "I think you should have a word with Tommy's parents," she said.

Willis looked up.

"We can't let him bully our son."

Willis gave an exasperated sigh. "I don't think that's a very good idea. Andy has to work these things out for himself. We can't fight his battles for him."

"But Tommy is a year older. It isn't fair."

Willis shook his head. "Life isn't fair," he said, "in case you hadn't noticed."

When *Lassie Come Home* was released as a movie that year Willis suggested they take their son to see a matinee.

"I'm not sure," said Trudy. "It might be too much for him."

"Too much?"

"Well, you remember what happened when we took him to see *Bambi*. It upset him. The only way we could get him to sleep was to bring him into our own bed. It was a week before things got back to normal."

"He has to get used to these things," Willis said. "Life has its ups and downs. It's only a movie, after all."

Willis prevailed, and *Lassie Come Home*, with its happy ending, had no discernible after-effects on Andy.

The same could not be said for *Treasure Island*. They'd begun reading Andy some of the classic stories at bedtime. Their son loved *The Adventures of Tom Sawyer*, and *Huckleberry Finn*, too, although it had its sad moments. But when they moved to the Robert Louis Stevenson classic, one passage from the first chapter triggered trouble:

> How that personage haunted my dreams, I need scarcely tell you. On stormy nights, when the wind shook the four corners of the house and the surf roared along the cove and up the cliffs, I would see him in a thousand forms, and with a thousand diabolical expressions. Now the leg would be cut off at the knee, now at the hip; now he was a monstrous kind of a creature who had never had

> but the one leg, and that in the middle of his body. To see him leap and run and pursue me over hedge and ditch was the worst of nightmares. And altogether I paid pretty dear for my monthly four-penny piece, in the shape of these abominable fancies.

This time Andy's eyes didn't flutter as Willis read, the usual indication that he was falling asleep. Eventually Willis tired and got up from the chair beside his son's bed.

"Leave the door open?" Andy called after his father.

Willis looked back. "Okay," he said, "but just a crack."

Later that night Willis was roused from a deep sleep when his son crawled in bed with them. "What's the matter?" he asked groggily.

"Nightmare," Trudy whispered.

The next day Willis complained to his wife. "We can't have him running into our bed every time he has a bad dream," he said.

"Yes, I know," replied Trudy, "but we can be a little more careful in our choice of bedtime stories."

However, this posed a difficulty, because Andy was no longer satisfied with children's books. He had developed a liking for adventure stories, but they required careful screening on Trudy's part. *Frankenstein* and *Dracula* were quickly ruled out, whereas *The Three Musketeers* and *The Scarlet Pimpernel* were acceptable. Then one evening Andy asked about *Robinson Crusoe.* He'd heard a teacher at school talking about it and was interested.

"What do you think?" Trudy asked Willis that evening after their son was asleep.

"Can't really hurt," said Willis. "It has a happy ending."

"But the story of a man shipwrecked on a desert island, all alone," said Trudy; "that might be scary for a seven-year-old."

"Well, he isn't alone for the whole story. Doesn't a native show up later?"

"Yes," said Trudy. "Crusoe calls him 'Friday,' because he was discovered on a Friday."

"I think it might be worth trying," Willis said. "It was Andy's suggestion, after all."

Robinson Crusoe proved a hit, but with a surprising result. "I liked the first part of the story," said Andy, "when Robinson is all alone and has to get along on his own, finding food and water, building a shelter. That was exciting."

"And the second part of the story?" Trudy asked.

"It was okay, but not as much fun as the first part."

"What do you make of that?" Trudy asked Willis later in the evening.

"I don't know," said Willis thoughtfully. "Maybe he liked the self-sufficiency in the first section."

"I guess so," said Trudy, not completely convinced.

Radio programs were another sensitive area. The after-school programs such as *The Lone Ranger*, *Gene Autry*, *Red Ryder*, *Superman*, and *Dick Tracy* were no problem, and a weekly wartime serial, *L for Lanky,* about the missions of a Lancaster bomber with a Canadian crew quickly became a favourite, as was *Captain Lawrence and the Secret Service Scouts.* But *Inner Sanctum*, *The Shadow*, and *Suspense* were ruled out, because they resulted in sleepless nights. At first Andy protested, but when it became apparent to him that listening to these mysteries was a recipe for a troubled night, he gave in.

Willis had to tread a fine line when it came to disputes with Trudy about what was appropriate for their son. He felt that her fragile state was often the motivating factor when it came time to establish rules for Andy. Some of her suggestions made perfect sense, but others, in Willis's opinion, were spawned by Trudy's own fears. Discussions about these matters became delicate. Willis sought to avoid the implication that his wife's unrealistic concerns were the product of her own mental problems. Rather than risk a confrontation, Willis would usually acquiesce on the trivial decisions, putting his foot down only on matters of principle. Trudy insisted that her son was creative, and naturally high-strung. Willis thought that Andy's fears were often the product of what he described as an "over-active imagination," and was not sure whether this condition was something he had inherited or the result of Trudy's own flights of fancy.

One habit Willis sought to break was their son's tendency to seek the sanctuary of their bed whenever he was troubled.

"He has to learn to tough it out," he argued.

"I know, I know," agreed Trudy. She didn't seem able to maintain that resolve, however, when his crying woke her in the middle of the night. Willis's difficulty arose from the fact that he was able to sleep soundly through just about anything, whereas Trudy was a light sleeper, aroused by the slightest noise. It was often morning before Willis realized he was sharing his bed with both his son and his wife.

Andy, for his part, was learning strategies to deal with his parents, particularly his mother. Trudy's mood swings were unpredictable, but her son quickly learned that almost any demonstration of creativity could revive her spirits. It might be a drawing, a song, or even a story about something that happened at school. It didn't always work, but when it did he felt a surge of gratification; he had managed to "make Mommy smile." His approach to his father was different. There was no need to please Daddy, who could be relied upon to solve problems, answer perplexing questions, and deal with difficult situations.

In many ways 1943 was a turning point in the war. Canadian troops landed in North Africa and the Soviets finally defeated the Germans after a long siege at Stalingrad. More and more stories began appearing about the brutal treatment of Jews under the Nazi regime. It was the year when Polish Jews in the Warsaw ghetto fought a heroic but futile battle against the occupying enemy. In the Battle of the Atlantic hundreds of German submarines were sunk with depth charges, as a result of improved detection. American troops, already heavily involved in Europe, took the fight to the Japanese by invading the Solomon Islands.

At home, Canadians were coping with food rationing. No one seriously complained. It was as if each family, when it sat down for dinner, was doing its part for the war effort. War Bonds were proving popular, and the top entertainers were travelling to military bases to stage shows for the troops. One of the most successful songs in the United States was Bing Crosby's version of "I'll Be Home for Christmas." Vera Lynn's song "You'll Never Know," whose first line was "You'll never know just how much I miss you," was a major hit in Britain.

Both Britain and the United States were broadcasting radio programs directly to their troops. Already, in 1932, the BBC had established an English-language short-wave network, the Empire Service, to beam news to the far-flung British colonies. In 1939, when war seemed inevitable, its name was changed to the BBC Overseas Service, and programs began to be broadcast in German. By 1942 the Service included transmissions in all of the major European languages. The year 1942 was also when the Voice of America was launched. At first it used BBC facilities to broadcast its message, but in 1943 work was begun on a powerful new VOA short-wave transmitter in Bethany, Ohio.

These developments were not lost on the federal politicians in power in Ottawa. In 1942 Prime Minister Mackenzie King promised that Canada would soon have its own short-wave transmission capabilities. In September an Order-in-Council was signed that created the CBC International Service. A site near Sackville, New Brunswick, was selected for the transmitter, but the wartime shortage of materials delayed the start of work. It would be another three years before Canada's voice could be heard abroad.

Willis read with interest the newspaper accounts of the CBC's plan for an International Service. He remembered his own work on the short-wave equipment at the Hornby transmitter. His job at the Research Council was still challenging, and he took immense pride in the contribution the Council was making to the war effort. There were recurring rumours among the staff about the development of a super-weapon. There had long been talk among physicists about the power potential of the atom, if only it could be split. Willis knew that this was no longer just talk.

Chapter 66

Ottawa, Montreal, and Sackville, N.B., 1943

In the fall, John Wheatley, the personnel director at the National Research Council, asked to see Willis. He reported shortly after arriving at work on a Tuesday morning in September.

"Have a seat, Willis," said Wheatley.

Willis sat down and lit a cigarette.

"I'm sure you've stayed in touch with your CBC friends," the personnel director said, glancing up from a file folder.

"From time to time," Willis admitted, worried that this interview might have something to do with security.

"Well, then," said Wheatley, "you probably know that they're setting up short-wave facilities to broadcast Canadian news to our troops overseas."

"I've read about it in the newspapers," said Willis.

"It seems they've encountered some technical problems, and they asked if we could help out. I went over your file, and you seem to be the ideal candidate. You've already worked for the Corporation, and probably know a good many of the engineers involved in the project."

"Could be."

"And I see that you helped install short-wave equipment at the transmitter in Toronto."

"Yes, I did," said Willis, "but we never activated it."

"Nevertheless, I think you might be able to help them out."

"I'd be glad to give them a hand."

"The CBC has chosen Montreal as the headquarters for the service. The studios will be there too, because both English- and French-speaking staff are available to host and produce programs. We'd like to send you to Montreal for talks with their people, then have you visit the transmitter site outside Sackville, New Brunswick, to assess the situation. You can report back to us and we'll take it from there."

Willis left the office with mixed feelings. He was reluctant to leave the work he was doing for the Council, but at the same time he knew that once the war was over he would be returning to the CBC. Perhaps this was an opportunity to reconnect with the Corporation and explore his post-war possibilities.

Trudy, as usual, was unhappy that he would be away. "How long will you be gone?" she asked.

"Probably a week or so."

"Can't they get someone else to go? You said your work here was essential."

"It is, but so is the CBC's need to get Canadian radio programs to our troops."

In Montreal Willis was briefed on the project by his old friend, Alphonse Ouimet. The following day he boarded a train to Sackville and began reading through the plans. Engineers had already selected a salt marsh just outside the town as the site for the towers. Its proximity to the east coast, its elevation, and the conductive powers of its soil made it an ideal location for short-wave broadcasting. It was clear that considerable work had to be done before the facility would be ready for transmission. There would be an opportunity for Willis to provide significant input if the Council decided they could spare him. Willis thought the project could be completed in six months if they were able to obtain the equipment they needed.

Willis checked into a hotel in Sackville, unpacked, showered, and went downstairs to the dining room to meet with the chief site engineer, Gordon McInstry, who welcomed him and pumped his hand enthusiastically. "Call me Mac," he said.

Mac was a ruddy-faced, clean-shaven man in his thirties. Over coffee he explained that the main impediment to progress had been a shortage of essential equipment. "I hope you have influence," Mac said, "because we'll need it if we're going to get on the air before the war is over."

"I'll do everything I can," Willis promised. He realized that pressure from the National Research Council might get things moving. "I'd like to see the site."

McInstry drove east from Sackville, and eventually they pulled over to the side of the dirt road. Bulrushes extended as far as the eye could see.

"These are the Tantramar Marshes," said McInstry, pointing east toward the horizon, "but we've filled in several acres to provide a solid foundation for the towers. We have some of the equipment, but we're missing some vital parts. Once we have them we can get busy. I have a team of construction workers ready. They can be here in a matter of days."

On the train back to Ottawa Willis began working on his report. He drew up a plan that included a timetable for the erection of towers, a tentative schedule for the completion of the transmitter building, and a target date for the first test transmission. When he returned to work at the NRC he sent one copy of his recommendations to the CBC International Service in Montreal and another to his supervisor at the Council.

In October he was summoned once again to the office of the personnel director. John Wheatley greeted him with a smile.

"You certainly impressed the folks in Montreal," he said. "They were pleased with your evaluation." He held up a copy of Willis's report. "I've had our people here go over your plans and they agree the timetable is reasonable. There's been considerable pressure from the Prime Minister's office to get this off the ground. He wants our boys in Europe to hear the voice of Canada."

"I agree with him. The Brits and Yanks have their short-wave services. We need our own."

"Well, I've talked to the head of your department here in Ottawa. He thinks he can spare you by the end of the month. We'd always planned to send you back to the CBC once the war ended, but this project has accelerated that prospect."

Willis wasn't sure where this was leading.

"We'd like to transfer you to the CBC in Montreal so that you can help out with the work on the International Service."

"Let me get this straight. They want to move me to Montreal, not back to Toronto?" He knew Trudy would not be pleased.

"Yes, that's their plan."

"Would I be returning to Toronto once the short-wave service is transmitting?"

"That would be up to the CBC," said Wheatley. "Well, what do you think?"

"Can I talk it over with my wife?" Willis asked.

"Certainly," the director said. "But there's a war on, and given the political pressure I'm not sure you have much choice."

"I see," said Willis.

As predicted, Trudy wasn't pleased. The idea of moving again, to another new city, filled her with fear.

"I can't speak a word of French, Willis. I'd be lost in Montreal. We don't know anyone there. And it would mean finding a new school for Andy."

"Calm down, Trudy. I've worked in Montreal, and it's a bilingual city. Just about everyone speaks both languages. And there are English communities. We can settle into one of those. Then, when I've finished my work for the International Service we'll probably be transferred back to Toronto."

"How long do you think that will take?"

"Well, I'm guessing, but if things go according to plan we should be wrapped up in a year or so. Who knows? Maybe the war will be over by then."

The next day Willis told Mr. Wheatley he'd talked things over with his wife and was ready for the transfer. "Fine," said the personnel manager. "There is a problem, though, and that's finding suitable housing in Montreal. You've worked there. Any preferences?"

Willis thought back to his time at Northern Electric and he remembered his visit to St. Lambert. He thought it would be a perfect fit for Trudy, a small English-speaking community not so different from her hometown. He offered his suggestion.

"I'll see what we can do," said Mr. Graham.

"If not St. Lambert, NDG would be our second choice."

"NDG?"

"Notre Dame de Grâce. It's another predominantly English area of the city."

In December word came through that the transfer had been approved. They had managed to find a home for Willis in St. Lambert, and a date was set for the move. They would leave Ottawa on the fifteenth of January. Their own furniture would have to be put into storage, because the house they'd be living in was furnished. It was a six-month lease. Once they were settled in St. Lambert they could look for another, more permanent place.

As usual, this arrangement didn't go down well with Trudy. "It's difficult enough pulling up stakes," she complained, "but then we'll have to move again?"

Willis had looked into the psychiatric services available in Montreal.

"Trudy, it's going to be difficult, but once we're settled we'll be able to get you the help you need. Montreal has some of the best doctors in the country, far superior to anything here in Ottawa, and even better than the ones in Toronto."

Trudy looked at her husband, and tears began to stream down her face. "Do you really think so?" she said. "I am so tired of this . . . this . . . postpartum depression, this anxiety."

Willis took her in his arms. "I'm sure you are, Trudy. There has to be an answer." In his own mind he was confident. His faith in science was rock-solid. There were no problems that couldn't be solved. It was just a matter of time. His father's life had been saved by the discovery of insulin. Penicillin was revolutionizing the treatment of infections. Willis believed that doctors would soon come up with something to cure his wife's troubled psyche as well.

He had done his homework. The Royal Victoria Hospital's newly opened psychiatric department, the Allan Memorial Institute, was showing considerable promise. Located in a converted Victorian mansion on the slopes of Mount Royal, separate from the hospital, it was affiliated with McGill University, the home of Canada's top medical school. The Allan featured an open-door policy for patients. No locked rooms, no padded cells. The person responsible for this revolutionary innovation was the Institute's first director, a psychiatrist of international stature named Ewen Cameron.

The man was a rising star. When the war was over he would be chosen to sit on a select panel in Nuremberg to judge the sanity of Nazi war criminals. Later he would co-found the World Psychiatric Association and become its president. Willis liked everything he read about Cameron. It seemed to him that the doctor was applying engineering principles to the field of psychiatry. No more hours on a couch going over a patient's past history. Cameron was determined to find a quick cure for mental illness, a magic bullet that would transform psychiatric treatment the way antibiotics had changed the treatment of infectious diseases. This aggressive approach soon gained the doctor a reputation as one of the world's leading psychiatrists. It would be more than thirty years before that same world would come to know the true nature of his work and leave Ewen Cameron's reputation in ruins.

But by then it would be too late for Trudy, too late for Willis.

References

Wilfrid Eggleston. *National Research in Canada: The N.R.C., 1916–1966.*
Clarke, Irwin and Company, 1978

W.E. Knowles Middleton. *Radar Development in Canada: The Radio Branch of the National Research Council of Canada 1939–1946.*
Wilfrid Laurier University Press, 1981.

Knowlton Nash. *The Microphone Wars: A History of Triumph and Betrayal at the CBC.*
McClelland and Stewart, 1994.

Arthur Siegel. *Radio Canada International: History and Development.*
Mosaic Press, 1997.

Eric Arthur Zimmerman. *In the Shadow of the Shield: The Development of Wireless Telegraphy and Radio Broadcasting in Kingston and Queen's University 1920–1957.*
Self-published, 1991

Manufactured By: RR Donnelley
Momence, IL USA
January, 2011